Global Justice
and Bioethics

Edited by Joseph Millum

and

Ezekiel J. Emanuel

OXFORD
UNIVERSITY PRESS

Oxford University Press, Inc., publishes works that further
Oxford University's objective of excellence
in research, scholarship, and education.

Oxford   New York
Auckland   Cape Town   Dar es Salaam   Hong Kong   Karachi
Kuala Lumpur   Madrid   Melbourne   Mexico City   Nairobi
New Delhi   Shanghai   Taipei   Toronto

With offices in
Argentina   Austria   Brazil   Chile   Czech Republic   France   Greece
Guatemala   Hungary   Italy   Japan   Poland   Portugal   Singapore
South Korea   Switzerland   Thailand   Turkey   Ukraine   Vietnam

Published by Oxford University Press, Inc.
198 Madison Avenue, New York, New York 10016
www.oup.com

Oxford is a registered trademark of Oxford University Press

Library of Congress Cataloging-in-Publication Data

Global justice and bioethics / edited by Joseph Millum and Ezekiel J. Emanuel.
   p. ; cm.
ISBN 978-0-19-537990-7 (alk. paper)
1. Medical ethics. 2. Globalization. 3. World health. I. Millum, Joseph. II. Emanuel, Ezekiel J., 1957-
[DNLM: 1. Bioethical Issues. 2. Internationality. 3. Social Justice. 4. Social Responsibility. WB 60]
R724.G5955 2012
174.2—dc23   2011024377

9 8 7 6 5 4 3 2 1

Printed in the United States of America
on acid-free paper

Dedicated to the staff, grantees, and trainees of the Fogarty International Center—working to promote global justice through science

ACKNOWLEDGMENTS

The editors would like to thank Bruce Agnew for editorial assistance, Peter Ohlin at Oxford University Press for his patience and encouragement, David Elliott for help with the cover design, and the faculty and staff of the Clinical Center Department of Bioethics at NIH for their support.

CONTENTS

CONTRIBUTORS

Allen Buchanan
Duke University

Tony Cole
Brunel University

Ezekiel J. Emanuel
University of Pennsylvania

Nir Eyal
Harvard University

Lisa Fuller
State University of New York
  at Albany

Robert E. Goodin
Australian National University
  and University of Essex

Robert O. Keohane
Princeton University

Joseph Millum
National Institutes of Health

Mathias Risse
Harvard University

Gopal Sreenivasan
Duke University

Alan Wertheimer
National Institutes of Health

Jonathan Wolff
University College London

CHAPTER 1

Introduction

Global Justice and Bioethics

JOSEPH MILLUM[*] AND EZEKIEL J. EMANUEL

Kosnatu and her sister have traveled all day from her village in rural Sierra Leone to the maternity referral hospital in Freetown. Kosnatu is pregnant and about to give birth. Following serious complications with the birth of her first child, a community health worker told her that she must have a cesarean section if she became pregnant again. The government recently introduced free health care for pregnant women, breast-feeding mothers, and children under five. Kosnatu's family is very poor; she is hoping that the government's promise will be honored and she will not be charged "fees." For now, they sit in a hallway waiting. There are many patients but very few hospitals, hospital beds, or qualified health care personnel.

Nguyen Van Long is an ex-heroin addict who lives in the outskirts of Ho Chi Minh City, Vietnam. Five years ago, he contracted HIV by sharing needles and since developed AIDS. Each month, Van Long goes to a government clinic where his condition is monitored and he collects the antiretroviral drugs that have halted the progression of his disease and allowed him to return to work. His doctor is now worried. Over the past two months Van Long's viral load has increased, and the doctor suspects that the virus is developing resistance. Second-line antiretrovirals exist

---

[*] The ideas and opinions expressed are the author's own. They do not represent any official position or policy of the National Institutes of Health, Public Health Service, or Department of Health and Human Services.

but are far too expensive for normal Vietnamese like Van Long. The doctor suggests that he make inquiries into a drug trial being conducted in a city hospital by a U.S. research group. It is rumored that the trial provides free treatment in exchange for research participation.

Each of these medical encounters can be described in terms of the interactions between individuals. But such descriptions, like the ones above, do not paint a complete picture. Kosnatu has to travel so far to get to a hospital because Sierra Leone's health care infrastructure is so weak. Government corruption and years of civil war have impeded development and mean that millions rely on subsistence farming to survive. The civil war was launched from neighboring Liberia and funded by international sales of diamonds from Sierra Leone's diamond fields. It was ended by intervention from regional powers, the United Nations, and finally the British army (the UK having once colonized the country). The government's announcement of free health care for pregnant women was possible only because of funding from the UK government and loans from the International Monetary Fund. Whether it will be sustained depends on them. A long wait for medical attention is inevitable in a country where there are so few health professionals; targeted recruiting and better wages mean that over 50% of the health care workers trained in Sierra Leone emigrate to neighboring countries, such as Ghana, and to the West.

Nguyen Van Long's generic antiretrovirals are imported from India, whose intellectual property laws were, until recently, substantially weaker than in the West. The cost of AIDS drugs has been brought down by competition from generics, and protracted negotiations between multinational pharmaceutical companies and civil society groups and politicians. Moreover, for thousands of Vietnamese, treatment for HIV/ AIDS has been possible only because of the funding provided by the U.S. government through its President's Emergency Plan for AIDS Relief (PEPFAR). The more recent generation of antiretrovirals needed by patients like Van Long are still protected by patent laws and therefore much more expensive than the first-line treatments. And though he might receive treatment by enrolling in a clinical trial, this depends on the research needs of international organizations that sponsor treatment trials, and it forces these organizations to decide whether and how to treat Van Long.

Kosnatu's and Van Long's situations only really make sense from a global perspective that takes in the structures that affect people's lives around the world. It is only from a global perspective that we can properly explain how these people's complicated medical encounters arise. And it is only from a global perspective that options for addressing their difficulties can be identified.

In this book, twelve scholars working at the intersection of philosophy, economics, and bioethics critically examine pressing global issues in medicine and ethics. The book is organized into three parts. The first part focuses on questions of ideal theory—that is, questions related to the ultimate global order at which we should be aiming. The writers in this part answer questions related to the normative significance of state boundaries for bioethics, rights to essential medicine, and the duty to ensure that everyone has access to health care. The last part of the book focuses on non-ideal theory—that is, questions about what the various actors engaged in biomedical research and care (individual researchers, research funders, foreign ministries, nongovernmental organizations, pharmaceutical companies, and others) ought to do in the face of an unjust world in which others do not do as they should. Here the problems include working out what medical researchers ought to do for research partici-pants and host communities that lack access to health care outside of clin-ical research; addressing cultural differences in providing health care to people who are very poor; and mobilizing consumers and investors to improve global health. Joining the two parts are two papers that address more theoretical issues concerning the nature of non-ideal theory and its relationship to ideal theory.

This introduction provides a brief overview of the burgeoning field of global justice and bioethics. We begin historically, trying to explain why the types of issues addressed by the contributors to this volume have become so pressing. For those readers unfamiliar with international polit-ical theory or bioethics, we then provide some basic factual and concep-tual background. Finally, we survey some of the key themes and problems that connect global justice to bioethics and analyze the strategies used in this book to address them.

BIOETHICS AND POLITICAL THEORY

Traditionally, the central subject matter of bioethics has been the ethics of individual interactions—between physicians and patients, and between researchers and research participants.[1] For the most part, this ethical analysis has also been relatively parochial, with U.S. bioethicists, for example, focusing on ethical problems arising within the context of U.S. clinical care and research. However, over the past decade or so, the purview of bioethics has greatly expanded.

The expansion of bioethics has at least two aspects. First, there is increasing recognition of the importance of the *systems* within which health care and research operate, which shape people's options and health care decisions, and which determine the resources available to them. Bioethicists now concern themselves with, for example, the development of health policy, the organization of health care providers, the institutions that govern medical research, and the social determinants of health. Though never wholly absent from bioethical discourse, these broader concerns are much more prominent than before. Systemic concerns like these naturally raise questions about the justice of the systems. Second, the reach of health systems is increasingly *international*. For example, increasing amounts of clinical research take place at multiple sites in multiple countries.[2] Faced with the disparities between the care that is available at home and the care received by their participants, Western researchers working in developing countries are dramatically forced to face the question of what they owe their participants during and after clinical trials. The levels of development assistance fall short of what is needed to combat health problems, raising questions about priority setting. Shortages of physicians, nurses, and pharmacists in developed countries, and recruitment of these providers from developing countries, raise questions about global distribution of health care personnel. With rapid travel, diseases are no longer localized but have global impact. And the implementation of the Trade-Related Aspects of Intellectual Property Rights (TRIPS) agreement, along with other bilateral and multilateral agreements, means that intellectual property protection is international—affecting the development and pricing of medical technologies and treatments everywhere.

Like bioethics, political philosophy has until recently had a rather domestic focus. Indeed, it is only a small exaggeration to describe the history of political philosophy as a history of theorizing about the state. The central problem for modern political philosophers has been working out the conditions under which the state's coercive power over its citizens can be justified. Indeed, even for Plato, the answer to the question "What is justice?" turned out to require a detour through a description of the political order of an ideal state.[3] One reason for this focus on the state is the perception that it is the source of many of the social conditions that determine the quality of people's lives. As John Rawls writes: "The basic structure [of society] is the primary subject of justice because its effects are so profound and present from the start."[4] But this justification for focusing solely on the state no longer seems viable. Recent growth in international communications, collaboration, finance, and commerce, and the institutions that regulate them, means that the life prospects of citizens from every country are dependent on events happening outside their country's borders. Such international effects are no more chosen by the people who experience them than the effects of national or local institutions; indeed, generally less so. Justice is a global concern.

The facts of globalization mean that a responsible bioethics must address problems of international scope. But the expansion of the scope of both theories of justice and the problems of bioethics into the global arena means that the concerns of the two now intersect to an unprecedented degree. Consequently, it is now impossible to engage with many of the most pressing problems of bioethics without also engaging with political philosophy (if, indeed, it ever was possible).

BACKGROUND: SOME KEY FACTS

A proper understanding of global bioethics and how considerations of justice relate to it is not possible without having some idea of the context in which bioethical issues of global scope arise, including the distribution of wealth around the world, the global burden of disease, and the key international political institutions.

Despite the massive increases in global productivity since the Industrial Revolution, the majority of the world's 6.8 billion people remain very poor.[5] More than 3 billion live on less than $2.50 a day, and over 80% of the world's population lives on less than $10 a day.[6] Though the amount of absolute poverty has fallen in the past three decades, the majority of this reduction is accounted for by China, where 600 million people have been brought out of poverty during its dramatic economic growth. In the rest of the world, the number of people living in absolute poverty—that is, suffering an absolute deprivation of their basic needs—has fallen only slightly.[7] Meanwhile the distance between the wealth of the richest and the poorest people in the world has dramatically increased. Disparities in wealth are reflected in disparities in consumption. For example, the wealthiest 20% of people account for 76.6% of consumption, while the poorest 20% account for just 1.5%, and 12% of humanity is responsible for 85% of human water use.[8]

It is often convenient to divide countries into rich and poor (or "developed" and "developing"), but such a simple division misses several important features. First, the differences in wealth between countries that are considered developing are immense. Mali has a gross domestic product (GDP) per capita of around $1,000, whereas Brazil's is closer to $10,000.[9] Second, disparities within countries can be very great. Though Equatorial Guinea had a GDP per capita of $30,000 in 2007—roughly the same as Italy—the majority of its people remain subsistence farmers, while a tiny ruling elite controls the country's substantial oil revenues. Consequently, Equatorial Guinea's average life expectancy is just 50 years (compared to Italy's 81) and nearly two thirds of its people lack access to clean water.[10]

Poverty and poor health frequently go hand in hand. People living in poorer countries tend to suffer greater morbidity, have lower life expectancies, and are less likely to be able to access modern health care. However, there are exceptions to all of these associations. For example, the high life expectancies found in relatively poor communities in countries such as China, Costa Rica, Cuba, Sri Lanka, and the Indian state of Kerala indicate that poverty *per se* need not lead to poor health. In fact, these places are characterized by their focus on the broad determinants of health of their citizens. For example, Kerala introduced universal free primary and secondary education in the early 20th century, has invested in maternal and child nutrition, provides universal access to free health care and family planning, and is culturally and politically committed to the empowerment of women. As a result its performance on key population health indicators—such as mortality rates for mothers, infants, and children, numbers of underweight children, and life expectancy—is dramatically better than other Indian states, even though it has a low per capita income even relative to India as a whole.[11]

As well as disparities between countries (and within them) in terms of wealth and health, there are significant differences in the sources of morbidity and mortality. While some noncommunicable diseases, such as heart disease and stroke, cancer, and mental illness, are severe health problems in almost all countries, the burden of infectious diseases is far greater in South Asia and sub-Saharan Africa. HIV/AIDS, malaria, tuberculosis, and diarrheal diseases kill millions of people in these regions but very few people in developed countries. On the other hand, diabetes is a growing problem in populations rich enough to overeat, and Alzheimer's and other dementias are a serious health burden in populations with long life expectancies.[12] There are also important differences in the causes of injuries and in environmental health hazards to which people are exposed. For example, the indoor air pollution caused by cooking with solid fuels such as dung or wood is responsible for 1.6 million deaths each year in developing countries.[13] Differences like these affect whether research into new health care interventions responds to the global disease burden. For example, since the purchasing power of the global poor is very low, there is little financial incentive for private companies to develop treatments for diseases that burden only them, nor to develop delivery mechanisms for treatments that are designed to operate in resource-poor environments.

In terms of institutions, the nation-state is still the most important political entity on the world stage; however, there are a number of significant transnational institutions. The UN is a political organization comprising 192 member states whose stated purpose is the promotion of international peace, cooperation, development and human rights.[14] The UN system also includes the World Health Organization (WHO), an agency of the UN devoted to international public health. The WHO conducts and supports research and monitoring, sets international health standards, and provides countries with technical support.[15] The World Trade Organization (WTO) deals with the regulation of trade between its 153 member states through the negotiation of binding agreements. Compliance with WTO trade agreements is backed by the threat of economic penalties. Other, mainly economic, transnational institutions also have profound impact. The European Union now comprises 27 countries, from Germany and Sweden to Romania and Cyprus. It has its own parliament, single market and currency, and an annual budget of over €120 billion.[16] The World Bank offers financial loans and credits and technical assistance to developing countries to fight poverty. In 2010 it provided over US$45 billion and is now involved in over 1,800 projects, ranging from microcredit to large infrastructure development.[17]

These global institutions are responsible for a number of global agreements that are of direct relevance to bioethics. The UN established the International Bill of Rights, which comprises the Universal Declaration of Human Rights, the International Covenant on Economic, Social and Cultural Rights, and the International Covenant on Civil and Political Rights. Among other entitlements, the International Bill of Rights includes a right to freedom from discrimination, to freedom of movement, to health, and to an adequate standard of living. The WTO now oversees some 60 trade-related agreements, perhaps the most famous of which is TRIPS. TRIPS aims to standardize world intellectual property protection at the high levels found in the United States, the European Union, and Japan. This process is expected to be complete in 2016, when all signatories, including the "least developed countries," are required to have implemented the WTO's intellectual property provisions. Intellectual property rules play an important role in determining which health care interventions get developed and how much they cost.

BACKGROUND: SOME KEY CONCEPTS

Political theorists who discuss justice are usually concerned with the evaluation of public policies or social institutions. This requires that they consider questions of *distributive justice*—how policies, institutions, or a social system as a whole distributes some set of benefits and burdens. For example, social institutions are responsible for the allocation of property, opportunities for work, political rights, and so forth. All of these allocations may be properly assessed to determine whether they are just. In analyzing theories of distributive justice, or when assessing the justice of a particular institution, there are several ways in which it can be helpful to conceive of justice. One way is to think of justice as equal treatment.[18] Different theories of justice can then be understood as different theories about what equal treatment consists of, and an institution can be criticized if it fails to treat people as equals. An alternative is to think of justice in terms of justifiability. An institution is then just if and only if its structure and its actions can be justified to each of the people it affects.[19]

Domestic and Global Justice

Discussions of justice used to be limited to the state. But we noted above that many of the reasons to be concerned about domestic distributive justice now seem to be reasons to be concerned about global distributive justice too. These include the imposition of institutions like the WTO. People have no more choice about their country's WTO membership than they do about most of its domestic institutions. The effects of global factors like international trade and finance on a country's economic growth and stability, as well as its people's life prospects, are immense.[20] One of the foundational questions we may ask about global justice is how morally important the boundaries of the state are. Philosophers have offered a continuum of possible answers to this question, ranging from a *cosmopolitan* denial that state boundaries have any intrinsic significance, to what we may call a *statist* view that the existence of the state is a necessary condition for requirements of distributive justice to apply.[21]

The idea of justifying principles of distribution to those they affect can be helpful when assessing particular global institutions rather than the global system as a whole. For example, property claims (over both physical and intellectual property) are now held against everyone in the world, and almost all the world's physical resources are owned. This suggests that the global institution of property ownership requires justification to all; it should have a form that everyone can accept.[22]

The borders of the state are often thought to make a normative difference: different duties are owed to fellow citizens than outsiders. But people often claim that other associations are morally significant, too, such as shared national, ethnic, or religious identities. Such groupings may be important because they ground special duties between people who share the group identity. Modern nationalists, like Will Kymlicka and David Miller, point to such aspects of identity to explain why we should care more about fellow nationals than other people, and why national groups have a claim to form their own state.[23] The question of whether associations like the nation make a normative difference is distinct and additional to the previous question about the normative significance of the political state. It is possible to downplay the relevance of state borders to distribution while still thinking that co-nationals should be preferred. Likewise, someone may think that the existence of a state transforms the normative relationships between its citizens while denying that this is connected to characteristics they have in common (other than co-citizenship).

Others argue that there are moral duties to respect the internal workings of other people's associations, such as families, religions, and nations. This respect might be instantiated by a presumption against intervention, or it might be argued that (in at least some instances) there is no impartial position from which to criticize a practice from another culture. Lisa Fuller considers some of the complexities of intervening in societies with very different value systems in her chapter on international nongovernmental organizations (INGOs).[24]

Ideal and Non-ideal Theory

The papers in this collection are categorized according to their focus on ideal or non-ideal theory. *Ideal theory* concerns hypothetical institutional arrangements that are just, known to be just, and whose requirements are largely complied with by those to whom they apply.[25] The central task for ideal theorists, therefore, is working out realizable conceptions of just institutional arrangements. Conversely, *non-ideal theory* deals with the obligations that arise either when institutional arrangements are not just or when some of the individuals subject to the institutions do not comply with them. These correspond to two branches of non-ideal theory: transitional theory and partial compliance theory. In this volume, the proper taxonomy of non-ideal theory is critically explored by Gopal Sreenivasan, and one aspect of the relationship between ideal and non-ideal theory is analyzed by Robert E. Goodin.[26] Here we make a further analytical point that may be helpful.

A binary classification into ideal and non-ideal theory is not always the most helpful way to think about specific policy questions or institutions. This is because when we evaluate actual or proposed institutions, we can idealize the context in which we consider them to varying degrees and along different axes. Suppose we are trying to answer the question of what rules should govern the pricing of medicines around the world. Given the vast disparities in purchasing power in the actual world, we might think it useful to address the question as a matter of transitional justice: what should the rules governing the pricing of medicines be, assuming the world's wealth and health disparities remain more or less as they are? Our answer to this question might tell us how the governments of rich industrialized countries should act, what the WTO ought to agree, and so forth. But the answer to this question is an idealization: it assumes full compliance with the demands of transitional justice, and so it is, as it were, a type of ideal non-ideal theory. We may also be interested in what an individual actor, such as the UK government, ought to do if other governments do not follow our recommendations about the right transitional system for pricing medicines. Our theorizing will then be non-ideal along two axes, since we will assume that the global distribution of wealth remains unjust *and* that not everyone is going to do what he or she ought; hence it will be a matter of both transitional theory and partial compliance.

A lot of the work that bioethicists do falls into this intermediate zone between ideal and non-ideal theory. Depending on the purpose of one's analysis, it may be very helpful to idealize along some, but not all, dimensions. Liam Murphy has argued that considerations of fairness constrain certain moral duties under conditions of partial compliance, so that they do not require more of us than we would be expected to do if everyone did his or her share.[27] Thus, according to Murphy, working out what each individual ought to do under conditions of full compliance might tell us what a fair division of duties is. Were this argument sound, it would suggest that we can work out our *actual* duties by means of analyzing our *ideal* duties. Sreenivasan takes precisely this route in considering what Canada's minimal obligations of global redistribution are.

PRESSING PROBLEMS OF GLOBAL JUSTICE AND BIOETHICS

Bioethical problems that can be resolved only by paying attention to questions of international justice are liable to arise whenever institutions with global reach affect health or health care. These institutions themselves may be the subject of ethical appraisal—for example, when international treaty obligations appear to conflict with a government's supplying cheap drugs through its public health system. In other cases, the institutions or their effects provide a backdrop of injustice against which bioethical issues arise—for example, when clinical researchers in a low-income country have to decide how to deal with research participants who do not have access to health care.

Table 1.1 lists a broad but not exhaustive sample of important international bioethical problems divided into four categories: clinical care, research, health policy, and theory. Of necessity, it is a simplification; for

Table 1.1. SELECTED PROBLEMS IN GLOBAL JUSTICE AND BIOETHICS

Subject	Problem	Key Question
Clinical care	Health tourism	Is it permissible for wealthy people to use the health systems of other countries for faster or cheaper care?
	Organ trafficking	Is international trade in organ transplantation permissible?
	Access to medicines	How should access to essential medicines for everyone be ensured?
Research	Responsiveness	Must health research be responsive to the needs of the community in which it is carried out?
	Benefit-sharing	How should the benefits of research be shared with the different groups who contribute to it?
	Standards of care	What standard of care should be offered to participants in trials when the prevailing care in their community is less than the global best?
	Ancillary care	What treatment ought researchers to give their participants over and above the treatment that is required for the scientific design and safety of their trial?
	Post-trial access	Should research participants and communities be guaranteed access to successful interventions after the trial?
Health policy	Parallel health systems	Should NGOs and private providers run health care systems independent of public health care systems?
	Intellectual property	How should the international intellectual property regime be structured as it applies to health interventions?
	"Brain drain"	Should health workers' migration be restricted?
	International disease threats	Who is responsible for monitoring and combating infectious diseases that move across national borders?
	Lifestyle exports	Do any obligations fall on countries or corporations who export unhealthy lifestyles?
Theory	Cultural variation	How should variation in cultures affect international research and health care provision?
	Priority setting	How should local and international priorities for research and health care be set?
	The right to health	What is the right to health and against whom is it held?
	Ideal and non-ideal theory	How should we move from the actual world to a world where there is justice in health?

example, some problems arise in more than one category, and answers to many questions have implications for other categories of questions.

Categorizing problems by subject area gives us some idea about the sheer breadth of the questions concerning global justice and bioethics. We can perhaps learn more, though, by examining the methods used to answer these questions. Looking at the contributions to this book, there appear to be three strategies. First, some writers take a top-down approach to a problem. They first work out a general view about an aspect of global justice and then apply that view to the particular case at hand. In this collection Mathias Risse uses this strategy.[28] He starts with a discussion of how private property, in general, is justified. He then adapts his conclusions to private intellectual property, and finally applies his thinking to the particular case of pharmaceuticals.

A second strategy attempts to circumvent the controversies about more general questions of justice while still making use of the theoretical apparatus that political theory provides. This strategy looks for common ground between the different theories of global justice and then draws conclusions from that common ground. If, for example, there is good reason to think that any plausible theory of justice will require that trade rules be written so that they do not make the global poorest any worse off, then this premise can be used in discussion of trade agreements without having to defend any particular theory of justice. Ezekiel J. Emanuel uses a version of this strategy in his chapter on how researchers should design clinical trials in developing countries.[29]

The third strategy is to bracket the institutional questions and focus on individual interactions. Arguments that adopt this route take the non-ideal nature of the world with its injustices as a given and consider what specific individuals ought to do under such circumstances. In his discussion of the obligations of medical researchers amid injustice or deprivation, Alan Wertheimer is not considering the question of how the institutions governing medical research should be structured, nor the question of how researchers should attempt to repair the social injustice in the societies in which they work; he is asking how the clinical research itself can be ethically carried out when the research participants and their communities are in such bad situations.

On a related note, it is worth mentioning the different actors whom writers may address. Sometimes they are addressing individual people, answering the question of how those particular moral agents ought to act. In his chapter, Nir Eyal develops the concept of global-health impact labels, which are certifications that individual, concerned citizens can set up themselves and that allow individual consumers to affect global health through their economic decisions.[30] He argues that individual consumers and investors ought to establish and make use of these labels. At other times, bioethicists' concerns lie more with institutional creation or reform. In discussing the basis of the global health duty, Jonathan Wolff is more concerned with the structure and actions of social institutions than with what any particular person does.[31] Exactly who one's target is may affect what argumentative strategies are available. For example, it is easier to set aside more fundamental questions of global justice if you are primarily addressing individual doctors or clinical researchers rather than health systems as a whole. It may also make a difference to the appropriate level of idealization one adopts. Recommending to an individual what he or she alone should do right now is the most non-ideal of non-ideal theory; recommendations about institutional reform, with their requirements that multiple actors coordinate, tend to be "more ideal."

THIS BOOK

A single book could not cover all the possible issues that arise in global justice and bioethics, and this book makes no attempt to be comprehensive. Instead, our contributors provide their perspectives on a wide range of important topics. In doing so, they move discussion forward along a number of axes, methodological as well as subject-specific. They also demonstrate how rigorous academic work can cross from the most abstract concerns of political theorists to the most practical concerns of bioethicists. In doing so, they avoid two problematic extremes of this kind of academic work: that political theory may be too abstract to be of practical use, and that bioethics may be insufficiently grounded in theory, and so may lack rigor. We hope that this volume will introduce some of the ideas and methods of political theory to bioethicists, and likewise introduce bioethics to some political theorists.

NOTES

1. Norman Daniels adds to these dyadic relationships the ethics of "Promethean challenges," by which he means the ethics of dealing with new medical technologies. Daniels N. (2006). Equity and population health: towards a broader bioethics agenda. *Hastings Center Report* 36(4):22.
2. Thiers FA, Sinskey AJ, Berndt ER. (2008). Trends in the globalization of clinical trials. *Nature Reviews Drug Discovery* 7:13–14.
3. Plato (2008). *The Republic* (trans. by B. Jowett). Fairfield, IA: First World Library.
4. Rawls J. (1999). *A Theory of Justice* (rev. ed.). Cambridge, MA: Harvard University Press, p. 7.
5. World population figure taken from U.S. Census Bureau. International Data Base (IDB). Available at: http://www.census.gov/ipc/www/idb/worldpopinfo.php.
6. Shah A. Poverty facts and stats. *Global Issues*, Updated: March 28, 2010. Available at: http://www.globalissues.org/article/26/poverty-facts-and-stats. These dollar amounts are derived from purchasing power parity (PPP): $10 represents the amount of goods that could be purchased in the United States for US$10.
7. Chen S, Ravallion M. (2008). *The Developing World Is Poorer Than We Thought, But No Less Successful in the Fight Against Poverty*. World Bank. Available at http://go.worldbank.org/BS90FWR1O0.

8. Shah, *op. cit.*
9. United Nations Development Programme. *Human Development Report 2009. Overcoming Barriers: Human Mobility and Development.* Available at: http://hdr.undp.org/en/reports/global/hdr2009/. GDP is a measure of a country's total economic output. These amounts are also in PPP.
10. United Nations Development Programme, *op. cit.*
11. Skolnik R. (2008). *Essentials of Global Health.* Sudbury, MA: Jones and Bartlett Publishers, pp. 35–37.
12. Lopez AD, Mathers DC, Ezzati M, Jamison DT, Murray CJL, eds. (2006). *Global Burden of Disease and Risk Factors.* New York: Oxford University Press.
13. World Health Organization. *Indoor Air Pollution and Health: Fact Sheet No. 292.* Available at: http://www.who.int/mediacentre/factsheets/fs292/en/index.html.
14. United Nations. *UN at a Glance.* Available at: http://www.un.org/en/aboutun/index.shtml.
15. World Health Organization. *About WHO.* Available at: http://www.who.int/about/en/.
16. European Commission. *The European Union Budget at a Glance.* Available at: http://ec.europa.eu/budget/budget_glance/index_en.htm.
17. World Bank. *About Us: Projects.* Available at: http://go.worldbank.org/M7ARDFNB60.
18. Kymlicka W. (2002). *Contemporary Political Philosophy: An Introduction* (2nd ed.). Oxford: Oxford University Press, pp. 3–5, following Dworkin R. (1977). *Taking Rights Seriously.* Cambridge, MA: Harvard University Press, pp. 179–183 and elsewhere. See also Sen A. (1992). *Inequality Reexamined.* Cambridge, MA: Harvard University Press.
19. Scanlon TM. (1999). *What We Owe to Each Other.* Cambridge, MA: Harvard University Press.
20. See Millum J. "Global Bioethics and Political Theory," Chapter 2.
21. *Statist* may have slightly misleading connotations, but there does not seem to be a better alternative. See Millum, *ibid.*, p20*n.*
22. See Risse M, "Is There a Human Right to Essential Pharmaceuticals?" Chapter 3.
23. Kymlicka W. *Multicultural Citizenship.* Oxford: Oxford University Press, 1995; Miller D. (1995). *On Nationality.* Oxford: Oxford University Press.
24. Fuller L, "INGO Health Programs in a Non-ideal World," Chapter 9.
25. Rawls, 1999, pp. 7–8.
26. Sreenivasan G, "What Is Non-ideal Theory?" Chapter 6; Goodin RE, "The Bioethics of Second-Best," Chapter 7.
27. Murphy L. (2000). *Moral Demands in Nonideal Theory.* New York: Oxford University Press, pp. 74–101.
28. Risse M, "Is There a Human Right to Essential Pharmaceuticals?" Chapter 3.
29. Emanuel E, "Global Justice and the 'Standard of Care' Debates," Chapter 8.
30. Eyal N, "Global-Health Impact Labels," Chapter 10.
31. Wolff J, "Global Justice and Health," Chapter 5.

PART ONE

Ideal Theory

CHAPTER 2

Global Bioethics and Political Theory

JOSEPH MILLUM[*]

INTRODUCTION

The world is very unequal. While a billion people live in luxury, billions more struggle with hunger, pollution, inadequate housing, unsafe working conditions, economic vulnerability, and poor health care. Many of the questions confronted by contemporary bioethicists are affected by or the product of these gross international inequalities. Some issues concern interactions that take place against the backdrop of inequality. For example, faced with the disparities between the care that is available at home and the care received by their participants, Western researchers working in developing countries are forced to consider whether they owe participants clinical care, whether there should be local access to the fruits of research, and so forth. Other questions concern the structures of systems that might themselves be analyzed as just or unjust. For example, the implementation of the Trade-Related Aspects of Intellectual Property Rights (TRIPS) agreement means that intellectual property (IP) protection is international, which affects the ability of everyone to access patent-protected medicines. And economic disparities between countries lead to "brain drains" of essential workers from the poor countries where they are most needed to rich countries, and allow better-off citizens of industrialized

* The ideas and opinions expressed are the author's own. They do not represent any official position or policy of the National Institutes of Health, Public Health Service, or Department of Health and Human Services.

countries to take advantage of less expensive health care available in developing country hospitals through health tourism.

Global inequality is, then, the source of a number of bioethical problems. Considerations of global justice therefore appear relevant to bioethics. However, exactly how we should apply the concept of justice beyond the borders of the nation-state remains a matter of fierce dispute among political theorists. Some—cosmopolitans—argue that exactly the same principles of justice apply internationally as they do domestically. Others with more statist leanings think that the requirements of justice beyond national borders are severely attenuated. It is therefore important to consider how these debates are relevant to bioethics. Can bioethicists ignore them? Or must they argue for a position on global justice before they can properly resolve their specific problems?

Most bioethicists who address questions to which global justice matters have not considered the significance of the disputes over the correct theory of global justice. Some restrict themselves to analyzing the morality of individual interactions, and so treat these international bioethical problems no differently than other cases in applied ethics. Alternatively, those bioethicists who take distributive justice seriously generally adopt or defend some version of cosmopolitanism. Consequently, the significance of the differences between theories of global justice for bioethics has been obscured.

In this paper, I consider when and how these differences are important. I argue that certain bioethical problems can be resolved without addressing disagreements about global justice. People with very different views about global justice can converge on the existence of a duty to aid the very badly off—those in absolute poverty—wherever they may be.[1] However, despite agreement on extreme cases, there should be disagreement over the extent of international obligations to those who are only relatively poor. Consequently, different theories of justice will diverge in their implications for a number of important problems in contemporary bioethics. I close by sketching in more detail two contemporary bioethical issues—concerning pharmaceutical patents and the health worker brain drain—and show how responses to them might be developed by cosmopolitan and statist liberals. These sketches demonstrate the relevance of specific theoretical claims about global justice to particular bioethical problems.

JUSTICE

Justice, says John Rawls, has as its primary subject matter "the basic structure of society, or more exactly, the way in which the major social institutions distribute fundamental rights and duties and determine the division of advantages from social cooperation."[2] Since social cooperation may generate burdens, too, we may also say that justice should be concerned with their distribution.[3] Thus, the subject matter of what has come to be known as "distributive justice" is primarily the analysis of social institutions, where "institution" can be understood very broadly to include both formalized and informal social practices. I include respect for human rights under this heading, too, since the realization of human rights is a matter of institutional organization.[4] Distributive justice should be distinguished from two other senses of justice that may be of interest to bioethicists. The first is *transactional justice*, which is concerned with the conditions under which individual interactions between people or institutions are permissible (for example, when considering whether a clinical trial hosted in a developing country would be exploitative). The second is *justice as requital*, which concerns what should be done when someone experiences a loss (for example, through injury as a result of a surgeon's error). I return briefly to these other senses of justice later in this chapter; here, I focus primarily on theories of international distributive justice. When I mention justice *simpliciter* I mean this sense of justice.

When analyzing theories of justice, we should distinguish *ideal* and *non-ideal* theory. According to Rawls, *ideal* theory concerns well-ordered institutional arrangements—that is, those institutional arrangements that are just, are known to be just, and with which individuals subject to the institutions willingly comply.[5] The central task for ideal theorists, therefore, is working out realizable conceptions of just institutional arrangements. *Non-ideal* theory, by contrast, deals with the obligations that arise either when institutional arrangements are not completely just, when the individuals subject to the institutions do not fully comply with them, or, commonly, both.[6]

Academic philosophers have most rigorously explored questions of ideal theory. Their answers tell us what a just state or a just global order would look like. However, our day-to-day concerns about justice are not just about the final goal at which people working for justice should aim. We also want to know what to do in the face of the practical problems that arise in our imperfect world. These are generally questions of non-ideal theory. For example, ethical questions about health worker migration (the brain drain) are premised on the massive economic inequality between countries. If the disparities between countries were not so large, movement between them would probably not be so unidirectional (i.e., from poor to rich), and people would therefore be unlikely to be so troubled about the question of its regulation. International bioethicists must therefore concern themselves with non-ideal theory, as well as ideal theory.

THE EXTENT OF INTERNATIONAL REQUIREMENTS
OF JUSTICE

A central question that divides theorists writing about global justice, and that affects most directly problems in international bioethics, concerns what the people and governments of rich countries owe to those outside their borders.[7] The possible answers are generally thought to range from "the same as they owe to similarly situated people inside their borders" to "nothing at all." Consequently, answering this question satisfactorily requires addressing both the scope and the content of the requirements of global justice. In this section, I sketch the spectrum of possible positions and explain how they might be defended. This allows me to show when and how disagreements about global justice should make a difference to answers to bioethical questions.

The views occupying the two extremes I label *cosmopolitan* and *statist*.[8] According to pure cosmopolitans, the principles of distributive justice that apply in the domestic sphere apply equally internationally.[9] Thus, if Rawls' difference principle were the right way to allocate primary goods within a country, there should also be a global difference principle allocating primary goods among all people in the world.[10] According to pure statists, principles of distributive justice apply only domestically. Between these two lie more moderate statist positions that acknowledge cross-border duties of justice but claim that they are weaker than domestic duties.

One need not be of any particular political creed to be a cosmopolitan or a statist. For example, liberal egalitarians could think that the principles they endorse require redistribution for equality across the globe; but equally, a libertarian could think that the same principles should apply to institutions in order to protect property rights and negative liberties around the world.[11] Nevertheless, the discussion of the international scope of principles of justice has been dominated by arguments between cosmopolitan liberals on the one side, and statist liberals and communitarians on the other.[12]

Cosmopolitanism

Cosmopolitan positions may be distinguished by the grounds that are thought to underlie them. *Humanitarian cosmopolitans* believe that duties of justice do not arise from associations like the state, but from characteristics of persons as such, independent of their relations to other persons. Any utilitarian theory of global justice should take this route, since utilitarians hold that the ultimate justification for principles of justice is their contribution to aggregate utility, and it is irrelevant in whom that utility is located.[13]

Political cosmopolitans accept that some characteristics of associations like those epitomized by the state are the grounds for the applicability of the requirements of justice, but argue that these characteristics are in fact found in international associations, too. For example, it might be thought that mutually beneficial cooperation between individuals is necessary and sufficient to ground requirements of justice in the distribution of the products of that cooperation. The extent and importance of global trade could then be cited as evidence that mutually beneficial cooperation extends well beyond national borders. This would imply that the requirements of justice extend beyond national borders, too.[14]

This latter strategy is used by Charles Beitz in *Political Theory and International Relations*. He argues that the global institutional order is analogous to a state in two key respects. First, there is an analogy between the arbitrariness of the natural distribution of talents between individuals and the arbitrariness of the natural distribution of resources between states.[15] As with natural talents, Beitz argues, social institutions should not allow these morally arbitrary differences to disadvantage people. He concludes:

> The underlying principle is that each person has an equal prima facie claim to a share of the total available resources, but departures from this initial standard could be justified (analogously to the operation of the difference principle) if the resulting inequalities were to the greatest benefit of those least advantaged by the inequality.[16] GLOBAL DIFFERENCE PRINCIPLE

Second, Beitz argues that substantive global economic interdependence exists in the form of international investment and trade, and that this interdependence "yields substantial aggregate benefits."[17] These international transactions are regulated by global institutions, as well as by informal practices of economic policy coordination. Thus, there exist cooperative interactions that generate benefits and burdens that must be distributed, and there exist institutions that already affect the distribution of those benefits and burdens. These are the very conditions, Beitz argues, that motivated Rawls to argue that the institutions that constitute the basic structure of a society must be designed in accordance with principles of justice. In exactly the same way, therefore, we can determine the principles that should govern the distribution of benefits and burdens. Beitz proposes a global "original position" constituted by individuals (not peoples, as Rawls later has it).[18] From this original position, ignorant of their nationality, Beitz argues, people would choose the same principles in the international case as in the national one, and for the same reasons.

Unlike humanitarian cosmopolitanism, the political strategy for justifying cosmopolitanism makes its truth contingent. In a world of autarkic states there would not be any duties of international justice. In the present world, presumably, if there are states or communities that are sufficiently separate from the rest of the international community, then they are not owed assistance as a matter of justice. To some, this may be disquieting: whether the billions of poor people in the world have a justice-based claim to assistance is dependent on whether they happen to be part of the global economy. Worse, it is the very poorest in the world who are most likely to be excluded from the global economy and therefore, on this account, most likely to be excluded from justice-based consideration.[19] To others, this fits with longstanding intuitions about the applicability of claims of justice.

Statism

We may divide statists into *nationalist statists* and *political statists*. Nationalist statists think that there is something morally important about nationality, independent of the state, that is sufficient to ground special duties toward co-nationals and that makes it permissible to prefer co-nationals. For some, this is because being part of a national culture is necessary for a person's identity as a moral being.[20] For others, it is because of the great value of national cultures.[21] Nationalists argue against cosmopolitanism on the grounds that it does not make room for the moral importance of national ties.

Nations and political states may, of course, come apart, since political borders are not always drawn around groups of people who share a national identity. I include nationalists as a form of statist because they privilege co-nationals (as political statists privilege co-citizens) rather than treating everyone the same, and because nationalists typically view an independent state as one of the desiderata for a nation.[22] Indeed, one of the primary motivations of more practically minded nationalist theorists is to defend the demands of some cultural or ethnic group for its own state.

Political statists argue that obligations of justice arise only when a state already exists. The associative relationships between people in the same state generate normative relationships between them that did not exist before. However, political statists deny that the current relationships between people in different states are sufficiently like being in a state to generate similar normative relationships. Thomas Nagel, for example, argues that the state is distinctive because of the coercive power that it wields over its citizens and because the actions of the state (e.g., in wielding this power) are taken in the name of its citizens. Neither of these characteristics, Nagel says, applies to existing international institutions.[23] Political statists therefore accept that the global order *could* be such that the requirements of justice were universal (if, for example, there were a world government). They simply deny that the requisite empirical conditions exist. They therefore reject the second strategy for defending cosmopolitanism outlined above.

Degrees of Statism

It is helpful to distinguish statist positions according to the degree of normative primacy they assign to the state. A *pure statist* would deny that there are any duties of justice that extend outside the borders of the state. However, few theorists seem willing to adopt such an extreme position. More are what I will call *strong statists*: they hold that the main locus of duties of justice is the state, but allow some very attenuated duties of justice outside its borders. For example, Rawls' Law of Peoples states the principles of justice that should govern the interactions of liberal states with other states. These principles do not include distributive principles, like those that apply to domestic institutions: Rawls does not think inequalities between states are the proper subject matter of justice. However, as well as constraints on how states may interact, Rawls acknowledges a duty of assistance to "other peoples living under unfavorable conditions that prevent their having a just or decent political and social regime" to help them reach a point where they can have a just or decent regime.[24] As I note in the following section, depending on how many people live under such unfavorable conditions, this duty may be quite stringent.

Although most discussion of the scope and content of requirements of global justice so far has contrasted cosmopolitan and strong statist positions, there is also room for intermediate positions. Such *moderate statist* positions would recognize some significant requirements of justice beyond the boundaries of the state, but would limit their nature or extent. Limitations on international duties could take different forms. For instance, there might be duties owed to co-citizens that must be fulfilled before duties to non-citizens take effect; it might be that what is owed to non-citizens is less than what is owed to co-citizens; or it might be that certain duties of justice apply both domestically and internationally and others apply only domestically.[25]

For example, one way to argue for a moderate statist position would be to show that some duties of justice apply in virtue of people's moral status and other duties require some special relationship between the people to whom the duties apply. Different duties of justice falling into the latter category could then be stratified according to the nature of these relationships. For example, purely economic transactions might be governed by principles of fairness for the distribution of the benefits and burdens of just those transactions. But political rights might remain with just the people living within the borders of a state, since only they would be rightfully subject to that state's coercive authority. Such moderate statist positions have not been much explored in the literature on global justice.[26]

ABSOLUTE POVERTY AND AGREEMENT
ON INTERNATIONAL DUTIES

Despite the large differences between theories concerning the scope and content of the requirements of global justice, there is still a substantial core of agreement among most commentators regarding certain duties that extend past national borders. In the case of people who are sufficiently badly off, almost everyone endorses normative principles that commit them to some duty of assistance incumbent on those who can help. Consequently, when it comes to bioethical problems that are primarily of relevance to this group, there is no need for extended arguments about global justice: agreement on a duty to help can be assumed.

This agreement applies clearly to the case of people living in conditions of what Peter Singer calls "absolute poverty."[27] Their deprivation is such that even their most basic needs (for nutritionally adequate food, clean water, shelter, basic health care, and so forth) are not met. According to the most recent figures from the World Bank, in 2005 approximately 1.4 billion people were living on US$1.25 a day or less,[28] the measure the World Bank now uses to judge absolute poverty.[29] Absolute poverty indicates both that the people suffering from it are very poor indeed, and also that their poverty is not simply relative to other people. There are also billions of people in the world who live in what I will call *only relative poverty*—people who are able to meet their basic needs, but who are very poor relative to other people in their community, country, or the world. I argue shortly that agreement on international duties of assistance disappears when we come to this group. Consequently, when it comes to bioethical questions that concern the only relatively poor, complete answers will likely require bioethicists to articulate and defend political views.

For many of the absolutely poor, there are many actors who could ameliorate their situation. National governments in all but the poorest countries could prioritize their spending differently and save many lives. Other governments, particularly those of the richest countries in the world, could provide directed aid that would help the poor, and could negotiate transnational economic policies that help instead of exacerbating global poverty. Even individuals now have the power to save the lives of distant others, through donations to nongovernmental organizations (NGOs) that provide essential services in very poor environments.[30]

Given the extent of the deprivation of the absolutely poor and the possibility of ameliorating at least some of it, we can see why people who have quite different political views are still likely to converge on the conclusion that governments and individuals have a duty to assist the absolutely poor.[31]

Consider first the different theories of international distributive justice. Most clearly, humanitarian cosmopolitans believe that all the absolutely poor have a right to assistance, since their claims to resources have an equal standing with the claims of everyone else. This claim is held against any and all governments that can meet it (and, perhaps, against nongovernmental institutions and individuals, too). For political cosmopolitans, the requirements of justice apply when people are involved in systems of cooperation that generate benefits and burdens, like the state. They believe that the existing international order—such as the global economic order that is given formal shape by institutions like the World Trade Organization (WTO)—is sufficient to generate obligations of justice with regard to those within it. As I noted above, this is liable to exclude people who are not part of the international economic system. Moreover, many of the absolutely poor are marginalized to the point that their contribution to the global economy is negligible. This might lead someone to the conclusion that the absolutely poor would therefore not have a justice-based claim to resources under the political cosmopolitan's rationale. But this would be a mistake. The great majority of the absolutely poor are still affected by the global economic system, even if they are not great contributors to it. For example, whether or not they have employment, training, or social security may be determined partly by international economic policies and agreements. Thus, though many of the absolutely poor may be excluded from beneficial participation in the global economic system, it is still a powerful determinant of their life prospects.

One does not have to be a liberal cosmopolitan to think that the existence of people in conditions of avoidable absolute poverty entails a duty to assist them.[32] Only pure statists deny that there are *any* international duties of justice.[33] So, for example, Rawls, a strong political statist, denies that we have duties to ameliorate the poverty of people in other states, even if they have much less than we do, so long as they are able to live under just or decent regimes. But, as I noted in the previous section, he does think that there is a duty of assistance to people who are unable to live under such regimes. This duty of assistance includes securing the "basic needs" of people in burdened societies.[34] All except the most extreme nationalist statists may agree. Though nationalists believe that there are reasons to favor co-nationals over non-nationals, most accept that there are *some* duties owed to people of different nationalities, and that those duties can take preference over helping co-nationals, when the needs of the other people are urgent enough. For example, David Miller writes:

> [W]e can agree that the existence of societies scoring very low on the HDI [Human Development Index, calculated by the United Nations Development Programme] is a global injustice without agreeing about why it is an injustice—whether by virtue of the inequality between rich and poor societies, or simply by virtue of the absolute level of deprivation experienced by most members of the poorest societies. Our moral responses to the global status quo are over-determined, and so we can agree in practice about what needs to be done most urgently to promote global justice without having to formulate explicitly the principles that lie behind this judgement.[35]

Even if it were argued that there are some people beyond national borders who fall outside the scope of distributive justice, there are other universally acknowledged moral duties. First, the very poor are frequently considered to be the victims of a global economic order that has been imposed upon them (mainly by rich country governments whose bargaining power allows them to shape international agreements in favor of their own perceived economic and geopolitical interests). For example, the structural adjustment programs instituted during the 1980s and 1990s by many developing-country governments, under pressure from global institutions like the World Bank and the International Monetary Fund, appear to have made many people in those countries worse off.[36] People may therefore be starving as a direct result of the actions of rich-country governments. If these empirical claims are correct, then those responsible owe assistance to their victims as a matter of compensatory justice.

Second, there are humanitarian duties, which are thought to apply just by virtue of the beneficiary of the duty having morally significant needs, and not because of some special relation between the duty bearer and the beneficiary.[37] Consider, for example, the duty of rescue, according to which we are obliged to provide urgently needed assistance to others when we can do so without great cost to ourselves. Though there is dispute over the extent of duties of rescue, including exactly how much we can be obliged to sacrifice for others and whether the duty changes in the face of others' non-compliance,[38] it is hard to see how someone could consistently believe in any duty of rescue but deny that the rich people and governments of the world have a duty to spend the relatively small amounts of money necessary to provide people with clean water or life-saving vaccinations.

Thus, the depth of the deprivation of the absolutely poor means that it is overdetermined that citizens and governments of rich countries have some duties to help them. Since everyone endorses at least one of the normative principles I just outlined, everyone is committed to a global duty of assistance to the absolutely poor.[39] The consensus that exists is therefore a function of the truly awful situation of the world's poor, rather than any theoretically interesting agreement between different theories of justice.

LACK OF CONSENSUS ON RELATIVELY POOR

This consensus is not available when we turn our attention to people who are very poor by the standards of rich countries, but not in such desperate straits as the absolutely poor. The justifications I just gave do not apply to them, or, at least, it is controversial whether they do. First, strong statists generally argue that relative poverty is the business of the state, and not the business of other states. For Rawls, for example, once a people has the resources to live under a "just or decent political and social regime," the duty of assistance cuts off.[40] Second, the principle of compensatory justice may not apply to many of those who are only relatively poor. Standards of living have improved for many people in the world over the past few decades, which makes it far from clear that they have been harmed by the international economic system.[41] Finally, the duty of rescue applies only when a great good can be provided to someone at a low cost to the provider (that is, when it is possible to execute an "easy rescue"). But as we move further from the satisfaction of basic needs to other things that might improve the well-being of the relatively poor, the extent to which they benefit is likely to decrease, while the cost of providing the benefit increases. For example, the cost of saving a human life by providing the traditional immunization program of childhood vaccinations in South Asia and Sub-Saharan Africa is approximately US$205.[42] By contrast, hip resurfacing surgery, which generally aims only at improved mobility and pain reduction and is mostly performed on elderly patients, costs around US$5,000 in private Indian hospitals.[43]

Compared to the average citizen of a developed country, billions of people in the world are only relatively poor. Were these people citizens of one of these developed countries, most statists (and cosmopolitans, of course) would agree that they deserved assistance as a matter of justice.

However, as I just argued, strong statists will generally not agree that those in other countries who are only relatively poor deserve assistance from developed countries. This, then, is the key difference between the implications of statist and cosmopolitan theories of global justice. It is a difference that has implications for how we should respond to a number of important bioethical problems. It demarcates the range of problems concerning which bioethicists cannot assume that there will be agreement about duties to people outside national borders, and therefore the area where fully addressing these problems requires engaging with the debate over global justice. In the section entitled "Two Theories of Justice and Two Bioethical Questions" I illustrate this by considering how different theories of justice apply to two specific problems—pharmaceutical patents and the brain drain. But before I can do this, I must say something about the notion of basic health care, because these bioethical problems relate to health care, and the inability to access basic health care is indicative of absolute poverty.

ABSOLUTE POVERTY AND BASIC HEALTH CARE

Whether someone is in absolute poverty depends on the resources to which he or she has access. The absolutely poor lack access to the resources that are necessary for leading a minimally decent life. Access to health care is one of these resources: without health care, someone's life prospects are likely to be severely diminished. I will call *basic health care* the level of health care someone must be able to access in order to escape absolute poverty.[44] People may be able to access basic health care because they have the personal wealth needed to pay for medical treatment or health insurance, or they may be able to access it because it is supplied to them through a public health care system. How they are able to access it is not important to my analysis here.[45]

Unlike, say, necessary nutrition, defining basic health care is complicated. Everyone has similar nutritional needs, and so we could confidently place an upper bound on the amount of food someone would need to be able to access in order to escape absolute poverty. But people have very different health needs. People with certain conditions may need treatment that is very expensive. For example, the average price of a possibly life-saving heart bypass is around US$20,000.[46] Are people who are unable to afford a heart bypass absolutely poor? This seems implausible: if we were to call people absolutely poor because they could not afford every medical intervention they might possibly need, this would detach the concept of absolute poverty from our intuitive understanding of what it means, and from the justifications that imply a duty to help people in absolute poverty. Thus, basic health care will include some, but by no means all, of the medical care that someone might need in order to lead a minimally decent life. Whether a particular intervention is included will presumably depend on whether its cost-effectiveness reaches a certain threshold.[47]

It follows that people in only relative poverty are nevertheless likely to lack health care for many conditions. For example, people in only relative poverty still might not have access to hepatitis A vaccines, treatment for deteriorating eyesight, heart bypass surgery, and so forth. Such treatments are not trivial: they are necessary for good health. However, they may not be necessary for someone to escape absolute poverty.

This analysis tells us how the conclusions I drew about agreement and disagreement between different political theorists should apply to the particular cases of medical research and health care provision. Agreement on the existence of a duty to aid those in absolute poverty leads to the conclusion that there is a universal obligation to encourage research and health care provision such that everyone is able to access basic health care. However, disagreement over what is owed to the only relatively poor will create disagreement about what medical research should be funded, what care should be provided by whom, and how the rules governing research and provision of care are constrained. For the strong statist, relative poverty matters within the state; outside the state it does not. Thus, according to a strong statist, people living within a comparatively rich country may have a justice-based claim that their government support research into diseases that affect them, and ensure that they can access existing treatments for these diseases. The same government will not have similar obligations to people living in relative poverty in other, poorer countries.

In the following section I explore this implication in greater detail, by examining two important questions of international health policy.

TWO THEORIES OF JUSTICE AND TWO
BIOETHICAL QUESTIONS

For the most part, bioethicists who have addressed questions involving international justice have taken one of two routes. First, some bracket the broader theoretical questions of distributive justice and make do with moral analysis on a more narrow scale. They discuss, for example, particular interactions between individuals, or appeal directly to the needs people have or the harms they perceive are being caused. In short, they do traditional applied ethics in the international sphere.[48] Second, those who argue from more foundational theories of justice to specific conclusions about policy or practice usually do so by assuming or defending some version of cosmopolitan liberalism.[49] Furthermore, these analyses frequently focus on the plight of the worst off (who are generally the absolutely poor). The result is that the implications of the differences between theories of global justice have often been obscured.[50]

I close by sketching two bioethical problems and showing how cosmopolitan and statist liberals are likely to develop responses to them. One concerns IP and its effects on access to medicines; the other concerns the brain drain of health care workers from developing to developed countries. For reasons of space these are no more than sketches; they are not intended as contributions to the debate about these specific issues, but to illustrate the importance of the differences between theories of global justice noted above.

Intellectual Property and Access to Medicines

Private research and development of health care products is primarily motivated by patent laws. A patent allows the inventor of a novel product or process to prevent others from making commercial use of that product or process for the lifetime of the patent (normally 20 years). Patents on medicines allow pharmaceutical companies to charge much more than the cost of production for the medicines and so recoup their substantial research and development expenditure (as well as making a tidy profit).

In 1994 the WTO established the TRIPS agreement, which will standardize IP laws across WTO states to the higher levels found in developed countries. The least developed countries have until 2016 to implement TRIPS, but its effects are already being felt. In India, the generic drug industry is now threatened by the 2005 Indian Patents Act implementing TRIPS and establishing a 20-year patent life for products and processes. Individual countries are also pressured into implementing IP laws that give even more protection than TRIPS ("TRIPS-plus" agreements), principally through bilateral free trade agreements with the United States or European Union.

The patent-based approach to incentivizing the development of new medicines has several drawbacks, which are exacerbated in the global environment. First, for those products that get developed, the prices set during the life of the patent are (in principle) determined by what will maximize revenue, not what will maximize the number of people using the product. This means that new medicines are frequently priced out of the reach of a substantial proportion of the people who would benefit from them. Second, the patent system affects the products that actually get developed. Since the central motivation of pharmaceutical companies is profit, companies tend to develop products on the basis of whether they can be sold, not their impact on health. This has at least two consequences: it encourages the development of expensive products rather than cheaper interventions that might benefit health more, and it means that the health problems for which products are developed will be the problems of people who have the money to buy them. Consequently, the current international IP system has been widely criticized for producing medicines that only the rich can afford, and that focus mainly on the problems of the rich.[51] It means that millions of people do not have access to urgently needed medicines, and the development of new interventions does not focus on the problems that are most urgent for the global poor.

Alternative ways to incentivize the development of important health interventions have been proposed, including alternative types of patent that reward inventors in proportion to the positive health effects of their interventions,[52] and rules that give greater power to governments to negotiate prices or overrule patents.[53] Arguments in favor of and against these various systems get couched in the language of social justice. I am not going to assess the merits of these proposals here: the essential point for my argument is that there are alternative ways in which a system could be set up, and therefore the system is amenable to moral assessment. Moreover, I now argue, different theories of global justice should diverge in their views about which system should be preferred.

Considering the previous arguments about research priorities and access to medicines, there should be some agreement among different theorists. Those arguments suggest that even for strong statists the international IP regime should not impede access to medicines that are required for basic health care.

But for the strong statist, when it comes to other interventions and conditions—those that apply to the development and marketing of treatments that are not part of basic health care, as I have defined it—whatever gets negotiated between governments stands, even if it is to the disadvantage of some and the advantage of others. Nagel, for example, denies that the sort of collective commercial engagement constituted by agreements overseen by the WTO is sufficient to trigger obligations of justice.[54] Instead, he argues, the agreements made may have whatever content the contracting parties agree upon: "contracts between sovereign states . . . are 'pure' contracts, and nothing guarantees the justice of their results."[55] Thus, for example, Nagel would say that so long as it does not impede the ability of British citizens to access health care, and so long as there are sufficient exceptions to allow people in developing countries to access truly basic health care, support by the UK government for TRIPS or "TRIPS-plus" IP regimes cannot be criticized on the ground of justice.

The strong statist line would be opposed by cosmopolitan liberals. This can be seen most clearly if we assume that the different parties agree on the grounds for IP protection (e.g., if we suppose that they agree that an instrumentalist justification is correct). Suppose, for example, that the strong statist and the cosmopolitan are agreed that the right system of IP over pharmaceutical products is whatever system maximizes the long-term health gains from medical treatments. The strong statist thinks that the health gains that count for making this evaluation are gains to people within a particular state, subject to the caveat just noted concerning basic health care. Thus, according to the strong statist, when the UK sets domestic IP laws or negotiates international laws, it should evaluate the health gains to people within the UK only (again subject to the caveat). The cosmopolitan, on the other hand, thinks that the gains that count for this evaluation are health gains to everyone in the world. Thus, when the UK sets domestic IP laws or negotiates international agreements, it should take into account the effects of the laws on everyone.

Different ways of setting up the international IP system will lead to different research being conducted, and so to different products coming to the market, at different prices. In other words, different IP systems will be better or worse at meeting particular sets of health research priorities. But the global health research priorities are very different than the domestic health research priorities of developed countries.[56] For example, of the estimated 241 million cases of malaria each year, 204 million take place in Africa and a mere 3 million in the Americas and Europe.[57] Thus, we can expect that the IP system for pharmaceuticals supported by a strong statist will be quite different from the system supported by a cosmopolitan.

It might be objected that the implications of the theoretical disagreement between cosmopolitans and strong statists are much less substantial than I have claimed. Suppose that the international IP system has a negative effect on people who are only relatively poor within a particular country. If it thereby precludes their government from fulfilling its *domestic* duties of justice, it can be condemned on those grounds. Thus, it might be thought, the strong statist will have to condemn the international IP system in much the same set of cases as the cosmopolitan.[58] In response, we should remind ourselves of what the requirements of domestic justice are: they demand, among other things, fairness in the distribution of resources within the state. But this says nothing about the amount of resources a state is able to distribute. Its international agreements may positively or negatively affect those resources, as may the actions of other states. But those agreements need not affect whether the resources within a state are distributed fairly between its citizens. Thus, the international IP system is unlikely to actually preclude a government from being domestically just, at least with regard to the only relatively poor, and therefore cannot be condemned on those grounds.

It is worth noting that moderate statists might also oppose the strong statist view regarding the permissible IP rules governing pharmaceutical patents. The WTO is a complex international organization, which has significant effects on people's lives and the power to enforce compliance with agreements.[59] Leaving the WTO is liable to come with substantial economic costs. For moderate statists, these characteristics may be sufficient to generate some obligations of justice, even if the WTO's requirements are not as stringent as the obligations generated by a fully fledged state. For example, moderate statists could argue on that basis that at least the economic and health gains from the international IP system should be shared fairly among the various parties affected by it.

The Health Care Worker Brain Drain

My second case concerns the brain drain: the systemic loss of skilled workers from one economic sector to another. In many developing countries, brain drains of health workers (such as doctors and nurses) are of grave concern. They take two forms: first, within-country brain drains from the public to the private sector; second, international brain drains, from poorer to richer countries. I focus here on the latter.

Huge numbers of health workers are economic migrants. For example, a 2006 WHO report estimated that 25% of the doctors and 5% of the nurses trained in Africa were currently employed in industrialized countries that are members of the Organization for Economic Cooperation and Development (OECD).[60] Frequently, these workers are actively recruited by the recipient country in order to make up for a perceived shortage of locally trained workers. The policies of recipient countries, specifically with regard to the training, recruitment, and immigration of health workers, thereby have large and predictable effects on whether health workers trained in developing countries migrate. This migration exacerbates existing inequalities in trained personnel. For example, the doctor–patient ratio in sub-Saharan Africa is now estimated at about 20 per 100,000, compared with a ratio of 220 per 100,000 in developed countries.[61]

The emigration of health workers is generally thought to negatively affect donor countries.[62] Though there is some gain through remittances home, transnational networking, and workers who eventually return better trained to their country of origin, these are more than negated by the lost investments in training and shortages of personnel. I consider the justice of policies related to health worker migration on the assumption that this empirical claim is correct.

Three key actors are involved in the health worker brain drain: the recipient country, the donor country, and the health workers themselves.[63] Each may assert claims. For example, donor countries may argue that they have the right to compensation from recipient countries or that they may require locally trained health workers to work domestically for some period after they qualify. Health workers may claim that they have both a right to the education they receive and the right to emigrate to any country that will accept them. Finally, recipient countries may assert their right to open their borders to whomever they choose.

What do the theories of international justice tell us about the duties of recipient countries, with regard to their recruiting policies, and the rights of donor countries?

First, as with the effects of IP rules, the effects of the brain drain on absolute poverty must be considered. Some minimum level of health care worker coverage is clearly necessary for everyone to have basic health care, and therefore for everyone to be brought out of absolute poverty. In countries where, for instance, the number of health care providers is so low that routine childhood vaccinations cannot be supplied, or where many people have no access to maternal health services, the lack of trained health care personnel perpetuates absolute poverty. For example, Chen et al. suggest a threshold of 2.5 health care workers per 1,000 for meeting targets of 80% coverage of measles immunizations and skilled attendants at birth.[64] WHO estimates that the average density of health workers in Africa is 2.3 per 1,000, as compared to 18.9 per 1,000 in Europe and 24.8 per 1,000 in the Americas.[65] While sub-Saharan Africa has only 3% of the world's health workers, it has 24% of the world's burden of diseases.[66] Again, theorists of all persuasions ought to agree on an international duty to prevent further loss of health care workers from absolutely impoverished areas and help them reach the minimal levels of coverage necessary to bring them out of absolute poverty. Some measures may have to be domestic, since the brain drain typically involves a loss of trained workers from rural to urban (and from public to private) environments within a country. But others can be international, since the international brain drain appears to exacerbate this loss. However, once the minimal level is met for a population, large disparities in health care workers will likely remain. The different theories of global justice will again diverge in their views about the appropriate response to those disparities.[67]

For the strong statist, above the minimal level of health care workers needed for the provision of basic health care, the additional health gains from having more trained doctors, nurses, and so forth will no longer be the concern of foreign countries and governments, even though people may have claims against their own governments to supply them.[68] Thus, strong statists are likely to see nothing wrong with hiring health workers from the developing world, so long as they come voluntarily and legally (according to immigration agreements) and do not violate any duties the hiring state has to give its citizens priority for employment.[69]

In contrast to strong statists, the cosmopolitan liberal will look for those emigration and immigration policies that would have the optimal impact on global health, independent of location. This would surely militate in favor of policies to restrict the migration of health workers from developing countries (insofar as that does not violate their human rights) or at least to heavily incentivize them to stay in the countries where they are trained. Moreover, the policy recommendations of the cosmopolitan should apply equally to both emigration and immigration: donor countries have an obligation to do what they ethically can to retain needed health care workers, and recipient countries have a duty to restrict health care workers' movement away from the places where they can do the most good. Given cosmopolitans' rejection of the normative significance of national boundaries, in an ideal world we might expect them to be most in favor of freedom of movement. Somewhat ironically, in the actual non-ideal world they should support restricting freedom of movement.

In fact, consistent cosmopolitans might be committed to even more extensive redistributive duties. As noted above, almost all of the countries in sub-Saharan Africa have massive burdens of preventable or treatable diseases, but inadequate health care infrastructure and far fewer health workers than they need. The good that any particular physician could do (with appropriate supportive funding) in sub-Saharan Africa is far greater than the good he or she could do in the UK (to take an arbitrary developed country). Now, on the cosmopolitan view, the UK should definitely be investing in training health workers in sub-Saharan Africa (as well as taking measures to reduce their emigration to the UK). But this will take time. In the short term, the greatest progress towards a just allocation of doctors would be for the UK to send substantial numbers of its doctors to countries in sub-Saharan Africa—that is, to encourage a "reverse brain drain." After all, at the moment they could do far more good there.

Naturally, such a proposal would generate all sorts of objections: it would wrong the doctors to send them abroad (but: salaries could simply be adjusted to provide appropriate incentives rather than coercing doctors, and they could be sent for limited tours of duty); it would not cure the underlying problem (but: it would naturally go alongside a radical cosmopolitan scheme of building domestic capacity); and so forth. My aim is not to defend the specifics of some such proposal. Instead, this is intended to illustrate how radical the practical implications of the cosmopolitan's theoretical commitments might be.[70]

CONCLUSIONS: THEORETICAL DISAGREEMENT AND MORAL PROGRESS

This chapter has given an overview of the current debate on theories of international justice, and shown how international justice is relevant to contemporary bioethical problems. I have argued for two specific conclusions.

First, some progress in international bioethics can be made in the face of uncertainty about the correct theory of global justice. A great deal of agreement exists on a minimum duty to provide aid to people living in conditions of absolute poverty. Such agreement can have implications for policies that can be enacted now. We should not exaggerate the differences between different views when it comes to the moral claims of people in desperate need.

Second, when it comes to the situation of the billions of people in the world who are only relatively poor, not absolutely poor, the differences between theories of justice make a difference. Consequently, to address comprehensively their distinctive problems in the international arena, bioethicists must work on issues across the spectrum of normative theory, from classical bioethics through to political theory.[71] Solutions to many of the most important problems in bioethics require simultaneously grappling with questions of international justice. The arguments just sketched concerning how cosmopolitan and statist liberals should approach IP and the health worker brain drain suggest that resolving these bioethical problems is likely to require committing to substantive claims about the correct theory of global justice.[72]

NOTES

1. For further work on this question see Sreenivasan G. (2002). International justice and health: a proposal. *Ethics & International Affairs* 16(2):81–90, and Wolff J, Chapter 4 in this volume.
2. Rawls J. (1971). *A Theory of Justice.* Cambridge, MA, and London: The Belknap Press of Harvard University Press, p. 7.

3. Cf. Beitz C. (1999). *Political Theory and International Relations.* Princeton, NJ: Princeton University Press, p. 131.
4. Pogge T. (2002). How should human rights be conceived? In Pogge T. *World Poverty and Human Rights.* Cambridge, UK: Polity Press, 52–70.
5. Rawls, 1971, pp. 8–9.
6. These correspond to two branches of non-ideal theory: *transitional theory* and *partial compliance theory.* See Sreenivasan, Chapter 6 in this volume, for an extended analysis of these categories.
7. Almost nothing has been written on the question of what other countries (i.e., those not among the rich countries of North America, Western Europe and East Asia) owe to people within and without their borders.
8. As I explain shortly, the latter set of positions includes both those who think that the existence of a state is what is morally significant and those who think that the existence of a nation (defined in virtue of shared characteristics of its members) is morally significant. I choose the term "statist" to describe both, rather than "nationalist," since the latter has more connotations that might lead us astray. Nonetheless, my use of the term "statist" must be distinguished from its use to describe political views that privilege the state over the people within it or that extol state intervention in a country's economy. I impute neither of these views to the statists I describe.
9. My way of defining cosmopolitanism therefore focuses on the conclusions that the theory of justice draws. This contrasts with views that focus on the reasons why the theory of justice draws its conclusions. So, for example, Charles Beitz defines cosmopolitanism as follows:

   At the deepest level, cosmopolitan liberalism regards the social world as composed of persons, not collectivities like societies or peoples, and insists that principles for the relations of societies should be based on a consideration of the fundamental interests of persons. (Beitz, C. (2000). Rawls's Law of Peoples. *Ethics* 110(4), p. 677)
   For alternative definitions see Pogge T. (2002). Cosmopolitanism and sovereignty. In Pogge T. *Human Poverty and Human Rights.* Cambridge, UK: Polity Press, p. 169, and Brock G. (2009). *Global Justice: A Cosmopolitan Account.* Oxford, UK: Oxford University Press, pp. 11–14.
10. The difference principle is Rawls' preferred principle for allocating primary goods. It states that social and economic inequalities are permissible only if they are to the greatest advantage of the least well-off members of society (for discussion, see Rawls, 1971, pp. 65–73).
11. Cf. Nagel T. (2005). The problem of global justice. *Philosophy and Public Affairs* 33:113–147: "I am leaving aside here the very important differences over what the universal foundation of cosmopolitan justice is. Cosmopolitans can be utilitarians, or liberal egalitarians, or even libertarian defenders of laissez faire, provided they think these moral standards of equal treatment apply in principle to our relations to all other persons, not just to our fellow citizens" (p. 119n).
12. Communitarians believe that the principles governing a just society must take into account the existing shared practices and understandings within cultures, whether by basing principles of justice on these shared practices and understandings or by taking account of them in working out the content of the requirements of justice.
13. See Goodin RE. (1995). *Utilitarianism as a Public Philosophy.* Cambridge, UK: Cambridge University Press, for an application of utilitarian thinking to social justice, including global justice.
14. Note that this view does not commit its proponents to the desirability of a global government: the types of institutions needed to treat everyone equally are determined by practical as well as normative considerations and these might favor a decentralized

global network of power centers rather than a single world government. Cf. Beitz, 1999, pp. 182–183, and Tan KC. (2000). *Toleration, Diversity, and Global Justice.* University Park, PA: Pennsylvania State University Press, pp. 100–102.

15. Compare Barry B. Humanity and justice in global perspective. In Pennock JR, Chapman JW. (1982). *Nomos XXIV: Ethics, Economics, and the Law.* New York: New York University Press, 219–252, who considers the implications of treating the world's resources as belonging to all of humanity.

16. Beitz, 1999, p. 141.

17. Beitz, 1999, p. 145. Though note that he does not require that it produce benefits to all in order for considerations of justice to apply, either in the domestic or the international case. Considerations of justice apply, according to Beitz, to all "institutions and practices (whether or not they are genuinely cooperative) in which social activity produces relative or absolute benefits or burdens that would not exist if the social activity did not take place" (p. 131).

18. Beitz, 1999, pp. 152–153. Rawls J. (1999). *The Law of Peoples with "The Idea of Public Reason Revisited."* Cambridge, MA, and London: Harvard University Press.

19. My thanks to Bob Goodin for emphasizing this point.

20. See, e.g., MacIntyre A. (1984). *After Virtue* (2nd ed.). Notre Dame, IN: Notre Dame University Press, and Walzer M. (1983). *Spheres of Justice.* New York: Basic Books.

21. See, e.g., Miller D. (1995). *On Nationality.* Oxford, UK: Oxford University Press, and Tamir Y. (1993). *Liberal Nationalism.* Princeton, NJ: Princeton University Press.

22. Cf. Barry B. (1998). Statism and nationalism: a cosmopolitan critique. In Shapiro I, Brilmayer L. *Nomos XLI: Global Justice,* pp. 20–21.

23. Nagel, 2005, p. 140.

24. Rawls, 1999, p. 37. Other strong statists include Nagel *op. cit.* and Michael Blake (e.g., Blake M. [2002]. Distributive justice, state coercion, and autonomy. *Philosophy and Public Affairs* 30(3):257–296).

25. We might expect there to be moderate nationalist positions, too. Nothing in the justifications given for nationalism entails that special duties to co-nationals must always, or even usually, trump other special or general duties, including duties owed to foreigners. Cf. Michael Blake, who writes:

    > But surely, if cultures are goods, then they may be evaluated as such; if they are morally important, they nonetheless may be outweighed in moral importance by other goods—such as the claims of non-members to, for instance, simply survive as human beings. Establishing the importance of a distinctive way of life does not establish that such a creature is more important than others' abilities to survive. (Blake M. International justice. In Zalta EN, ed. *Stanford Encyclopedia of Philosophy* (Winter 2008 edition). (URL: http://plato.stanford.edu/archives/win2008/entries/international-justice/).

26. Though see Cohen J, Sabel C. (2006). Extra rempublicam nulla justitia? *Philosophy & Public Affairs* 34(2):147–175.

27. Singer P. (1993). *Practical Ethics.* Cambridge UK: Cambridge University Press, pp. 218–220. The term comes from Robert McNamara, who describes absolute poverty as "a condition of life so characterized by malnutrition, illiteracy, disease, squalid surroundings, high infant mortality and low life expectancy as to be beneath any reasonable definition of human decency" (quoted in Singer, *Practical Ethics,* p. 219).

28. Chen S, Ravallion M. (Aug. 1, 2008). *The Developing World is Poorer than We Thought, But No Less Successful in the Fight Against Poverty.* World Bank Policy Research Working Paper Series No. 4703. (Available at SSRN: http://ssrn.com/abstract=1259575). The figure of $1.25 a day represents the amount of goods that could be purchased in the United States in 1993 for US$1.25 (rather than the amount that could be purchased in any particular country for US$1.25).

29. The World Bank's method of measuring poverty has been criticized (see, e.g., Reddy S, Pogge T. (2002). How Not to Count the Poor. Unpublished Manuscript. Available on www.socialanalysis.org, who argue that there are both conceptual and empirical flaws in the World Bank's methods) and, in any case, the concept it seeks to measure may not map exactly onto the concept of absolute poverty I am using. However, this figure does give some idea of the massive extent of absolute poverty.

30. Singer P. (2009). *The Life You Can Save: Acting Now to End World Poverty.* London: Picador.

31. Note that I am not arguing that most people believe that there is such an international duty of assistance. I am arguing that given the facts, and given the normative principles to which most people (including political theorists) are committed, they ought also to endorse an international duty of assistance.

32. Gopal Sreenivasan argues along similar lines in order to show that the OECD countries ought to allocate 1% of GDP to improving the health of the worst off. He concludes: "Any plausible and complete ideal of international distributive justice, I suggest, will at least require better-off states to transfer *one percent of their gross domestic product* (GDP) to worse-off states" (Sreenivasan, 2002, p. 83; italics in original).

33. Though libertarians could accept that there are international duties of justice, but argue that they do not require the provision of assistance to anyone, just as they do not in the domestic context.

34. Rawls, 1999, pp. 106 and 114–116. See, also, Nagel, 2005, p. 118, and Blake, 2002, pp. 258–260.

35. Miller D. (2005). Against global egalitarianism. *Journal of Ethics* 9, pp. 63–64.

36. See, e.g., Shah A. Structural adjustment—a major cause of poverty. *Global Issues*, Updated Oct. 29, 2008. (Accessed Jan. 25, 2010. http://www.globalissues.org/article/3/structural-adjustment-a-major-cause-of-poverty); and George S. (May 1993). The debt boomerang: the Third World debt crisis is doing you harm—whether you live in the North or the South. *New Internationalist*, p. 243.

37. These duties are sometimes called duties of humanity, duties of beneficence, or duties of charity.

38. See, e.g., Murphy LB. (1993). The demands of beneficence. *Philosophy and Public Affairs* 22(4):267–292.

39. Wouldn't the different justifications for a duty to aid pick out slightly different sets of very poor people and prescribe slightly different duties of aid to them? Almost certainly. But there will also be many difficulties in measuring people's needs, working out what would address them, comparing the expected impact of different policies or institutional set-ups intended to help them, and so forth. These other sources of imprecision are likely to swamp the differences between the various sets, so that it will generally be unhelpful to separate them.

40. Rawls, 1999, p. 111.

41. Chen & Ravallion, 2008.

42. Brenzel L, et al. (2006). Vaccine-preventable diseases. Chapter 20 in Jamison et al., *Disease Control Priorities in Developing Countries* (2nd ed.), Washington DC and New York: The World Bank and Oxford University Press, p. 401. This includes the cost of providing the vaccinations and the probability that a child will die from one of the infectious diseases vaccinated against if he or she is not vaccinated.

43. Mudur G. (2004). Hospitals in India woo foreign patients. *BMJ* 328:1338. The price is substantially greater in developed countries.

44. This is a stipulative definition for the purposes of this paper. The set of interventions included in basic health care need not correspond to other lists of basic health care interventions, such as WHO's list of essential medicines (see http://www.who.int/topics/essential_medicines/en/).

45. Though it may make a difference to how secure their access to basic health care is, which might then be relevant to assessing their level of poverty.
46. This is the price in the United States. The prices of operations vary widely from country to country. For example, the price of bypass surgery in Canada appears to be about half of that in the United States (see Eisenberg MJ et al. [2005]. Outcomes and cost of coronary artery bypass graft surgery in the United States and Canada. *Arch Intern Med* 165:1506–1513). Nevertheless, my point should be clear.
47. James Dwyer makes essentially the same point:
    "Basic needs" is a vague concept, of course. Sometimes we know whether something is a basic need, but sometimes we do not; and in my own view, there is no easy, context-free way to specify basic needs, especially those that concern health care. I may "need" a very expensive, rare, and technical treatment (extracorporeal oxygenation, for example) in order to live. From an individual point of view, this treatment may seem like a basic need because it is the only way to sustain life. But from a social point of view—taking into account population needs, cost-effectiveness, and opportunity costs—we may not want to count this need as basic. (Dwyer J. (2007). What's wrong with the global migration of health care professionals? Individual rights and international justice. *Hastings Center Report* 37(5), p. 39).
48. For this style of argument applied to the brain drain see, e.g., Hooper CR. (2008). Adding insult to injury: the healthcare brain drain. *Journal of Medical Ethics* 34:684–687. On intellectual property and access to medicines see Resnik D. (2001). Developing drugs for the developing world: An economic, legal, moral and political dilemma. *Developing World Bioethics* 1:11–32; and Schüklenk U, Ashcroft RE. (2001). Affordable access to essential medication in developing countries: conflicts between ethical and economic imperatives. *Journal of Medicine and Philosophy* 27(2):179–195.
49. On the brain drain see, e.g., Gostin LO. (2008). The international migration and recruitment of nurses: human rights and global justice. *JAMA* 299(15):1827–1829, who assumes a particular (and apparently cosmopolitan) view of global justice and in any case focuses his analysis on the worst off (those whose right to health has not been fulfilled). Thomas Pogge argues for what he describes as a cosmopolitan view (Pogge T. [1992]. Cosmopolitanism and sovereignty. *Ethics* 103(1):48–75) and has developed a detailed proposal for incentivizing the development of medicines on that basis (see, e.g., Pogge T. [2005]. Human rights and global health: a research program. *Metaphilosophy* 36:182–209).
50. A notable exception is Norman Daniels, who cites the two examples I discuss to illustrate the need for bioethicists to address broader questions of global justice (Daniels N. [2006]. Equity and population health: towards a broader bioethics agenda. *Hastings Center Report* 36(4):29–34).
51. Pogge, 2005, *op. cit.*
52. Pogge, 2005, *op. cit.*
53. For example, greater use of compulsory licensing for public health emergencies, whose legality is affirmed in the Doha Declaration on the TRIPS Agreement and Public Health, WT/M/DEC/W/2, Nov. 14, 2001. (Available at http://www.who.int/medicines/areas/policy/tripshealth.pdf. Accessed Oct. 8, 2009.)
54. Nagel, 2005, pp. 142–143. Though he does acknowledge that we might have a duty to restrict the content of international agreements for humanitarian reasons.
55. Nagel, 2005, p. 141.
56. For example, one important difference is that infectious diseases still account for a much higher proportion of the projected burden of disease in 2030 for low-income countries

than for high-income countries. This suggests that their current domestic research priorities should likewise differ. See Mathers CD, Loncar D. (2006). Projections of global mortality and burden of disease from 2002 to 2030. *PLoS Medicine* 3(11):e442.

57. World Health Organization. *The Global Burden of Disease: 2004 Update*. (Available at: http://www.who.int/topics/global_burden_of_disease/en/. Accessed Sept. 28, 2009.) The malaria disparity is actually much greater than this suggests, since the great majority of the American cases occur in Latin America.

58. My thanks to Bob Goodin for pressing this objection.

59. World Trade Organization. (2007). Chapter 3: Settling disputes. In: *WTO. Understanding the WTO*. Geneva: World Trade Organization, pp. 55–61.

60. World Health Organization. (2006). *World Health Report 2006—Working Together for Health*. Geneva: WHO.

61. Hamilton K, Yau J. (2004). *The Global Tug-of-War for Health Care Workers*. Washington, DC: Migration Policy Institute. (Available at: http://www.migrationinformation.org/feature/print.cfm?ID=271. Accessed Oct. 8, 2009.)

62. Kuehn BM. (2007). Global shortage of health workers, brain drain stress developing countries. *JAMA* 298:16:1853–1855; Johnson J. (2005). Stopping Africa's medical brain drain. *BMJ* 331:2–3; World Health Organization. (2006). *Working Together for Health: The World Health Report 2006*. Geneva: WHO, pp. 97–104.

63. There is also an important collective action problem when we come to consider what policies individual developed countries should follow in the face of expected continuing recruitment by other developed countries.

64. Chen L, Evans T, Anand S, Boufford JI, Brown H, Chowdhury M, et al. (2004). Human resources for health: overcoming the crisis. *Lancet* 364:1984–1990. These are not arbitrarily chosen figures but derive from the Millennium Development Goals. I take them, therefore, as a rough proxy for what might be necessary to lift a population out of absolute poverty. As before, however, nothing in my argument turns on this particular figure being correct; it must merely be the case that there is some level of health care workers needed to bring a population out of absolute poverty, that this level is not universally met, but that many populations who are relatively poor do meet it.

65. World Health Organization, 2006, p. 5.

66. World Health Organization. (April 2006). *The Global Shortage of Health Workers and its Impact, 2006*. Fact sheet No. 302. (Available at: http://www.who.int/mediacentre/factsheets/fs302/en/index.html.)

67. There may be room even for a little more consensus than this, on the grounds of compensatory justice. Daniels argues that one cause of the brain drain is the Structural Reform Programs imposed by the International Monetary Fund and the World Bank in the 1980s, which devastated public health care provision in developing countries and thereby drove many health workers out of the public sector (Daniels, 2006, p. 33). This suggests that there could be a consensus view according to which when global institutions cause the conditions for brain drains, they commit some *prima facie* wrong. But this seems at best to be an indirect argument for restrictions on immigration and targeted hiring; it would be more plausible to argue that the actions of the IMF and World Bank should be altered and/or directly compensated for.

68. Dwyer suggests that cosmopolitans and political statists would agree about medical migration, because the brain drain tends to undermine social justice in donor countries (Dwyer, 2007, pp. 40–42). But except when we are dealing with basic needs, this seems false: the donor countries are responsible for distributing the human health care resources that they have in a just way, and differences in the amount of human health care resources between countries are not the concern of justice for the strong statist.

69. Again, we might look for a moderate statist position on the brain drain. However, there is much less of a case for claiming that a global institution governing health worker migration exists than in the case of IP protection: there is no obvious analogue to the WTO here.

70. In this respect, it is interesting to note the content of the 2003 Memorandum of Understanding between the Government of the United Kingdom of Great Britain and Northern Ireland and the Government of the Republic of South Africa on the reciprocal Educational Exchange of Healthcare Concepts and Personnel, which resulted in the training of South African nurses in the UK and senior nurses from the UK working in South African hospitals as mentors (Nullis-Kapp C. Efforts under way to stem "brain drain" of doctors and nurses. *WHO Bulletin* 2005;83:2).

71. Cf. Daniels, 2006, pp. 33–34.

72. For helpful comments on previous drafts of this chapter I would like to thank Danielle Bromwich, Kirstin Borgerson, Bob Goodin, Marika Warren, and Alan Wertheimer.

CHAPTER 3

Is There a Human Right to Essential Pharmaceuticals?

The Global Common, the Intellectual Common, and the Possibility of Private Intellectual Property

MATHIAS RISSE

Every human being has a right to vital pharmaceuticals—not in the sense that he or she can claim the right to medications that are not yet available, but in the sense that access to pharmaceuticals must not be limited by overly strong private intellectual property (IP) rights. (The meaning of what counts as "overly strong" will be developed in this study.) My argument in support of such a human right draws on foundational considerations about IP. My analysis is driven by exploring parallels between a Global Common and an Intellectual Common, to both of which all of humanity would have symmetrical ownership rights.

INTRODUCTION

To the extent that this is a matter of law, one can make a case that there is a human right to vital pharmaceuticals such as those on the World Heath Organization's (WHO's) list of essential medicines. Yet while lawyers explore questions such as what such a right amounts to, and whether it conflicts with the Trade-Related Aspects of IP Rights (TRIPS) agreement, our concern is to explore whether philosophical approaches to human rights deliver a right to pharmaceuticals—specifically, whether there is such a right according to a conception I have recently offered. This approach regards human rights as membership rights in the global order, where one source for such rights is humanity's collectively owning the earth.[1]

In a nutshell: Yes, there is a human right to essential pharmaceuticals. Yet we must be careful in assessing precisely what kind of right this is. My argument does not deliver a right that anybody receive pharmaceuticals that are not yet available.[2] Instead, I show that it is owed to people across the world that IP generally and vital pharmaceuticals in particular not be regulated in a way that acknowledges far-reaching private IP rights (rights that go beyond what we need to compensate inventors and set incentives for future inventions)—especially at the global level. In that sense, there indeed is a human right to pharmaceuticals.

My conclusions follow from reflections on the possibility of private IP, rather than the more customary appeal to foundational rights to welfare or health (care). Therefore much of what I argue should be useful even to those who reject my conception of human rights.[3]

The question of what support there is for a right to vital pharmaceuticals arises separately for each conception of human rights. Addressing it within the confines of my conception should be of particular interest because considerations of collective ownership play an important role in both that conception and the debate about the possibility of private IP. While reflections about a Global Common are at the core of my conception, reflections about an Intellectual Common inform debates about IP. An exploration of how these approaches interact underlies the subsequent discussion in this chapter.

Since I have to introduce my conception of human rights in this study and bring it to bear on our question about essential pharmaceuticals, my argument will be complex, and will bring together apparently rather disparate themes.[4] So it should be useful to provide an overview of what is to come. I proceed in three steps. First I establish the link between human rights and collective ownership. Individuals possess a set of natural rights that characterize their status as co-owners. The existence of states puts these rights in jeopardy, and a set of associative rights must ensure that states preserve these natural rights.[5] These associative rights provide us with a conception of human rights, such rights being membership rights in the global order. The standpoint of collective ownership serves as one source of these rights.

Although this view may strike many as missing the point of human rights, it has its virtues: it rests on foundations that should be universally acceptable; it can show why the language of "rights" rather than goals is aptly used in the context of human *rights*; and it entails a global responsibility for these rights. That humanity collectively owns the earth was a predominant idea in 17th-century political philosophy: Grotius, Pufendorf, Locke, and others debated how to capture this status and the conditions under which parts of the Global Common could be privatized.[6] Although their views were religiously based, we can revitalize this concept non-theologically. Such revitalization is sensible in light of the problems of global reach that have recently preoccupied us and that can be addressed from the standpoint of collective ownership of the earth.

One area where an idea of collective ownership has had an impact is IP, which leads to the second step of my argument. Just as there is a Global Common, so ideas may form an Intellectual Common. Those who "have" ideas are then not inventors or creators, but explorers or discoverers. Whatever claims there may be to control the use of ideas would have to be assessed in light of the principle that such ideas originally belong to a Common. This approach draws on Locke's discussion in Chapter V of the *Second Treatise*, a discussion so influential in theorizing property that Drahos assigns it "totemic status."[7] Locke merges his account of collective ownership with a labor-based ("mixing") approach to privatization. Since many commentators have thought Locke's ideas transfer readily to IP, those ideas have inspired a particular approach to IP in terms of an Intellectual Common. Rather than Locke, however, my effort to revitalize the standpoint of collective ownership draws on the work of Hugo Grotius.

Like no other work in the philosophy of international relations, Grotius' *De Jure Belli ac Pacis Libri Tres* (DJB), *Three Books on the Law of War and Peace*, published in 1625, makes ownership of the earth central to the relations among both individuals and political entities. Grotius' concern is with the "differences of those who do not acknowledge one common Civil Right whereby [these differences] may and ought to be decided" (I.1.I), differences Grotius seeks to adjudicate non-parochially. By making collective ownership central, Grotius formulates a version of what one may call global public reason. I develop the first step of my argument from a Grotian standpoint, although it matters little whether we choose a Grotian or, say, a Lockean starting point. I take a Grotian starting point for the second step too, but then the Grotian origins do matter.

Grotius is best known for his views on the freedom of the seas. Throughout his writings, he argues in different ways that the seas cannot be owned. Assessing his arguments leads to reflections about the conditions under which anything that is originally collectively owned can be privatized. While Grotius' reasons no longer fit the case of the seas, they do fit the case of IP, as long as we can classify the objects of IP as belonging to an Intellectual Common. What Grotius says about the sea helps make a case against private IP rights beyond what is needed to compensate inventors or perhaps set incentives for future inventions. Strikingly, a similar argument is available if there is no Intellectual Common. What emerges, and what the second step of my argument mainly provides, is a general case against the possibility of private IP beyond compensation and incentive setting wherever IP is regulated—a case that in no way depends on whether there in fact is an Intellectual Common.[8]

The third step connects the first two. In the first step we saw that collective ownership was one source from which to derive human rights *qua* membership rights in the global order. But collective ownership is not the only source from which we can derive global membership rights. Other such sources include global interconnectedness, enlightened self-interest, and additional moral reasons, such as appeals to human needs, reasons that must be acknowledged at the global level and that are tied to global obligations. Crucially, any argument seeking to show that X is a human right in this sense must show that the matter is indeed of genuinely *global* concern and is appropriately captured as a right.[9]

The second step itself requires that wherever IP is regulated, private IP is acceptable only within the limits sketched. This result does not depend on my conception of human rights. Yet putting the first two steps together, we also find that there is a human right to vital pharmaceuticals, in the sense that such pharmaceuticals ought to be regulated at the global level and that *private* rights to control access to pharmaceuticals should be constrained as before. To make this point within the confines of my conception, we need to establish that the regulation of a sub-domain of IP—one that includes pharmaceuticals—is of genuinely global concern. To that end, I argue that the second step lets us conclude that the constraints on the regulation of that sub-domain offer the sort of moral reason that is of genuinely global concern. This reasoning then delivers the conclusion that, indeed, there is a human right to essential pharmaceuticals.[10]

Drawing on Grotius rather than Locke for a foundational discussion of IP has considerable consequences that are worth exploring independently of whether one wishes to adopt my conception of human rights. Locke introduced an account of collective ownership of the earth, and then explained how things can be legitimately privatized. He did not take much interest (if any) in the possibility that loomed so large in Grotius, namely that there was good reason for co-owners not to accept privatization of parts of the collectively owned earth (especially the seas). Yet the realm of ideas is analogous to precisely *those* parts of the earth for which there is good reason not to subject them to appropriation at the exclusion of others. Since Locke did not pay attention to the idea that parts of the earth should reasonably be left unappropriated, this analogy cannot be drawn in Locke (or in any event, he does not himself create room for it). So we can expect a Grotian approach to deliver more restrictive private IP rights than a Lockean approach.

COLLECTIVE OWNERSHIP

To introduce the concept of collective ownership, let me touch on a few themes from Grotius. Grotius offers this account of collective ownership of the earth:[11]

> Almighty God at the creation, and again after the Deluge, gave to Mankind in general a Dominion over Things of this inferior World. All Things, as Justin has it, were at first common, and all the World had, as it were, but one Patrimony. From hence it was, that every Man converted what he would to his own Use, and consumed whatever was to be consumed; and such a Use of the Right common to all Men did at that time supply the Place of Property, for no Man could justly take from another, what he had thus first taken to himself; which is well illustrated by that Simile of Cicero, Tho' the Theatre is common for any Body that comes, yet the Place that every one sits in is properly his own. And this State of Things must have continued till now, had Men persisted in their primitive Simplicity, or lived together in perfect Friendship. (DJB, II.2.II.1)

God's gift can rightfully be put to use without any agreement. But this only works under primitive conditions, and does not even include a right to recover things left behind. Agreement is needed to create further-reaching rights, at least according to the account in *De Jure Belli ac Pacis*. Still, God's gift makes clear that the earth is *for* the use of human beings. As Buckle puts it, "in using the world for their own ends, human beings are not strangers (or trespassers) on a foreign soil. They are at home."[12]

Once primitive conditions have been left behind, property arrangements are conventional. To be adequate, these conventions must recognize the fact that the earth was originally given to humankind collectively. One implication of this point is the postulation of a "right of necessity;" for, as Grotius states:

> in a case of Absolute Necessity, that antient Right of using Things, as if they still remained common, must revive, and be in full Force: For in all Laws of human Institution, and consequently, in that of Property too, such cases seem to be excepted. (DJB, II.2.VI.2)

This right does not derive from charity (II.2.VI.4). Instead, it restricts private property rights as they could have been intended, or in any event, their legitimate scope. After all, in addition to his account of the divine gift, Grotius also offers an account of natural rights that include "the Abstaining from that which is another's, and the Restitution of what we have of another's, or of the Profit we have made by it, the Obligation of fulfilling Promises, the Reparation of a Damage done through our own Default, and the Merit of Punishment among Men." Society was formed for the protection of what a person can properly regard as his own, in Latin, the *suum* (DJB, I.2.I.5), and a sphere of what is properly ours exists prior to actual property arrangements. Whereas Hobbes thought the most basic insight one could make uncontroversially was that everybody had a right to self-preservation, Grotius started with a number of laws of nature which spell out what individuals have a right to in ways meant *to be reasonable for everybody*. Grotius is guided by certain ideas of human solidarity, and an understanding of humanity as susceptible, in principle, to moral motivation.

Some limitations to property are not rights of necessity but general restrictions of what may be claimed under any conditions. Other people may avail themselves of innocent profits (e.g., sail on our rivers), or demand free passage (even when trading with third parties, II.2.XI–XIII), rights that if denied can be claimed by force (II.2.XIII.3). People may rest ashore to recover from a journey, even build "a little Cottage" (II.2.XV.2), and seek "a fixed Abode" (II.2.XVI.2) if they are persecuted at home, assuming they abide by local laws. Products must be sold at reasonable prices if they are not needed by the producers (II.2.XIX). Even the right to marriage ought not to be denied, women apparently being part of the common stock (II.2.XXI). All these rights are owed to all, not just to a selected few (II.2.XXII). What these cases make clear is that the collective ownership status of earth, in conjunction with the additional natural rights Grotius postulates, puts considerable limitations on the possibility of privatization.

EGALITARIAN OWNERSHIP

While Grotius took the biblical concept of the earth as a divine gift, like Locke he held that this view should be acceptable even if humankind had never received that revelation. Indeed, the view that the earth originally belongs to humankind collectively is plausible without religious input. Philosophically, we have much to gain by developing the idea that humanity collectively owns the earth, since this status affects what people can do with portions of the planet. Among other things, this idea generates constraints on what immigration policies to adopt,[13] and leads to a conception of human rights. Two points are obvious enough: first, the resources of the earth are valuable and necessary for any human activities to unfold; and second, those resources have come into existence without any human effort. These points must be considered when individual accomplishments are used to justify property rights strong enough to determine use across generations.[14]

Egalitarian Ownership is the view that the earth originally belongs to humankind collectively: all humans, no matter when and where they are born, must have *some* sort of symmetrical claim to it. ("Original" ownership does not connote time but is a moral status.) This is the most plausible view of original ownership, because of the two points above: that the existence of resources is nobody's accomplishment, whereas those resources are needed for any human activities to unfold. Egalitarian Ownership is detached from the complex set of rights and duties civil law delineates under the heading of property law.[15] At this level of abstraction from conventions and codes that themselves have to be assessed in relation to views on original ownership, all Egalitarian Ownership states is that all humans have a symmetrical claim to original resources. One may say that the term "ownership" is misleading here, but I use it because of its connection to the familiar, thicker notions of ownership in civil law; and we are, after all, concerned with what sorts of claims individuals have to resources. To be sure, the considerations motivating Egalitarian Ownership speak to raw materials only, not to what human beings have *made* of those materials. The distinction between what "is just there" and what has been shaped by humans is blurred, say, for land that human beings have wrested from the sea, or for natural gas harnessed from garbage deposits. But by and large, we understand well enough the idea of what exists without human interference.[16]

We must now assess different *conceptions* of Egalitarian Ownership. Such conceptions differ in how they treat the symmetry of claims individuals have to original resources. There are roughly four types of ownership status. An entity may have: *no* ownership; *joint* ownership—ownership directed by collective preferences; *common* ownership—in which the entity belongs to several individuals, each equally entitled to using it within constraints; and *private* ownership. Common ownership is a right to use something that does not exclude others from also using it. If the Boston Common were held in *common* ownership when it was used for cattle, each person's use could be constrained by a limit on the number of cattle that each could bring—a condition motivated by respect for co-owners and concern for avoiding avoid the Tragedy of the Commons.

Yet if the Common were held in *joint* ownership, each individual use would be subject to the approval of each co-owner. Joint ownership ascribes to each co-owner property rights as extensive as rights of private ownership, except that others hold the same rights: each co-owner must be satisfied on each form of use.

So there are various interpretations of Egalitarian Ownership: resources can be jointly owned or commonly owned, or each person can have private ownership of an equal share of resources (or a value equivalent). These conceptions carve out a pre-institutional space of natural rights that constrain property conventions which in turn regulate what natural rights leave open. I submit that Common Ownership is the most plausible conception of Egalitarian Ownership.[17] While I cannot offer a complete argument for this proposal here, I offer elaboration on what common ownership means, what it entails, and why it should be preferred to the other conceptions as an interpretation of Egalitarian Ownership.[18]

The core idea of common ownership is that all co-owners ought to have an equal opportunity to satisfy their needs to the extent that this turns on obtaining collectively owned resources. This formulation, first, emphasizes an equality of status. Second, it points out that this equality of status concerns opportunities to satisfy needs—although there is no sense in which each co-owner would be entitled to an equal share of what is collectively owned, let alone to the support of others in getting such a share. Third, it is concerned with those needs that can be satisfied with resources that are collectively owned (that is, nothing at all is said about anything to which the original intuitions motivating Egalitarian Ownership do not apply).

To put this in Hohfeldian rights terminology, common ownership rights must minimally include liberty rights accompanied by what Hart calls a "protective perimeter" of claim rights.[19,20] To have a liberty right to X is to be free of any duty not to X. Common ownership rights must include at least rights of that sort; that is, co-owners are under no duty to refrain from using any resources. But the symmetry of claims postulated by Egalitarian Ownership demands more than liberty rights. In light of the intuitions supporting Egalitarian Ownership, to count as an interpretation of it, Common Ownership must guarantee minimal access to resources—that is, it must impose duties to refrain from interference with certain forms of use of resources. Therefore, we must add that protective perimeter of claim rights to the liberty rights. We obtain enough mileage from the original intuitions to require that common ownership rights (for Common Ownership to serve as an interpretation of Egalitarian Ownership) be conceived of in sufficientarian terms, in the sense that no co-owner should interfere with actions of others if they serve to satisfy basic needs. These intuitions cannot be pressed beyond that, in a way in which other conceptions of Egalitarian Ownership do (namely, those according to which each person holds an equal share of the collective owned resources, or according to which each act of use requires approval by all co-owners).

Yet we do have to add one more right. We must also make sure individuals can maintain their co-ownership status under more complex arrangements. A necessary condition for the acceptability of such arrangements is that the core purpose of the original rights can still be met. That core purpose is to make sure co-owners have the opportunity to meet their basic needs. In Hohfeldian terminology, co-owners have an *immunity* from living under political and economic arrangements that interfere with the ability of those subject to them having such opportunities.

COMMON OWNERSHIP AND THE STATE

Although humanity owns the earth collectively, and although the high seas and Antarctica are treated as a Global Common,[21] the remaining land is covered by states. The imposition of a system of states that divide the world's resources needs to be reconciled with Common Ownership, on two grounds. First, each state imposes a complex system of political and economic relationships that determines which, if any, original resources individuals have access to. Second, a system of states imposes a system of ownership in which groups may claim group-specific, collective ownership of certain regions. Thus co-owners are excluded from exercising rights with regard to much of what is collectively owned. So regardless of its moral virtues or prudential advantages, a state system generates two problems for co-owners: it exposes them to the *ex ante* risks and *ex post* reality of finding themselves in conditions where their moral status as co-owners can be exercised at most in rudimentary ways if at all; and it allows them only limited exit options (if any) if they find themselves with an abusive government. By virtue of the concentrations of power that it represents, a state system has the power to violate the rights of co-owners both by undermining their opportunities to satisfy their basic needs and by impeding their ability to relocate.

It is under these conditions that we must ask what to make of the immunity that individuals have from living under political and economic arrangements that interfere with their opportunities to satisfy their basic needs. The relevant arrangement to which individuals are subject in this case is not merely the state in which they live, but the system of states *per se.* Each state, by virtue of its immediate access to individuals' bodies and assets, might deprive them of such opportunities. But so, crucially, might other states, by refusing them entry if they cannot satisfy their basic needs where they live.[22]

Common ownership rights are natural and pre-institutional. Once institutions are founded, guarantees must be given to co-owners that institutional power will not be used to violate their status. Since such a violation is threatened by the system of states *per se,* such guarantees assume the form of moral demands against that system of states. Responsibilities that arise in this manner must be allocated at the level of the state system, as collective responsibilities, rather than resting exclusively with individual states and then only with regard to their members. The rights thus derived guarantee that individuals' status as equal co-owners, be preserved, regardless of what particular property arrangements hold across the globe. I explain below why it makes sense to think of these rights as *membership rights in the global order* and why such rights in turn offer a plausible (alas non-standard) conception of *human rights.* But before doing so, let me return to Grotius and use his account to introduce the subject of IP.[23]

PROPERTY AND THE LAW OF THE SEA

One question that naturally arises from Grotius' account of collective ownership (*mutatis mutandis* for all such accounts) is whether *all* of the divine gift can be occupied, and thus put to a kind of use that excludes others. Grotius famously responded negatively, arguing that the *sea* could not and should not be so occupied. His reasoning is of interest to us since it bears on questions about IP. Grotius' work contains two accounts of private property, and so two ways of generating the question of whether *everything* can be appropriated. According to *De Jure Belli,* the original common property is divided ever more, in response to changing socioeconomic arrangements (II.2.II.5). People realize that adjustments are necessary, make agreements to that effect, or accept them tacitly. First occupancy decides who gets to privatize what.[24] At the beginning of DJB II.2.III, following his views about privatization, Grotius explains that the sea is excluded from privatization because it is big enough for everybody's use. This emphasis relates to his point in II.2.II, that the arrangement of common use served *the same purpose* as the subsequent introduction of private property. For the sea, no new property regime was *needed* to ensure that arrangements under different conditions serve the original purpose.

The earlier *Mare Liberum* (ML) also explains the further development of the *suum*—that is, what every person can properly regard as his or her own. At the earliest stage, again, there merely exists a right to use, and hence the *suum* is accordingly limited. But *Mare Liberum* does not turn on agreements to explain what happens next and thus how the *suum* is extended. Grotius distinguishes between two stages of private acquisition (ML, pp. 22f). First, acts of use create special relationships between things and certain individuals. Sometimes *use* amounts to consumption, thus to *abuse*: the apple I eat is no longer left for others to use similarly. Other things are made worse by being used. A form of private ownership is then inseparable from use. At the second stage, Grotius explains that something similar also occurs in other cases. The passage speaks of "a certain reason" (the Latin word being *ratio*). The value of assigning objects to specific people is realized: the *ratio* for such assignments was that occupation often changes objects of use (think of cultivating land). Instead of compacts *modifying* common use, private ownership arises through *natural extensions* of use.

Grotius next assesses the limitation of appropriation, especially regarding the seas (p. 24). The centrality of occupation as a basis of private ownership becomes clear again.[25] One reason why Grotius rejects the idea that some people can lay claims to the seas here is that the seas *cannot* be occupied. The mechanism that explains which individual would be the owner at the exclusion of others does not apply to the seas. Even if occupation were possible, it would be wickedness because the gains for occupiers would not depend on excluding others. There appears to be a tension between *De Jure Belli* and *Mare Liberum,* as one of them but not the other gives an important role to convention. Yet the point in both is that the earth belongs to everybody, but that it is left to the will of humans to develop this gift. Precisely how particular arrangements come about is inessential, as long as the changes continue to make sense of the original situation of equity, and the changes are reasonable adjustments to new circumstances given Grotius' starting points.[26]

Behind Grotius' reasoning, we can construct a conservative principle of occupation: unless there is a good reason to exclude people from parts of the earth, they should not be excluded. Although consistent with occupation at the exclusion of others under certain conditions, collective ownership also imposes obstacles to it. Collective ownership creates a reference point from which departures must be justified. (Recall the Grotian right of necessity and the inherent limitations to privatization.) The founding of political communities is a good reason for exclusion, and thus one way to meet the presumption against privatization: Grotius takes no particular issue with the existence of states. Still, the burden of proof is on those who wish to legitimize occupation at the exclusion of others. As Grotius saw it, that burden could not be met for the sea.[27]

FREEDOM OF THE SEA AND PRODUCTS OF THE MIND

Nowadays the sea can be monitored by air and water, so differences between the ability to occupy land and water are a matter of degree. Nor does it still hold that use by any one party leaves intact what others could do with the sea: that much is true for ships traveling through, but not for fishing and seabed exploitation. Writing in the late 19th century, Henry Sidgwick realized that Grotius' argument had expired with regard to fisheries,[28] and concern about over-fishing has only increased since then. Complete freedom of the seas would no longer be called for on Grotius' own terms.

Yet Grotius' reasoning also bears on a rather different domain: the *products of the mind*, such as scientific, musical, literary, or other artistic works and inventions, and also images, names, symbols, or design patterns. These products are subject to IP law, which, among other things, includes patents, copyrights, and trademarks. Transferring Grotius' reflections on the sea to the domain of IP, as I will now argue, entails restrictions on private rights in that domain. That is, IP law should compensate inventors and may set incentives, but should acknowledge no further benefits for inventors. Although I continue to talk about *Grotius*, these arguments are also available in the secularization of his account.

Let us explore how Grotius' arguments against private appropriation of the sea carry over to this case. To begin with, just as Grotius points out that use of the sea is consistent with everybody else's use of it, Thomas Jefferson classically makes this point about IP:

> If nature has made any one thing less susceptible than others of exclusive property, it is the action of the thinking power called an idea. . . . Its peculiar character . . . is that no one possesses it the less, because every other possesses the whole of it. That ideas should be freely spread from one to another over the globe, for the moral and mutual instruction of man, and improvement of his condition, seems to have been . . . designed by nature. . . . Society may give an exclusive right to the profits arising from them, as an encouragement . . . to pursue ideas which may produce utility, but this may or may not be done, according to the will and convenience of the society, without claim or complaints by anybody.[29]

There is a point to having private property in things like apples since only one person can make certain kinds of use of them. As Grotius insists with regard to the sea and Jefferson with regard to ideas, however, there is no such point in having private property rights in either of these. Crucially, certainly in the case of ideas, gains for occupiers do not depend on excluding others, if we talk about the actual use of ideas, rather than profits accrued *from the exclusion*.[30]

Second, *Mare Liberum* also argues for the freedom of the seas by appeal to its relevance for trade and so establishes that *everybody benefits* from leaving the seas free:

> For even that ocean wherewith God hath compassed the earth is navigable on every side round about, and the settled or extraordinary blasts of wind, not always blowing from the same quarter, and sometimes from every quarter, do they not sufficiently signify that nature hath granted a passage from all nations unto all? (ML, p. 10; see also pp. 49, 51)

Similarly, not only does use of ideas by some not subtract anything from their usefulness for others, it actually adds to it. Such use of ideas increases the overall amount of intellectual activity, which then in turn inspires yet more such activities and thereby increases the availability of whatever benefits such activities may have. *Everybody* benefits from a situation in which ideas are left unappropriated (given, in particular, that anybody's use of them does not interfere with everybody else's use). But only a few will benefit if the appropriation of ideas is protected by social and legal norms. Of course, were we to change IP arrangements *now*, some would indeed be made worse off by such changes—namely, those who so far had been allowed to appropriate ideas—so not everybody would benefit from these changes. Yet what I have argued does hold from an *ex ante* standpoint in which no IP arrangements have been made yet and from which we must assess what sort of private rights to IP (if any) there should be.

Let us proceed to the third point Grotius makes about the seas. The seas *cannot be occupied*, says Grotius. In certain, straightforward ways of understanding what it is to occupy something, that is true of ideas. One can keep ideas secret, or distract people from them, but one cannot do anything to an idea that keeps it from being independently grasped by others. One cannot do anything to an idea parallel to how, in the case of land, "the beginning of Possession is joining Body to Body" (DJB, II.8.VI). A body A's being joined to a body B decreases the space for a body C to be joined to B. Such joining might either affect the object itself in a way that would make it impossible for others to join the object in the same way, or else such joining would create a situation where others could join the object in the same way only by violating more basic moral rights of the person who did the original joining (say, because she needs to be pushed away). But a mind's grasping an idea decreases no other mind's capacity to do the same. A mind's grasping an idea does not affect the idea itself in any way that would make it impossible for others to grasp that same idea, nor does a mind's grasping an idea create a situation where others could do the same only if they violated more basic moral rights of the original grasper.

One might object that one can indeed "occupy" ideas in the sense that there could be (and in fact are) norms of intellectual ownership, such as patent law and copyright law. Alas, "occupation" of ideas is possible *only* through the acceptance of such norms—norms that require all people other than the holder of the respective private property right to renounce the option of making use of ideas, although their making such use of ideas could occur consistently and simultaneously with everybody else's making that same use of ideas. This observation then raises the question of why anybody ought to accept such norms, a question that, in turn, takes us back to the other two considerations against privatization we already discussed.

Yet, crucially, these parallels do not constrain private IP rights in quite the way in which Grotius' reflections restricted ownership of the seas, at least not without additional argument. In the case of the seas, I first argued that the earth was collectively owned, which created a presumption against privatization that must be overcome. In a second step I argued that, for the seas, this presumption cannot be overcome: no new property arrangements are needed, sensible, or otherwise acceptable for the seas. For IP, certain considerations *would* support limitations on privatization *were* there a presumption against privatization. A ready way of arguing for such a presumption is to show that there is an Intellectual Common in the same way in which there is a Global Common. A straightforward way of arguing that, in turn, is to defend a kind of realism about intellectual products.

Such realism denies that scientific, musical, literary, or other artistic works are literally "products" of the mind. Instead, they exist outside the realm of either material or mental objects. They belong to a (Fregean or Popperian) "third realm" of non-mental super-sensible entities, distinct from both the sensible external world and the internal world of consciousness. Alleged "products" of the mind would be such products only in the sense that a conscious mind can discover them. There would be no invention, no refinement, or any other *contribution* to these entities.[31] This view delivers a presumption against privatizing elements of this third realm, for objects in that realm exist prior to any human activities. In a second step we could add the considerations against privatization we extracted from Grotius' discussion of the sea, to show that this presumption is hard to overcome.

To be sure, this presumption *can* be overcome. First, individuals may fairly claim *compensation* for investments in making ideas accessible, compensation that might take into account the particular opportunity costs of the relevant individuals. Second, consistent with this argument for limitations on private IP rights is for societies to *set incentives* to stimulate creativity. Yet in a next step I argue that this presumption excludes, or anyway offers heavy resistance to, considerations supportive of benefits or private IP rights for inventors for reasons other than compensation and incentive-setting.[32] I used a guarded formulation ("excludes, or anyway offers heavy resistance") because it is hard to establish conclusively that only fairness-based compensation and a society's right to set incentives can overcome the presumption against private IP rights according to the view we are at present assuming. Perhaps there are considerations I am unaware of, and disagreement may remain with regard to the relative strength of different considerations, more than we can sort out now. Yet in the next section we will be able to establish this conclusion with regard to considerations commonly entertained in the literature.

Acknowledging compensation and incentive-setting as reasons for creating private IP rights, we still leave open much potential for disagreement about how far-reaching are the rights these considerations create— a point that we will not further pursue here, but need to acknowledge. Notice the following articulation of this point by Judge Frank Easterbrook:

> A patent gives the inventor the right to exclude competition for 20 years, and thus to collect an enhanced price for that period. Is 20 years too long, too short, or just right? No one knows. A copyright lasts the life of the author plus an additional period that Congress keeps increasing in response to producers' lobbying. What is the right length of a copyright? No one knows. A trademark lasts forever (or at least for as long as the product is made, and the name does not become generic in the public's mind). A trade secret (such as the formula for Coca-Cola, or the source code of a computer program) lasts as long as the developer can keep the secret. Are these durations optimal? No one knows. How much use, and by whom, should be permitted without compensation under the fair use doctrine? No one knows.[33]

ARGUMENTS FOR IP RIGHTS

Let us look at some arguments for more extensive rights to private property than can be based only on reasonable compensation and a society's right to set incentives, assuming an Intellectual Common. To begin with, there is the argument that protecting inventions does not make anybody worse off. Those inventions, as argued for instance by Robert Nozick, would not exist without the inventor.[34] Waldron replies that one may well be made worse off by inventions.[35] Suppose I am dying of a disease for which there is no cure. Suppose somebody finds a cure, but it is inaccessible to me. Then I am not merely dying, but I am dying knowing that I could be cured. Yet while Waldron's reply goes a long way towards answering Nozick, the issue is moot if there is an Intellectual Common. Consider the following excerpt from a 1907 textbook that Waldron quotes to illustrate Nozick's view (a view that captures the opposite attitude from Jefferson's above):

> It is as though in some magical way he [a patent holder] had caused springs of water to flow in the desert or loam to cover barren mountains or fertile islands to rise from the bottom of the sea. His gains consist in something which no one loses, even while he enjoys them.[36]

If indeed there is an Intellectual Common, the person who made the water flow hit on something with regard to which there exists a presumption against privatization. He should be compensated, but he cannot demand rewards based on the (now *irrelevant*) fact that nobody is worse off.

Yet this rebuttal may seem implausible, and Nozick's argument plausible, when achievements are made that seem utterly disconnected from the societal state of knowledge and ability. To make this point, Becker (1993) refers to the Borges story "Pierre Menard, Author of the Quixote," a story about someone who devotes his life to rewriting (reinventing) *Don Quixote* from scratch. Menard seeks to mimic Cervantes' mindset at the time of writing his masterpiece and to reproduce it, not from memory but indeed *from scratch*. The reason why this is absurd—in ways in which it is not absurd that Scott and Amundsen simultaneously raced towards the South Pole, or that Newton and Leibniz invented the basic ideas of calculus at roughly the same time—is that Cervantes' artistic achievements seem so essentially tied to the functioning of *his* mind that even somebody who knows precisely what he knew would write a different novel.[37]

To defend the view that, still, there should be no private rights beyond compensation and incentive-setting, one may insist that *anybody* who makes a discovery benefits from the labors of predecessors, no matter how big a leap to the invention. One may also say the usefulness of, or appreciation for, the invention is determined by a social context.[38] But the main reply to the point raised by the Borges story continues to be that if indeed there is an Intellectual Common, there will be more or less demanding *discoveries*, but there will simply be no *inventions*. This point is not rebutted by the ludicrous nature of efforts to recreate *Quixote*, any more than we could rebut the idea that Edmund Hillary and Tenzing Norgay reached the summit of Mount Everest first and in that sense *discovered* it, because nobody else will be able to do so again. (Were this the case: suppose a disease permanently damaged the physical potential of human beings the day after their success.) If there indeed is an Intellectual Common as sketched, some discoverers may be held in great awe and their abilities might be so immense that, due to their high opportunity costs, they could ask for fairly high compensation—but they could make no inventors' claims and thereby could not press for private property rights beyond what we have already granted them.

Similar considerations apply to Child's (1990) argument that there are infinitely many ideas: no inventor ever makes the stock of ideas smaller.[39] On that view, inventors should have rights beyond compensation and incentive-setting since they do no harm "removing" ideas from that pool. They bring something into our world in a way that makes nobody worse but some better off.[40] But again, these matters are moot if there is an Intellectual Common of the sort we are currently presupposing. The question of whether "removing" ideas from the common stock would diminish the size of that stock and thereby make anybody worse off would then have no bearing on our inquiry. To bestow additional plausibility on that contention, recall Waldron's appeal to think about property rights from the standpoint of those who are supposed to comply with them.[41] If there is an Intellectual Common, we need not appeal to social value to make it reasonable to resist compliance with private rights (rights to rewards, that is, that go beyond what is covered by compensation and incentive-setting). The metaphysical status of ideas renders an expectation of such compliance unreasonable.

CONSTRAINING IP RIGHTS

We needed strong assumptions to endorse the presumption against privatization that was central to these arguments. Realism about abstract objects, although controversial, is not outlandish. Yet we have not merely assumed that basic ideas, foundational themes, literary motives, or basic plots are elements of the realm of non-mental supersensory objects, but that the objects of patents and copyrights themselves—that is, finished scientific inventions, completed copyrighted poems, particular drawings, etc.—are such elements. We have made an assumption of real-ism about entities that bear a producer's distinct touch. Much patent law has been concerned "with the meaning and the characteristics of inven-tiveness and creativity, seeking to identify the locus of true innovation,"[42] and this strong realism makes these efforts look peculiar.[43] A weaker form would place only basic components into the third realm, but this would also weaken the presumption against privatization. In any event, to arrive at a plausible position, we must not base our argument on such strong realism about intellectual products. I will now argue for something much stronger than the conclusion that private IP rights are limited under the assumption of strong realism about intellectual products—namely, that we should restrict private IP rights to compensation and incentive-setting *regardless* of whether we endorse such strong realism.

Consider a characterization of intellectual products that overempha-sizes the subjective aspect, mirroring the way our earlier characterization overemphasized the objective aspect. According to this characterization, intellectual products are not discovered, but invented and created. There is no Fregean or Popperian third realm, no Intellectual Common, no pre-sumption against privatization.[44] We cannot even state that we instead have a presumption in favor of *privatization* because there is no starting point with regard to which anything could be *privatized*. But we now have a presumption in favor of private property rights, potentially much beyond what compensation and incentive-setting license.

But crucially, and perhaps somewhat surprisingly, what we above identified as the three Grotian considerations against privatization reenter. These considerations were: that ideas cannot actually be occu-pied in the same sense in which, say, land can be occupied; that the gains for users of ideas do not depend on excluding others; and that leaving ideas unappropriated benefits everybody. Above, these considerations ensured that the presumption against privatization could generally not be overcome. (The exceptions were fairness-based compensation and consequentialist considerations in favor of incentives for invention.) Now the considerations against privatization reenter by limiting the *extent* of rights for which in this case there is a presumption. These consid-erations again ensure that we consider the standpoint of those expected to comply with IP law. Both above and here, these considerations require that we should limit private property rights to what we can obtain via appeals to fairness and incentive-setting, although they enter in rather different ways.

So far we have operated with caricature views on the ontology of the objects of IP. The realist account unduly eliminates the contribution of human creativity, whereas the anti-realist account overstates the role of individual minds. This suggests an intermediate view, which according to Shiffrin would:

> locate only the subject matter and materials of intellectual products in the commons, for example, facts, concepts, ideas, propositions, literary themes, musical themes, and values. Authors discover these things and their interconnections. They make them publicly accessible by expressing them, often, in unique ways.[45,46]

As Shiffrin also remarks, the proper characterization of the metaphysical nature of the objects of IP law may not be in terms of such a view "in between" the extreme views. Instead, the proper view might be a domain-specific hybrid that holds that the appropriate characterization depends on the sort of intellectual product we are talking about. In any case, the argument for limiting private rights to compensation and incentive-setting applies across the board. As we have seen, the same results follow for IP regulation regardless of whether we have a third realm of ideas or ideas are human creations. So we can state the main result of this discussion of the possibility of private IP rights as follows: The ontological status of particular intellectual products will have to be characterized *to some extent* in terms of components readily placed into a third realm, and *to some extent* in terms of human creativity. (One of these extents may be vanishing.) So *to the extent* that we must appeal to something in that third realm, the considerations used for that case apply; *to the extent* that we are talking about products of the human mind, the considerations given in that case apply. Either way the arguments generate the same constraints on private rights. Therefore, these constraints apply to the whole range of IP.[47]

By way of comparison, consider Shiffrin's strategy.[48] Shiffrin argues that we can dispense with an appeal to a third realm of intellectual products (with regard to which all human beings would presumably be related symmetrically) to support the idea of an Intellectual Common. Instead, she postulates an Intellectual Common even on the very subjective view of the ontology of intellectual products: "Creations could become part of the common—available equally to all—when their nature did not require exclusive use, to symbolize the equal moral status of individuals." Shiffrin uses what I call the considerations against privatization to postulate an Intellectual Common regardless of ontological facts about intellectual products. Drawing on Locke, she argues:

> Locke's writings do not directly develop the foundations of the common property presumption. But there is reason to favor the second understanding. It, unlike the first, reflects the themes that initially animate Locke: the emphasis on equality, the connection between equality and common ownership, and reasoning about

property in light of its nature—that is, in light of what is necessary to make full and robust use of it. The qualities of IP strongly engage these Lockean themes—especially the facts that exclusive use is generally unnecessary for its proper use and that, to the contrary, its full exploitation commonly depends on nonexclusive use. These features generate moral reasons to regard intellectual products as part of the intellectual common, even if they are pure authorial creations. [The "first" view mentioned in this passage is the one that stipulates an independent existence of intellectual products; the "second" is Shiffrin's as just sketched.]

I have argued that the considerations against privatization play different roles depending on the metaphysical status of ideas (in one case making clear that a presumption against privatization can only be overcome for certain cases, in the other case delineating the extent of private property rights), but that we arrive at the same constraints on IP regulation regardless of whether there is an Intellectual Common. Shiffrin seeks to establish an Intellectual Common independently of metaphysical ideas about the objects of IP. I think this slightly mischaracterizes the work done by the considerations against privatization. But that is a minor disagreement.

GLOBAL IMPACT

Let me summarize what I have argued so far. In a first step, I have explained how natural ownership rights give rise to associative rights at the level of the state system *per se*. If these rights are preserved, the imposition of a system of states is acceptable to co-owners. From Grotius we took the idea of the earth being collectively owned by humanity, and that collective ownership delivers a presumption against privatization. We returned to Grotius for the second step of our discussion. We looked at his exploration of the question of whether all collectively owned space was open to private appropriation. Although we saw that Grotius' discussion of the seas is of dubious plausibility nowadays, his reasoning carries over to IP. Our main conclusion was that IP should be regulated such that only limited rights to private IP are acknowledged, determined by considerations of compensation and incentives only. We obtained that result independently of the ontological status of the objects of IP rights, although that status mattered for the characterization of the precise reasoning behind the respective conclusions.

The third step is to bring these trains of thought together to show that there is a *human right* to vital pharmaceuticals, as follows: The argument in the first step does some work toward the development of a particular conception of human rights; the argument in the second step helps us make sense of a human right to essential pharmaceuticals within that conception.

Let me elaborate, first, on the sense in which we have done work toward a conception of human rights. According to the conception in question, human rights are *membership rights in the global political and economic order.* The global order is the system of states that covers most of the land masses of the earth, as well as the network of organizations that, without constituting an actual government, provides for what has come to be called "global governance." These membership rights are partially derived from the concept of collective ownership as explained above. But as we will see, there are other sources too from which membership rights can be derived.

Being a member of that order means to live on the territory covered by it and to be subject to those bits of this interlocking system of jurisdictions that apply to one's situation. Nothing more is meant by "membership." By now all human beings are members in this sense. A condition for the existence of rights held within that order is that there is enough structure to the global order to render that very term applicable, as well as an accompanying capacity for coordinated action. And, indeed, there is enough structure because of the existence of organizations that are designed for, and in fact do concern themselves with, global problem-solving. Think of the population of the world as being contained in one large set, and of the global order as captured by relations among members of that set. All citizens of a given country stand in one such relation; all persons whose countries are in the WTO in another, etc. Membership rights in this order will be rights individuals hold *qua* members of this set with these structures imposed, where these internal differentiations matter when it comes to responsibilities.[49]

Such a conception of human rights, by virtue of resorting to features of an *empirically contingent* but *relatively abiding* world order, makes the applicability of human rights contingent. Individuals across history have not always held them, nor would those rights apply to a colony of humans on the back side of the moon should such a community be discovered. This does not mean one could do with such people as one pleases, but they would not be members of the global order. The main advantages of the present account—which I think are virtues considerable enough to compensate for the counterintuitive nature and implications of the connections between human rights, membership in the global order, and collective ownership—are that it readily makes clear why talk about "rights" is appropriate here (rather than talk about values or goals), and that it provides a non-parochial grounding of human rights in plausible starting points, which unlike, say, Griffin's view,[50] does not require inquiries into the nature of personhood, autonomy, or agency.

In light of the sheer relevance of the kind of thing whose ownership is at stake for all human purposes, one cannot simply reject this account as missing the point of "human" rights because it does not focus on providing content to the term "human." At any rate, ideas about universality that feed into our understanding of human rights are sufficiently complex to make it implausible that there will be a single philosophically most successful conception of human rights. Different conceptions capture different aspects of our common ideas about human rights. The conception offered here cannot plausibly exhaust all we want to say about human rights.[51]

The defining feature of human rights in this view is that they are important moral demands against authority as it applies to individuals in their immediate environment and that they are at the same time also *matters of urgent global concern.*[52] To argue that X is a human right, what is required as a first, preliminary step is that X be shown to be a matter of importance in the affected agents' immediate environment, and then, second, that a genuinely global urgent concern can be established. In the process it must also become clear why it is appropriate to talk about a *right* in a given case. (I do not mean to distinguish here between importance and urgency, but merely to use two different terms for the two stages of this argument.) Again, there is no reason to think that only if common ownership rights need to be protected can something become of global concern in a manner that renders the language of rights appropriate. On the contrary, a strength of this conception is that it can accommodate a range of reasons why certain matters should concern the world as a whole.

Additional sources might be substantive or procedural. Since little of the details will matter later, I will be brief.[53] As far as substantive sources are concerned, one can argue in at least three ways that something should be of global concern. First, one might argue that this is so on the basis of *mutual enlightened self-interest.* We are dealing with a moral notion here, of course, but considerations of self-interest enter by way of demonstrating the global urgency. For example, it may be necessary or conducive for the preservation of peace that authority is exercised in certain ways, based on the idea that unchecked governmental authority will be abusive vis-à-vis others or will create negative externalities (refugees, etc.). Second, one may argue that something is of global concern because there is a shared causal responsibility for the matter at hand that arises out of global interconnectedness.

A third substantive source, a collective category, involves moral considerations that do not turn on interconnectedness. The ownership standpoint could be enlisted here too, but we have discussed it. Such considerations include a natural duty of aid (which would have to be acknowledged independently and does not turn on particular features of the global order), as well as possible duties of rectification (where it would have to be shown that the global order *per se* owes the rectification). Such considerations can also include an appeal to basic human needs and their moral significance. In each case where such an appeal to a moral consideration is made, it will be crucial to offer an argument for why something is of urgent *global* concern—and this is precisely the argument I make below with regard to a right to vital pharmaceuticals.

The second kind of source is procedural. One way in which concerns can become common within a certain political structure, and one way in which they can become membership rights within that structure, is for them to come to be widely regarded as such, as a result of an authoritative process. Proposed "human rights" may receive support from any or all of these sources, and the strength of support arising from each source may vary. Not all sources from which membership rights can be derived apply equally to all individuals. Moreover, a critical discourse can occur if a proposed right fails to receive support from all different sources.

DERIVING A RIGHT TO VITAL PHARMACEUTICALS

So is there a human right to vital pharmaceuticals? To make that case, we need to show first that access to pharmaceuticals is a matter of importance in the affected agents' immediate environment, and then, second, to establish an urgent global concern. I take it that the preliminary step is met for essential pharmaceuticals.[54] Possibly, my conception of human rights also offers other resources to do so, but I seek to establish an urgent global concern through considerations of IP.

At first sight, the second step of the argument seems to speak against there being such a concern. That second step provides us with a moral standpoint for assessing IP regulation whenever such regulation takes place. What that standpoint does, however, is merely to clarify what *sort* of reason bears on the determination of private IP rights. This is compatible with much variation. (Recall Easterbrook's view.) Compensation is compensation for materials used, but also for time invested. Then the question becomes how highly societies should value an inventor's time, compared to the time of others. Also, some societies may care more about fostering innovation than others. (Recall the concluding part of the Jefferson quote.) One may say such regulation ought to be left to domestic law because no mandates follow from this view other than constraints on the kinds of reasons that can be used, and because countries would thus be able to shape their comparative advantage. But if IP explicitly ought *not* to be regulated globally, we find no human right to vital pharmaceuticals in this conception *if the goal is to establish one* via *considerations of IP.*

Alas, nowadays, there *is* international IP regulation, notably through TRIPS. To that agreement the conclusions of the second step apply. We have shown that wherever IP is regulated, it has to abide by certain constraints. This conclusion leads to a critical assessment of TRIPS: as private rights ought to be limited to compensation and incentives, and as it is implausible that either of these would impose obligations on very poor countries, TRIPS should not impose burdens on them. The wealth differential between rich and poor is so large that compensation and incentives for the pharmaceutical industry are not likely to depend on markets in very poor countries. An economic analysis would still need to be added, but we obtain this result from reflections on IP without saying anything about human rights.[55]

But can we also derive a *human right* to vital pharmaceuticals by enlisting considerations of IP, despite the reasoning to the contrary presented earlier? We can, if we can demonstrate an urgent concern at the global level with the regulation of a particular *sub-domain* of IP that includes essential pharmaceuticals—regardless of existing treaties, such as TRIPS, that may or may not have come about because of such a concern. Then the argument of the second step imposes constraints on the regulation of that sub-domain. To this end, we can offer two arguments. According to the first, the conclusion that the regulation of vital pharmaceuticals is a matter of urgent global concern emerges in the following three steps.

First, it follows from our reflections on IP that for a certain domain of such property, there is a presumption that indeed it should be regulated globally. This domain includes those ideas with regard to which it does make sense to speak of a discovery and an Intellectual Common. In that domain, any two human beings are symmetrically located with regard to ideas. There is then a *prima facie* case for global regulation parallel to the case for a global approach to the use of the three-dimensional space of the earth (with regard to which any two human beings are also located symmetrically), namely to preserve this symmetry of all human beings with regard to the respective common. Second, pharmaceuticals are in this domain because they draw on physiochemical properties of molecules. Such properties are among the most plausible entities for which it makes sense to say they were discovered and exist in an Intellectual Common.

These properties are what they are regardless of human activities, and regardless of whether it took ingenuity, effort, or serendipity to uncover them. And third, among ideas for which international regulation is plausible (i.e., those that are in an Intellectual Common), those that are immediately relevant to basic human needs make for an especially plausible and urgent case, and this includes vital pharmaceuticals.

There is therefore an urgent global concern with the regulation of vital pharmaceuticals. Using this conclusion, we see that there is indeed a human right to essential pharmaceuticals. It is owed to people across the world to regulate IP in vital pharmaceuticals at the global level in the first place, and to ensure that IP in vital pharmaceuticals not be regulated in a way that acknowledges far-reaching private IP rights. We can readily tie the availability of essential medicines to the protection of basic human needs, and thus to that particular source of membership rights in the global order. However, to show that in this manner we obtain a human right according to our conception, we needed to establish a *genuinely global concern* specifically for the availability of such medications. This is what the argument we now presented adds.[56] This argument also rebuts the objection that draws on the earlier considerations against global regulation of IP, although that objection continues to apply to those *domains* of IP not captured by the argument just offered.

Alas, these gains come at a price: this first argument makes a rather serious ontological commitment to realism about the content of the driving ideas behind pharmaceutical patents, a commitment that is essential to the argument because the symmetry of any two individuals with regard to these ideas depends on it. It is welcome news, therefore, that there is a second (independent) way of identifying a reason for there being an urgent global concern with the regulation of vital pharmaceuticals. This second argument does not turn on any ontological commitments about the subjects of IP law. Recall that our conclusion in the second part of this study was that IP law ought to abide by certain constraints—namely, it ought to acknowledge only compensation and incentive-setting as acceptable bases for private IP rights. We saw above that this generates a critical assessment of TRIPS. More generally, then, we can formulate an implication of the main result of the second part of our argument, as follows: There ought to be no IP regulation—*in particular* at the global level—that grants rights that go beyond what we can obtain by way of compensation and incentive-setting.

The first argument we just presented identified the symmetry of any two individuals with regard to the underlying ideas of a certain domain of IP as a reason for there being an urgent global concern with the regulation of that domain (and was based on particular, and controversial, ontological commitments). The second argument recasts the consider-ations against privatization that we extracted from Grotius' reflections on the freedom of the seas as a reason against property regulations of a certain sort at the global level. (Recall that these considerations were: that ideas cannot actually be occupied in the same sense in which, say, land can be occupied; that the gains for users of ideas do not depend on excluding others; and that use of ideas by others often adds to their usefulness for any given user of these ideas.)

So whereas the first argument identifies a reason *in favor of* regulation of a certain sort at the global level, the second argument identifies a reason *against* regulation of a certain sort. Put differently, the first argument formulates a positive case for regulation of vital pharmaceuticals at the global level and *then,* in a second stage, brings to bear the considerations we extracted from Grotius to formulate limitations on private property rights; however, the second argument dispenses with that positive case, and thus with the first stage of that first argument, and makes its case only in terms of the Grotian considerations against privatization.

Because there indeed is regulation at the global level now, and because global interconnectedness makes it inevitable that some property regime or other is in place at the global level regardless of whether there is one particular treaty governing that regime (such as TRIPS),[57] both of these arguments have the same implications. That is, neither generates a claim against anybody actually to invent vital medications that are not yet available. But both arguments lead to the conclusion that it is owed to people across the world that IP generally, and vital pharmaceuticals in particular, not be regulated in a way that acknowledges far-reaching private IP rights—*especially* (but not only) at the global level. In that sense, there indeed is a human right to pharmaceuticals: a right against constraints on access to pharmaceuticals on behalf of overblown private IP rights.

NOTES

1. Many thanks to audiences at the Harvard Kennedy School, the University of Utah, the Harvard School of Public Health, the Humboldt University in Berlin, King's College London, University College London, and the Johns Hopkins University, where I presented this material in 2008 and 2009. Thanks also to Norman Daniels, Nir Eyal, Axel Gosseries, Joseph Millum, and Dan Wikler for helpful exchanges. Hestermeyer (2007) sums up the state of the art in the legal debate; see also Toebes (1999). Sell (2003) discusses the background to TRIPS and the history of intellectual property arrangements. See Maskus and Reichman (2005a) on general developments in intellectual property protection at the global level. The WHO's lists of essential medicines is at http://www.who.int/topics/essential_medicines/en. For my conception of human rights, see Risse (2009a, 2009b).

I take the *concept* of human rights to refer to rights that are invariant with respect to local conventions, institutions, culture, or religion. I take it that the focus of the human rights language is on abuses committed by those in positions of authority: of two otherwise identical acts one may be a human rights violation, but not the other, depending on whether they can be interpreted as abusing authority. (On this, see Pogge, 2002, pp. 57 ff.) A *conception* of human rights adds contents to the concept, and consists of four elements: first, a list of these rights; second, an account of the basis on which individuals have them (i.e., of what features turn individuals into rights holders); third, an account of why that list has that particular composition—that is, a principle or process that generates it; and fourth, an account of who has to do what to realize these rights— that is, an account of the corresponding obligations. The universality captured by the concept of human rights and our human rights practices render it implausible that there is a single philosophically most sensible conception of human rights. As I explain below, my conception has some virtues, but also shortcomings, in terms of its ability to capture human rights discourse.

2. The argument of this study can interact effectively with work on incentive-setting for the pharmaceutical industry; see Kremer and Glennerster (2004), Love and Hubbard (2007), and Pogge (2008a, 2008b).

3. What should be of general interest independently of this particular conception of human rights is (a) the parallel between the Global Common and the Intellectual Common, and (b) my argument for constraints on the extent of private intellectual property rights that in turn ends up being *independent* of any ontological characterization of the objects of intellectual property law (in terms of an Intellectual Common or any other manner) and thus holds at considerable generality.

4. I develop this conception of human rights as part of an overall theory of global justice in my forthcoming book *On Global Justice*, to appear with Princeton University Press in 2012. See also Risse (2009a, 2009b).

5. "Natural" rights are rights that we can justify without reference to conventions or institutions that hold within or among groups, as well as without any reference to any transactions, such as promises or contracts. "Associative" rights are rights for whose justification we must appeal to the rights holders' belonging to a particular association.

6. See Buckle (1991) and Tuck (1999) for these discussions.

7. Drahos, 1996, p. 41.

8. Reed (2006) applies Grotian ideas to the gene pool (using the analogy to the sea), but seems to believe these ideas are of use only to Christians. Much of the second part of this essay engages with Shiffrin (2001), who has explored related considerations within Locke's approach to property. For introductions to the philosophical concerns behind intellectual property law, see Shiffrin (2007). See also Kuflik (1989) and Fisher (2001). For a discussion of these issues specifically with regard to patents, see Sterckx (2005). Lessig (2004) argues for very limited copyrights, which is kindred in spirit to the present argument. But as should become clear, my approach speaks more to patents than to copyrights. Ashcroft (2005) gives an argument in support of states' rights to expropriate private property in cases of public health emergencies, and Schueklenk and Ashcroft (2002) argue for a similar conclusion on broadly consequentialist grounds.

9. Arguments that proceed exclusively by an appeal to based human needs face a difficult boundary problem. One aspect of this boundary problem is that it is hard to say how much of a sacrifice is required to make sure others can satisfy their basic needs. Another aspect is that it is hard to draw immediate inferences from statements about, say, needs to the conclusion that, as matter of justice, very specific things need to be done, and that responsibility for them somehow lies at the global level. Such arguments are strongest when they do not rely exclusively on claims of need but integrate other considerations.

10. Buckley and O Tuama (2005) and De George (2005) deal with pricing issues in the pharmaceutical industry. Questions there concern what profits would be justified given what risks and difficulties that industry faces, but also given how much they benefit from public subsidies. For a perspective on these matters skeptical of the pharmaceutical industry, see Angell (2004). See also Cohen et al. (2006). Other important issues in this literature include the question of what areas this industry should invest in, possibly at the exclusion of others; and how to market products. One important question is also to what extent regulation of IP is causal to a lack of access to medication, and what remedies there might be. Kremer and Glennerster (2004) and Pogge (2008a, 2008b) make proposals for how to change the incentives of the pharmaceutical industry to make medications available to the poor. Maskus and Reichman (2005b) argue that TRIPS has given rise to a transnational system of innovation that could produce powerful incentives to innovate for the benefit of mankind, if developed properly. A crucial question about TRIPS is whether there is any sense in which the regulation proposed by that agreement is in the "enlightened" self-interest of developing countries.

11. I quote from DJB in the customary way. For instance, "II.2.II.1" means second volume; second book; second chapter; first section. The 2005 Liberty Fund edition is especially accessible. I also deal with Grotius' earlier work, *Mare Liberum, Free Sea*, which is part of a much larger work, *De Jure Praedae Commentarius, Commentary on the Law of Prize and Booty*, which, however, became available in full only in the 19th century (see the 2006 Liberty Fund edition). For *Mare Liberum*, I quote the pages from the 2004 Liberty Fund edition.

12. Buckle, 1991, p. 95.

13. See Blake and Risse (2007, 2009).

14. Much has been written on foundations of property; see Becker (1977), Reeve (1986), or Ryan (1987).

15. Honore, 1961.

16. A more difficult question is under what conditions man-made products, including improvements of original resources, should no longer be accompanied by special entitlements of those who made them or their offspring. See Blake and Risse (2009) for discussion. Egalitarian Ownership formulates a standing demand on all groups that occupy parts of the earth to inhabit the earth in a manner that respects this symmetrical status of individuals with regard to resources. That Egalitarian Ownership operates in this way should be intelligible and acceptable even within cultures where individuals are not seen as property owners. Nothing about Egalitarian Ownership precludes such cultures from being acceptable to their members even if they do not treat individuals themselves as property holders. Yet even cultures that do not see individuals themselves as property holders must indeed be acceptable to those who live in them, especially because all individuals have symmetrical claims to original resources.

17. In capital letters, "Joint Ownership" and "Common Ownership" are names of interpretations of Egalitarian Ownership and hence views about ownership of the earth, whereas in small letters "joint ownership" and "common ownership" are general forms of ownership of anything. I continue to say that humanity "collectively" owns the earth if the precise form of ownership does not matter.

18. Risse (2005) offers supportive arguments, showing why other conceptions are problematic. I develop all of this at length in my forthcoming book *On Global Justice*. See also Risse (2009a).

19. Hart, 1982, p. 171.

20. For the Hohfeld terminology, see Jones (1994), Chapter 1; Edmundson (2004), Chapter 5; Wenar (2005).

21. See Malanczuk (1997), pp. 149f and pp. 184ff. Outer space is also treated in this way.

22. (1) One might say that generally individuals who are threatened where they live do not have the opportunity to travel to another state to ask for entry, and therefore such states would not contribute to their predicament but merely fail to come to their aid. At any rate, they would only contribute to the predicament of those who make it to their borders and are turned away. But this under-describes the extent to which a state system based on self-determination and inviolability of territory contributes to such predicaments. Were our world no longer committed to such principles and were rich states more inclined to admit people who arrive at their borders because, say, their ability to make a living is threatened in their country of origin, organizations would spring up that specialize in making sure such individuals get to wealthier destinations. Such individuals would pay for these services by pledging future income, or charitable organizations would do this job. (2) One might also object as follows: Suppose we own a boat together and somebody interferes with your use of it. This would not mean I have to stop that interference or else let you use the boat when I am entitled to using it. But this is a wrong comparison. If we own the boat in common and this situation arises, *and we are all in the boat*, then you would have to give me refuge on your side, at least as long as the boat does not thereby turn over.

23. My view does not presuppose that individuals "participate" in the global order. Even secluded tribes possess human rights. They are co-owners of the earth and are constrained by the imposition of the state system even if they do not actually feel the constraints. In the case of such tribes there presumably are unusually strong reasons to set aside enforcement of human rights. Yet if by any chance humans are discovered on the back side of the moon, the considerations explored in this study would not apply to them. That does not mean one can do with them as one pleases. But as they would not be members of the global order, these considerations would not bear on their moral status.

24. See DJB, II.2.II.5, but also II.3.1 and II.3.IV.1, and in II.8.VI.

25. On the importance of occupation, see ML, p. 24, p. 34. On p. 116 we read that if things cannot remain common, they become the property of the first taker, both because the uncertainly of ownership could not otherwise be avoided, and also because it was equitable that a premium be put upon diligence.

26. See also Buckle (1991), p. 43.

27. For the more recent development of the law of the seas, see Malanczuk (1997), Chapter 12.

28. Sidgewick, 2005, p. 228.

29. Jefferson, "The Invention of Elevators" (Letter, 1813); quoted in Shiffrin (2001, p. 138).

30. Wilson (forthcoming) argues against the possibility of an intrinsic moral right to own intellectual property, and does so largely by way of reference to the fact that any one person's use of ideas does not exclude others from making the same use of them (i.e., their non-rivalrousness). Based on that idea, he argues that nobody will be harmed by being denied the possibility of creating an IP right.

31. Gottlob Frege's 1918 essay "Der Gedanke: Eine Logische Untersuchung" ("The Thought: A Logical Investigation") is a *locus classicus* for this view, although I am, for the sake of the argument, offering an extreme version of it. See Gideon Rosen's entry on abstract objects at the online *Stanford Encyclopedia of Philosophy*, http://plato.stanford.edu/entries/abstract-objects/. For a general background discussion of abstract objects and questions of their existence, see Burgess and Rosen (1997). Important to mention is also Karl Popper's theory of reality, which distinguishes among three worlds: World 1 is the world of physical objects and events; World 2 is the world of mental objects and events; and World 3 is the world of the products of the human mind. Cf. for instance Popper (1972).

32. This presumption has this effect regardless of which larger strategy of arguing for stronger private intellectual property one may choose, such as an approach in terms of natural law,

in terms of a hypothetical contract between inventors and society, general considerations of distributive justice, or in terms of rewards for contributions made.

33. Easterbrook, 2001, p. 406.
34. Nozick, 1974, pp. 178–182.
35. Waldron, 1993b.
36. John Bates Clark, *Essentials of Economic Theory* (1907), pp. 360–361; in Waldron (1993b, p. 866). A similar attitude is expressed in Bainbridge (1992): "The basic reason for intellectual property law is that a man should own what he produces, that is, what he brings into being. If what he produces can be taken from him, he is no better off than a slave. Intellectual property is, therefore, the most basic form of property because a man uses nothing to produce it other than his mind" (p. 17).
37. Becker (1993) also offers an illustration from a scientific context of the phenomenon that achievements are sometimes entirely disconnected from a societal state of knowledge and ability. He refers to the following statement of the mathematician Mark Kac: "'[T]here are two kinds of geniuses, the ordinary and the magicians. An ordinary genius is a fellow that you and I would be just as good as, if we were only many times better.' But for the second kind, 'even after we understand what they have done, the process by which they have done it is completely dark'" (Becker, 1993, p. 617, note 22).
38. See Hettinger (1989) on those two considerations.
39. Child, 1990, pp. 573–600.
40. See also Moore (1997).
41. Waldron, 1993b, pp. 841–887.
42. Lachlan, 2005, p. 107.
43. For the idea of invention, see Kneale (1955).
44. Paine (1991), which is a response to Hettinger (1989), captures the competing approaches to intellectual property very well: "We may begin thinking about information rights, as Hettinger does, by treating all ideas as part of a common pool and then deciding whether and how to allocate to individuals rights to items in the pool. Within this framework, ideas are conceived on the model of tangible property. Just as, in the absence of social institutions, we enter the world with no particular relationships to its tangible assets or natural resources, we have no particular claim on the world's ideas. In this scheme, as Hettinger asserts, the 'burden of justification is very much on those who would restrict the maximal use of intellectual objects' (p. 20). Alternatively, we may begin, as I do, by thinking of ideas in relation to their originators, who may or may not share their ideas with specific others or contribute them to the common pool. This approach treats ideas as central to personality, and the social world individuals construct of themselves. Ideas are not, in the first instance, freely available natural resources. They originate with people, and it is the connections among people, their ideas, and their relationships with others that provide a baseline for discussing rights in ideas. Within this conception, the burden of justification is on those who would argue for disclosure obligations and general access to ideas" (p 49).
45. Shiffrin, 2001, p. 159.
46. I am indebted to Shiffrin's article for this part of my argument. Shiffrin's concern is with Lockean approaches to intellectual property. She argues that the presumption against privatization that comes from the idea of original collective ownership has often been underestimated in Lockean accounts of private intellectual property. In her view, Locke's "mixing" account of privatization does not provide a foundation for privatization *per se*, but creates a way of assessing which individuals would be allowed to occupy something. Given a presumption against privatization, Lockean accounts of property deliver considerable constraints on the possibility of private intellectual property. For opposing understandings on the Lockean approach to property, see Hughes (1988), Becker (1993), Child (1990), and Moore (1997).

47. To use a mathematical analogy, we have offered an argument for two extreme cases, and now have argued that the same argument also holds for the intermediate cases that can be understood as convex combinations of the extreme cases.

48. Shiffrin, 2001, p. 164.

49. One worry that may arise here is that such rights will not apply to everybody. (What of North Korea?) Note two things. First, membership, as explained above, does not depend on the participation of one's country in political and economic activities of the global order. The existence of organizations of global reach is important only to fend off the objection that there is not enough structure to render talk of membership rights meaningful. Second, more importantly, the task of establishing whether membership rights hold for everybody falls to the discussion of the different sources from which these rights can be obtained. Rights that can be derived from collective ownership do apply to everybody, but others indeed may not.

50. Griffin, 2008.

51. (1) Cohen (2006) proposes that human rights have three features: they are universal and owed by every political society to everybody; they are requirements of political morality whose force does not depend on their expression in enforceable law; and they are especially urgent requirements. Any more particular account, says Cohen, has to meet these constraints, as well as two methodological assumptions: fidelity to the major human rights documents, so that a substantial range of these rights is accounted for; and open-endedness (we can argue in support of additional rights). These criteria do not entail commitments with regard to a range of questions about such rights. It is the function of a conception of human rights to provide a fuller set of answers to such questions. For instance, accepting these criteria does not imply that human rights must be understood as protecting essential features of personhood, though it is consistent with such an approach. A different way of adding detail to these criteria is to think of "human" rights as rights individuals hold *qua* members of the global and political order that *ipso facto* but contingently includes *everybody*. That is what the present conception does. (2) A conception that understands human rights as membership rights in the global order must be distinguished from Cohen's (2006) conception in terms of membership rights in political society. Cohen's notion of membership is that "a person's good is to be taken into account by the political society's basic institutions: to be treated as a member is to have one's good given due consideration, both in the processes of arriving at authoritative collective decisions and in the content of those decisions" (Cohen, 2006, p. 237f). Human rights then are rights individuals hold in their respective communities to ensure inclusion. In the conception I defend, rights that ensure inclusion in political communities will be among those that are the global order's responsibility, but this is so via an additional argument. For individuals everywhere to have a claim to something vis-à-vis their respective community does not suffice for this to be a claim of urgent global concern, in the sense that violations somewhere should be of serious concern to people everywhere or to global institutions. The difference between these two kinds of membership captures an ambiguity that permeates human rights talk, namely whether such rights in the first instance apply to each individual, or else are of global relevance. If one endorses the first stance, the question becomes why others far away should care; if the second, the question becomes how much of what is of fundamental importance to individuals can be incorporated.

52. I stated at the beginning that I took this reference to authority to be part of the concept of human rights.

53. I discuss these themes in more detail in my forthcoming book, *On Global Justice*.

54. For the connection between health and social justice, see Hofrichter (2003) and Wilkinson (1996); see also Chapter 6 in Barry (2005), Waldron (1993a), and Daniels (1985).

55. (1) This argument does not engage with the moral acceptability of TRIPS in ways internal to considerations of compensation and incentive-setting. For instance, Maskus and Reichman (2005b) argue that TRIPS has given rise to a transnational system of innovation that could produce powerful incentives to innovate for the benefit of humankind, if developed properly. Usual arguments supportive of strengthening IP protections even in developing countries include the ability to build local research and development; to attract technology transfers; and to attract foreign direct investment. The economist Joan Robinson once spoke of the "paradox of patents:" "The justification of the patent system is that by slowing down the diffusion of technical progress it ensures that there will be more progress to diffuse" (Robinson, 1958, p. 87, quoted in Streckx, 2005, p. 197). As far as TRIPS is concerned, the question is whether the long-term effects in terms of "more progress to diffuse" are sufficiently great to warrant the short-term costs not merely in terms of "slowing down the diffusion of technical progress," but also in terms of the more or less direct consequences of this slowdown (such as hampered access to medications). (2) One might say one concern behind TRIPS is to undermine certain possibilities of drug smuggling that benefits from the existence of countries without strong patent protection. However, I take it that the arguments in this study (in particular the argument for there being a human right to essential pharmaceuticals we are about to present) entail that different policies must be implemented to solve that problem, rather than the introduction of overly strong private IP rights that—crucially—require a plausibility independently of their effect on the smuggling of pharmaceuticals.

56. One might think we could obtain this result by a plain appeal to needs. But, again, there is a considerable boundary problem that arises for attempts to derive rights and obligations from the idea of basic human needs. It is easy to raise doubts about the claim that rights and obligations that are derived in this way are indeed of global concern ("Why would this particular claim not be exhausted by responsibilities that hold within certain shared political and economic structures?"). Since it thereby addresses such doubts, I think my argument is stronger for enlisting metaphysical considerations about the status of particular ideas (here, biochemical properties of molecules).

57. This is also the response (on this second argument) to the objection articulated earlier that the second part of our argument actually entails that there ought to be no regulation of IP at the global level. There is a sense in which this is a weaker argument than the first in support of a human right to pharmaceuticals in the manner to be identified at the end of this paragraph—that is, there is a philosophical gain from the ontological assumption that backs that first argument: The first argument generates a strong *pro tanto* reason in support of global regulation of intellectual property, one that has to be set aside in light of political, economic, or technological obstacles. As opposed to that, the second argument only generates a prohibition on particular forms of IP regimes at the global level. This prohibition applies only once there *de facto* is some global IP regime or other, in light of increasing global economic interconnectedness. So the strong premises of the first argument do make a difference in the strength of the conclusion that we can derive. But under those conditions just mentioned, then, which are the conditions under which we now live, these arguments do lead to the same conclusion.

LITERATURE

Angell, Marcia. (2004). *The Truth About the Drug Companies. How They Deceive Us and What To Do About it.* New York: Random House

Ashcroft, Richard. (2005). Access to essential medicines: A Hobbesian social contract approach. *Developing World Bioethics* 5(2):122–141

Bainbridge, David. (1992). *Intellectual Property*. London: Longman

Barry, Brian. (2005). *Why Social Justice Matters*. Cambridge: Polity

Becker, Lawrence. (1977). *Property Rights: Philosophical Foundations*. Boston: Routledge and Kegan Paul

Becker, Lawrence. (1993). Deserving to own intellectual property. *Chicago-Kent Law Review* 68:609–629

Blake, Michael, and Mathias Risse. (2009). Immigration and Original Ownership of the Earth." *Notre Dame Journal of Law, Ethics, and Public Policy* 23(1) (special issue on immigration):133–167

Blake, Michael, and Mathias Risse. (2007). Migration, territoriality, and culture. In J. Ryberg, T. Petersen, and C. Wolf, eds. *New Waves in Applied Ethics*. Ashgate Publishers

Buckle, Stephen. (1991). *Natural Law and the Theory of Property. Grotius to Hume*. Oxford: Clarendon

Buckley, Joan, and Seamus O Tuama. (2005). International pricing and distribution of therapeutic pharmaceuticals: an ethical minefield. *Business Ethics: A European Review* 14:127–141

Burgess, John, and Gideon Rosen. (1997). *A Subject with No Object. Strategies for Nominalistic Interpretation of Mathematics*. Oxford: Clarendon

Child, James. (1990). The moral foundations of intangible property. *The Monist* 73:573–600

Cohen, Jillian, Patricia Illingworth, and Udo Schüklenk, eds. (2006). *The Power of Pills: Social, Ethical, and Legal Issues in Drug Development, Marketing, and Pricing*. London: Pluto Press

Cohen, Joshua. (2006). Is there a human right to democracy? In Christine Sypnowich, ed. *The Egalitarian Conscience: Essays in Honor of G. A. Cohen*. Oxford: Oxford University Press, pp. 226–249

Daniels, Norman. (1985). *Just Health Care*. Cambridge: Cambridge University Press

De George, Robert. (2005). Intellectual property and pharmaceutical drugs: an ethical analysis. *Business Ethics Quarterly* 15:549–575

Drahos, Peter. (1996). *A Philosophy of Intellectual Property*. Aldershot: Dartmouth

Easterbrook, Frank. (2001). Who decides the extent of rights in intellectual property? In Rochelle Dreyfuss, Diane Zimmerman, and Harry First, eds. *Expanding the Boundaries of Intellectual Property*. Oxford: Oxford University Press: pp. 405–415

Edmundson, William. (2004). *An Introduction to Rights*. Cambridge: Cambridge University Press

Fisher, William. (2001). Theories of intellectual property. In S. Munzer, ed. *New Essays in the Legal and Political Theory of Property*. Cambridge: Cambridge University Press

Frege, Gottlob. (1918). Der Gedanke. Eine Logische Untersuchung. In *Beiträge zur Philosophie des Deutschen Idealismus* I: pp. 58–77

Griffin, James. (2008). *On Human Rights*. Oxford: Oxford University Press

Grotius, Hugo. (2004). *The Free Sea*. Edited and with an Introduction by David Armitage. Indianapolis: Liberty Fund

Grotius, Hugo. (2005). *The Rights of War and Peace*. Edited with an Introduction by Richard Tuck. Indianapolis: Liberty Fund

Grotius, Hugo. (2006). *Commentary on the Law of Prize and Booty*. Edited and with an Introduction by Martine Julia van Ittersum. Indianapolis: Liberty Fund

Hart, H. L. A. (1982). *Essay on Bentham*. Oxford: Oxford University Press

Hestermeyer, Holger. (2006). *Human Rights and the WTO. The Case of Patents and Access to Medicine*. Oxford: Oxford University Press

Hettinger, Edwin. (1989). Justifying intellectual property. *Philosophy and Public Affairs* 18:31–52

Hofrichter, Richard, ed. (2003). *Health and Social Justice: Politics, Ideology, and Inequity in the Distribution of Disease*. San Francisco: Jossey-Bass

Honore, A. M. (1961). *Ownership*. In *Making Law Bind: Essays Legal and Philosophical*. Oxford: Clarendon

Hughes, Justin. (1988). The philosophy of intellectual property. *Georgetown Law Journal* 77:287–366

Jones, Peter. (1994). *Rights*. London: Macmillan

Kneale, W. C. (1955). The idea of invention. *Proceedings of the British Academy* 41:85–108.

Kremer, Michael, and Rachel Glennerster. (2004). *Strong Medicine*. Princeton: Princeton University Press

Kuflik, Arthur. (1989). Moral foundations of intellectual property rights. *Chapter 11 in* Vivian Weil and John Snapper, eds. *Owning Scientific and Technical Information: Value and Ethical Issues*. New Brunswick: Rutgers University Press

Lachlan, James. (2005). A neuropsychological analysis of the law of obviousness. In Peter Drahos, ed. *Death of Patents*. London: Queen Mary Intellectual Property Research Institute, pp. *67–109*

Lessig, Lawrence. (2004). *Free Culture. The Nature and Future of Creativity*. New York: Penguin

Locke, John. (1988). *Two Treatises of Government*. Ed. by Peter Laslett. Cambridge: Cambridge University Press

Love, James, and Tim Hubbard. (2007). The big idea: prizes to stimulate R&D for new medicines. *Chicago-Kent Law Review* 82(3):1519–1554

Malanczuk, Peter. (1997). *Akehurst's Modern Introduction to International Law* (7th ed.). London: Routledge

Maskus, Keith, and Jerome Reichman. (2005a). *International Public Goods and Transfer of Technology Under a Globalized Intellectual Property Regime*. Cambridge: Cambridge University Press

Maskus, Keith, and Jerome Reichman. (2005b). The globalization of private knowledge goods and the privatization of global public goods. In Maskus, Keith, and Jerome Reichman. *International Public Goods and Transfer of Technology Under a Globalized Intellectual Property Regime*. Cambridge: Cambridge University Press

Moore, Adam. (1997). Towards a Lockean theory of intellectual property. In Adam Moore, ed. *Intellectual Property: Moral, Legal, and International Dilemmas*. New York: Rowman and Littlefield

Nozick, Robert. (1974). *Anarchy, State, and Utopia*. Cambridge: Harvard University Press

Paine, Lynne Sharp. (1991). Trade secrets and the justification of intellectual property: a comment on Hettinger. *Philosophy and Public Affairs* 20:247–263

Pogge, Thomas. (2002). *World Poverty and Human Rights*. Cambridge: Blackwell

Pogge, Thomas. (2008a). *World Poverty and Human Rights* (2nd ed.). Oxford: Polity

Pogge, Thomas. (2008b). Access to medicines. *Public Health Ethics* 1 (2):73–82

Popper, Karl. (1972). *Objective Knowledge: An Evolutionary Approach*. Oxford: Oxford University Press

Reed, Esther. (2006). Property rights, genes, and common good. *Journal of Religious Ethics* 34(1):41–67

Reeve, Andrew. (1986). *Property*. Atlantic Highlands: Humanities Press

Risse, Mathias. (2005). How does the global order harm the poor? *Philosophy and Public Affairs* 33(4):349–376

Risse, Mathias. (2009a). Common Ownership of the Earth as a non-parochial standpoint: a contingent derivation of human rights. *European Journal of Philosophy* 17(2):277–304

Risse, Mathias. (2009b). A right to work? A right to leisure? Labor rights as human rights. *Journal of Law and Ethics of Human Rights* 3(1):1–41

Robinson, Joan. (1958). *The Accumulation of Capital*. Homewood: Irwin

Ryan, Alan. (1987). *Property*. Milton Keynes: Open University Press

Schueklenk, Udo, and Richard Ashcroft. (2002). Affordable access to medication in developing countries: conflicts between ethical and economic imperatives. *Journal of Medicine and Philosophy* 27(2):179–195

Sell, Susan. (2003). *Private Power, Public Law: The Globalization of Intellectual Property Rights*. Cambridge: Cambridge University Press

Shiffrin, Seana. (2001). Lockean arguments for private intellectual property. In S. Munzer, ed. *New Essays in the Legal and Political Theory of Property*. Cambridge: Cambridge University Press

Shiffrin, Seana. (2007). Intellectual property. Chapter 36 in Robert Goodin, Philip Pettit, and Thomas Pogge, eds. *A Companion to Contemporary Political Philosophy*. Oxford: Blackwell

Sidgwick, Henry. (2005). *The Elements of Politics*. London: Elibron Classics

Sterckx, Sigrid. (2005). The ethics of patenting—uneasy justifications. In Peter Drahos, ed. *Death of Patents*. London: Queen Mary Intellectual Property Research Institute, pp. 175–212

Toebes, Brigit. (1999). *The Right to Health as a Human Right in International Law*. Antwerpen: Intersentia

Tuck, Richard. (1999). *The Rights of War and Peace*. Oxford: Oxford University Press

Waldron, Jeremy. (1993a). Social citizenship and the defense of welfare provisions. In Jeremy Waldron, ed. *Liberal Rights: Collected Papers 1981–1991*. Cambridge: Cambridge University Press

Waldron, Jeremy. (1993b). From authors to copiers: individual rights and social values in intellectual property. *Chicago-Kent Law Review* 68:841–887

Wenar, Leif. (2005). The nature of rights. *Philosophy and Public Affairs* 33:223–253

Wilkinson, Richard. (1996). *Unhealthy Societies. The Inflictions of Inequality*. London: Routledge

Wilson, James. (forthcoming). Could there be a right to own intellectual property? In *Law and Philosophy*

CHAPTER 4

༄

Global Justice and Health

The Basis of the Global Health Duty

JONATHAN WOLFF

INTRODUCTION

Global health is a growing industry, especially, though not exclusively, in the universities. Institutes of Global Health are being founded almost by the month, and funding no doubt is being diverted from other sources towards global health, or at least towards global health research, with actual impact on global health yet to be assessed. Yet why should developed nations take such an interest in global health? Indeed, why should they take any interest at all? Underlying this question is a whole host of others: What do we mean by global health? How do we define it, and measure it? What priority should health have in comparison to other urgent social concerns? What are the statistical trends underlying global health, and what are their causes? What detailed strategies are available for improving global health, or as it is also put, reducing the global burden of disease? To some degree, approaches to some of these latter questions will be implicit in what follows. But my focus in this paper is why it should be thought that the peoples of wealthy nations, or their governments on their behalf, should take measures to improve health in the developing world. For short, I shall call this *the question of the basis of the global health duty*.

In explaining why governments and citizens of developed nations should take on the global health duty, some writers have tried to show that, among other reasons, it is in the self-interest of the developed world to take a global perspective on health.[1,2] Others assume that the impulse

78

to help those in distant lands is a humanitarian one, which may or may not be the same as arguing that it is based on a duty of charity.[3] Finally, ethicists sometimes argue that the developed world has duties of justice to the peoples of the developing world.[4,5] The contrast between these different approaches can be brought out by considering what would be said if the developed nations failed to act to advance global health. On the self-interest argument, such a failure could be said to be foolish, perhaps seriously so, in that an opportunity to advance, or defend, our interests would have been forsaken, but there would be no violation of any sort of moral duties to the peoples of the developing world. The humanitarian argument supposes that the peoples of the developed world do have duties to advance the health of those elsewhere on the globe, at least in some cases, and that they can rightly be criticized if they do not do enough. But the humanitarian argument does not contend that people in the developing world have any *right* to assistance.[6] Individuals may have been wronged through neglect, but no rights would have been violated, on such a view. However, once the claim is made that rights will be violated if assistance is not forthcoming, the argument has shifted to one of justice: that we neglect or violate people's rights by failing to help.   RIGHTS

One obvious response is that all three arguments conclude that a concern for global health is required: so there is little point in debating whether the "true basis" for action is self-interest, charity, or justice. Alternatively, or in addition, it might be said that the global health duty is so urgent, and so obvious, that there is no need to discuss its foundation. I am sympathetic to both these points.[7] Nevertheless, there is good reason for pursuing the theoretical question. For one thing, it may well be that different types of programs, in different countries, have different justifications. Second, getting clear about the moral nature of the basis of our concern will have implications for the extent and pervasiveness of the duties. For example, a violation of rights is generally regarded as much more serious than a failure of self-interest or even of humanitarian action, and therefore is more demanding. Equally, understanding the moral basis can help determine when enough has been done. Duties of charity, arguably, run out when a certain absolute level has been achieved.[8,9] Duties of justice, and even concerns of self-interest, may have a different, more permanent, nature. Hence to get a clear idea both of the foundation and extent of duties, it is necessary to explore the conditions under which the different types of arguments may apply.

To put the point in the words of health activist and former international president of *Médecins Sans Frontières* (Doctors without Borders) James Orbinski, in a speech to the United Nations Security Council in 2000, "Language matters . . . it determines how a problem is framed and defines the range of possible solutions."[10] My first thesis in this paper is that Orbinski is right. Different ways of conceptualizing the global health duty will justify different forms of action. My second thesis is that of the various ways of pursuing justice-based arguments, the human right to health approach has a great deal to be said for it, both in terms of its theoretical basis and the types of action it mandates, and that therefore the issue of a *right* to health deserves further development. While I hope to demonstrate the first thesis in detail, I have to confess that my main contribution to arguing for the human right to health approach is to try to state it clearly, and to distinguish it from other approaches. A more substantial defense will not be attempted here.

JUSTICE AND GLOBAL HEALTH

Let us start with arguments from justice. The debate within the philosophical literature on the topic of global justice more generally can be approached by starting with some "common sense" reflections on the duties of justice we have to one another. It is commonly, though not universally, thought that citizens living within one country have duties of justice to offer extensive assistance to each other (if only through taxation and redistribution), but that those duties of justice stop at the borders of the political community. So in the case of health care, for example, all developed countries accept that at least some of the health care costs of certain disadvantaged groups—such as the elderly or the unemployed—should fall on the public purse to be met from general taxation. Again it is commonly, although not universally, thought that it would be unjust if this were not the case; that all citizens have some rights to health care; and that others within the society have a duty to contribute to the expense of such care, at least if the people concerned cannot afford to pay through no fault of their own. The claim here concerns moral rights, rather than legal rights. Legal rights are a matter of what the law demands; moral rights are rights one has independently of the law, and, indeed, the law can be criticized for not recognizing moral rights.

The common sense view, however, is that there is something different about the rights and duties that arise within a country and those that go beyond the country's borders. Whatever moral duties we think we have to pay for the health care needs of others in our own country, we would not generally think that we have a similar moral duty to pay for the health care needs of people in another country of a similar level of development. For example, people in the UK no doubt think that the health care needs of Canadians, at least when at home in Canada, is a matter for Canadians alone. On such a view, at least at first sight, it appears that political borders have moral significance: duties and rights change, or even disappear, at the customs post.

This common sense position does not go unchallenged, of course. One possible argument is to deny that we have duties of justice to offer assistance even to compatriots; this would be to adopt a libertarian position where the only rights individuals have are rights to non-interference, unless they have voluntarily undertaken more extensive duties or have violated the rights of others and now owe restitution or compensation.[11] The opposite response is to argue in the other direction; that not only do we have extensive duties of assistance to fellow citizens, but such duties also transcend national borders. On such a view, national borders exist for pragmatic, political, reasons, but have no moral significance in themselves. This is known as the cosmopolitan view: that the rights we have and the duties we owe to each other are universal in the sense that they extend to all other human beings, wherever they are located on earth. In its most radical form, the cosmopolitan view is that we owe exactly the same duties to all people in the world as we do to our compatriots.[12,13] Libertarianism and cosmopolitanism challenge the common sense view. However, it is also possible to argue that the common sense position is more or less correct: that we have duties of justice to assist our fellow citizens, but strictly such duties do stop at national boundaries. This does not mean that we should do nothing for people living in other countries, but rather that justice does not require us to do so, and therefore if we do have any duties across borders, they are duties of charity or humanity, rather than justice. This is commonly called a nationalist position.[14]

It is beyond the scope of this discussion to explore whether the libertarian, cosmopolitan, or nationalist view is correct. Rather we can ask what follows, on each view, for the question of whether the global health duty is a duty of justice. On the radical cosmopolitan view, the matter becomes rather simple. Because there is no moral difference between my relationship to my compatriots and my relationship to those in distant lands, then if I have a duty to assist those close to me, I equally have a duty to assist those far away.[15] Of course, in practice, what any of us can do is limited, and there are sensible reasons to prefer to act in areas where we can be more sure that we can make a difference. But the moral urgency of a claim on us depends on its extent, not its location, and on this view if we accept the duty to assist our fellow citizens in our own country, we equally have the duty to assist our fellow citizens of the world, wherever they are.

The basic idea behind the cosmopolitan position, then, is that differences of political allegiance and distance are morally arbitrary and cannot justify differential treatment. The point has been presented powerfully by Peter Singer, who compares one's duty to rescue a drowning child from a shallow pond a few feet away with saving the life of a malnourished child in a far-off country through sending a small amount of money. Singer suggests that it cannot make a moral difference that one child is a few feet away and the other many thousands of miles away.[16] Of course, issues of effectiveness and uncertainty affect what might be the most prudent way of directing my efforts, but the underlying moral issue is the same, suggests Singer, in both cases.

Although Singer does not state the argument in the language of justice, it is easy to see how the cosmopolitan justice argument can be mounted on exactly the same terms. If I have duties of justice to care about the health of those around me, then, as considerations of distance are irrelevant, I have the same duties of justice to care for the health of all people in the world. And so, for example, if I were to campaign for more money to be spent on research on drugs to cure conditions to which my family and neighbors are vulnerable, I must also campaign for research into conditions that do not occur in the geographical region in which I live. I should be happy to see the tax the government collects from me dispersed to wherever in the world it can do most good. As a start, it is commonly argued, justice requires that the countries of the developed world should significantly increase the amount of aid they send to the developing world to be spent on health-related projects.[17]

However, even if cosmopolitanism is morally correct, the full cosmopolitan position can seem an overly idealist, perhaps utopian, view, with little practical purchase. We live in a non-cosmopolitan world, where a form of the nationalist position is taken for granted. Nationalism entails something approaching libertarianism in respect of those who live beyond our borders: such people have a right that we do not interfere with their lives, but they have no right that we offer assistance. Or rather they have no right that we offer assistance unless we have contracted to do so, or have previously engaged, or currently engage, in behavior that violates their right of non-interference. If we have violated rights through our behavior, or have otherwise acted in ways that have harmed them, then we may owe duties of rectification or of corrective justice.

This introduces a powerful new possibility. The claim that the global health duty is one of justice does not, therefore, necessarily rest on cosmopolitan foundations. Even the libertarian and the nationalist should accept the duty insofar as it is shown to rest on claims of corrective justice.[18] Hence it is vital to explore the arguments that could lead to the conclusion that we do indeed owe the peoples of the developing world corrective justice, to rectify past or present violations of rights or other harms.

CORRECTIVE JUSTICE AND GLOBAL HEALTH

One argument often made is that developed countries owe duties of just reparation to their former colonies. This is promising, although somewhat problematic. In most cases, those who are now alive played no role in the colonial process, nor did today's citizens of former colonies suffer under colonial rule. Hence, it will be asked, how can there currently be duties of justice to repair injustices between members of former generations? One argument is that people alive today still benefit from the colonial legacy and so owe corrective duties to those who continue to suffer harm. The world order created by colonialism persists and is to the benefit of citizens in the developed world. Another argument is that we can take countries, as distinct from individual people, as the units of moral argument and responsibility, and the former colonial powers continue to owe duties of reparation to their former colonies. These arguments are, of course, controversial, and in this paper it is not my task to try to settle the issues.[19] Rather, I want to look at the types of arguments with which we can debate whether or not current generations in the developed world have moral duties, including duties of justice, to attend to the health of those in the developing world.

As we saw, the cosmopolitan view supposes that our common humanity provides strong reasons for thinking that we have duties of justice to all people on earth. The nationalist and the libertarian reject cosmopolitanism, but do accept that there are duties of justice to correct past rights violations, including international violations. One possible way of making such a case is the argument from the colonial legacy, which caused endemic and persistent poverty for the masses, which in turn has led to significantly adverse health effects. If this is correct, then even those who reject the cosmopolitan view can recognize that there are substantial justice-based duties to those who live in former colonies.

I will just run through quickly some other arguments that have been used to press the case that there are corrective duties of justice. Again, I don't want to assess these arguments. A second argument, after the argument from colonial legacy, is that the trade policies pursued by companies or individuals operating out of the developed world have done much to create the health problems that now need to be redressed. So, for example, the aggressive development of overseas markets for cigarettes, sugary soft drinks, and highly processed food, and the sometimes inappropriate promotion of powdered baby milk, and so on, have created a range of health problems. Even worse, perhaps, is the arms trade, fueling civil wars, as well as the world markets for drugs and sex. But arguably, even more significant has been the banking and debt policy enforced by some of the most respectable of our institutions, including international, government-backed, financial organizations such as the International Monetary Fund.[20,21]

We also cannot ignore a further source of harm; the recruitment of health professionals from the developing world to the developed world, wrecking attempts to build strong health services and draining human and financial resources from developing countries.[22,23] Finally, a further argument concerns highly exploitative labor markets, in which workers in developing nations produce cheap goods for the developed world for poor pay and in conditions that are detrimental to their health. Accordingly, if the truth is as sometimes alleged, arguments of corrective justice are available to support the global health duty.

Note, though, that if the argument for the global health duty is to be based on corrective justice, then any global health program must be related, in some way, to a previous injustice, and the purpose of the program will be to correct, or otherwise make up for, that injustice. So if, as suggested, the health brain drain constitutes an injustice, then justice demands either a change in practice and compensation for past wrongs or an agreed policy that puts future dealings on some sort of equal footing, involving, perhaps, significant transfer payments between governments. Similar arguments apply to unjust trade or banking policies that may have weakened health or health care systems in the developing world. Corrective justice would call for repair together, often, with compensation.

By contrast, if a health problem, such as an endemic disease or the consequence of a natural disaster, had nothing to do with the previous unjust actions of countries in the developed world, then there is no direct argument from corrective justice that can be used to support the global health duty.[24] This is confirmation of Orbinski's claim that "language matters," insofar as he means that language reflects moral reasoning. For example, there is great concern about such matters as the pricing of drugs and research on medicine for poor people's diseases.[25] Yet one will need to assume more than libertarian or nationalist premises if one wishes to argue that current drug pricing and research priorities amount to an injustice for which corrective justice is owed.

THE HUMAN RIGHT TO HEALTH

So far, I have not discussed human rights as specified in the Universal Declaration of Human Rights and related conventions. If the doctrine of universal human rights can be sustained, then it may be possible to argue that individuals throughout the world have positive rights for assistance, without endorsing the cosmopolitan view that state boundaries are morally arbitrary and all people of the world have exactly the same rights and duties to each other, wherever they live,[26] or relying on arguments from corrective justice. Nevertheless, insofar as it is correct to identify the provision of rights with justice, the human rights approach should be classified as a justice-based position.

In relation to health, Article 25 of the Universal Declaration of Human Rights provides a right to medical care, which was extended in the 1978 Alma-Ata declaration to a right to the highest attainable standard of health. There is increasing interest in exploring the idea of a human right to health.[27,28] The advantages of using the idea of a human right to health are clear: human rights are an internationally respected currency, backed by 60 years of institution-building and an enforcement mechanism, and ever-growing in influence. Yet both the purpose of human rights discourse and the implications of claiming that there is a human right to health remain to be explored in more detail. One worry, for example, relates to the distinction made earlier between moral rights and legal rights. Human rights, as recorded in conventions and ratified by governments, are clearly legal rights. The question is whether they are also moral rights; if they are not, it is hard to see how they can be relevant to the question of how to provide moral foundations for the global health duty. We will return to this vital question shortly.

To see the more general difficulties with arguments from human rights, it is worth beginning with skepticism about human rights, or at least about extending the notion of human rights beyond a small category of the most essential rights. There are at least four reasons for such skepticism. One is a worry about "rights inflation." If too many items are claimed to be "human rights," then the currency of human rights will be devalued and this will have a detrimental effect on even the most impor-tant human rights, such as the right not to be detained without trial. Second, rights inflation leads to the increasing possibility that human rights will conflict with each other, with no obvious method for resolving such conflicts.[29] Third, there is a concern that the language and institu-tions of human rights encourage a legalistic, confrontational culture, which sets people aside from their governments and can be counterpro-ductive in achieving desirable aims. Finally, there is a logical argument that it is incoherent to claim that a person has a right unless it is possible to identify a duty holder in respect of that right. In the case of the claimed "human right to health" it is not obvious who the duty holder will be.[30]

However, while important, these arguments do not seem decisive. The first, inflation, and third, legalism, are reasons for exercising caution, but fall short of showing that it is wrong to argue in favor of a human right to health. The second, that there could be conflicts that are hard to resolve, is a problem faced by almost every plausible approach to ethics, and so is not a distinctive difficulty for human rights theory. The fourth, the problem of the duty bearer, has two possible responses. One is to deny that it is necessary to identify a duty holder in order to claim a human right. The other is to suggest that it is possible to identify a duty holder in this particular case. It is interesting to pursue this second line, for there are two plausible candidates for the primary duty holder, which correspond to two quite different understandings of human rights.

On one view, the important feature of human rights is their universality, and as a result the duty holder is the international community as a whole. It is everyone's responsibility that each person's human rights are met. Of course, for practical purposes it may be necessary to devolve responsibilities to national governments, but this is by way of delegated responsibility from the international community. This view sees human rights as strongly related to an earlier philosophical tradition in which individuals were said to have natural rights, derived either from God's will or from reason.[31] On such a view individuals hold human rights against all other individuals, and human rights would exist even in an anarchist world without governments.

A second view suggests that the primary duty holder is not the international community but the government of the country in which the human rights holder resides. The role of the international community, in this view, is to take steps to encourage and assist national governments in carrying out their duties, and perhaps to pick up the duty in extreme cases of failure; but this is a secondary responsibility, not the primary responsibility of the immediate duty holder, the government of the country involved. Such a view emphasizes the idea of human rights as embodied in the practices of international law, and plays down the importance of placing accounts of human rights within any particular philosophical tradition, or of supposing that each human right needs to be given a philosophical foundation.[32]

The first understanding of human rights appears very similar to the cosmopolitan position, and therefore I will not devote further attention to it here. The second "international law" version, however, adds something new and, I think, very promising to the debate, and so it is this on which I will concentrate here. First, I need to say more about the moral foundations of such a view of human rights. Within the contemporary philosophical literature, there have been numerous attempts to provide a foundation for human rights, with many theorists claiming to have devised a justification that is in some way superior to those offered by others.[33] To some degree, however, this repeats a debate that took place 60 years ago, in the context of drafting the Universal Declaration. For then, as now, there was much greater agreement on the content of human rights than on the moral foundations for them. Of course, theorists may disagree about the exact formulation of some of the rights, and even about the inclusion of some, but the convergence on doctrine is remarkable, given the divergence on foundations.

Jacques Maritain, who helped provide background material for the convention at which the Universal Declaration was drafted, commented:

> During one of the meetings of the French National Commission of UNESCO at which the Rights of Man were being discussed, someone was astonished that certain proponents of violently opposed ideologies had agreed on the draft of a list of rights. Yes, they replied, we agree on these rights, *provided we are not asked why*. (emphasis in original)[34]

If the point was not clear enough, Maritain titled the section that contains this discussion "Men Mutually Opposed in their Theoretical Conceptions Can Come to a Merely Practical Agreement Regarding a List of Human Rights."[35] Unfortunately, the point is perhaps spoilt by the inclusion of the words "merely practical." To use Rawlsian terminology, we might argue that the Universal Declaration of Human Rights is a superb example of an overlapping consensus, in which each person can sign up to a political doctrine for his or her own moral reasons.[36] While people with different moral perspectives can agree on the political doctrine—that is, on the human rights that must be recognized—and can justify it on the basis of their own moral reasons, none of the moral perspectives has a privileged place as providing the core foundation for the doctrine. Different people will find their own justifications. In sum, then, on this view human rights have moral foundations, but no *particular* moral foundation. This, perhaps, explains the appeal of human rights doctrine within a broadly liberal framework.

The obvious concern here, of course, is to repeat Orbinski's point "language matters," or, more precisely, *foundations* matter. Different foundations may well produce different accounts of human rights, and so any claimed overlapping consensus is illusory. Indeed, Griffin devotes a paper to the argument that the Universal Declaration does not reflect the best philosophical understanding of human rights.[37] More generally we should expect differences between those who wish to base human rights on the protection of human agency and those, say, who wish to defend rights on the basis of human well-being.

The criticism is clearly important, and there is no purpose in denying that there can be disagreement about the content of human rights. Yet we need to keep such disagreement in perspective. First, we should note that human rights law is in this respect no different from other areas of law, be it family law, commercial law, or even criminal law. No legal doctrine can be complete; all legal doctrines need case law to settle unclarities, gaps, and tensions. Some pragmatic compromise and accommodation will be unavoidable in the attempt to develop a definitive account of human rights.

In this case, although philosophical difference is the cause of some disagreements, it is not the cause of all. So, for example, it is unlikely that the question of what should count as "accessible" health care, or what should be on the list of essential medicines, could be settled by philosophical argument about the foundations of human rights. Even if there were complete agreement on foundations, we would still have to face up to disagreements over specific matters. But let us restrict ourselves to cases where philosophical disagreement leads to a dispute about the content, or even the existence, of a human right. What should we do? Philosophers will produce whatever arguments they can muster, but it is unrealistic to think that if the disagreement is entrenched, there is any prospect of settling it to the satisfaction of the contending parties. In consequence, although language matters, there comes a point where philosophers will have to hand over their disagreements to international human rights lawyers.

This observation brings us to a further important feature of the "international law" understanding of human rights, which is that the international law understanding integrates philosophical theory with actual human rights practices, of conventions, declarations, and international courts. To do so has an interesting and radical consequence, for it provides a different "shape" to more familiar cosmopolitan or humanitarian arguments for thinking about the global health duty. Within international law, the standard responses to violations of human rights are, first, for both outsiders and insiders to criticize a regime for failing to meet the human rights of its citizens, and second, for other countries to put pressure on a regime, ranging from diplomatic to military in the most extreme cases, to change the behavior of the government (or even to change the government) so that human rights are respected once more. In other words, on this understanding, a human right is claimed against a particular government, which has the duty to meet the right. The international community is not directly expected to meet the claim itself, but has the second-order duty to help enforce the duty of the national government.

If the international law account of human rights is accepted, then conceiving of health as a human right changes the duties of the international community essentially from aid worker to policeman, with diplomats and lawyers, rather than doctors and nurses, the active agents. Consider the cholera epidemic in Zimbabwe in 2008–09. There seems little doubt that this was the result of a failure of government—specifically of governance and of public services, especially sanitation.[38,39] If there is a human right to health, then it is conceivable that it was violated in this case. However, an *international law* response to a human rights violation would not have been to send in emergency medical teams to treat cholera victims, and engineers to repair sanitation facilities (even though there were good humanitarian and perhaps cosmopolitan reasons for doing these things, too). Rather, if human rights are grounded in international law, the response would have been to put pressure on the government to do these things itself, with technical and financial assistance where necessary and desirable, but as a partner with the national government rather than an independent actor. Now, this may or may not be a welcome reorientation of the debate, but it shows that the international law perspective of human rights, if taken seriously, seems very different from a cosmopolitan foundation for the global health duty, because it appears to justify quite different actions. We might say that the logical structure of human rights grounded in international law mirrors one particular causal structure of action and intervention, whereas the logical structure of cosmopolitanism mirrors a quite different causal structure. This underlines the importance of trying to be clear about the foundation of the duty, at least in the case of global health.

In summary, we have looked at three different attempts to argue that the global health duty is one of justice. The first is the cosmopolitan view, that our duties of justice do not stop at national boundaries. The second is the "corrective justice" view, that the peoples of the less-developed world require assistance to overcome the harms that were caused through their interaction with the developed world. The third is the human rights view. All have their merits, and limitations, but we should note that the arguments are not exclusive, and that it may well be important to make room for all of them in a more complex view.[40]

THE HUMANITARIAN GLOBAL HEALTH DUTY

If the arguments from justice are not accepted, or if further support is desired, what is left? Shortly we will discuss arguments from self-interest. But first we must look at arguments from humanitarian concern, or as it might be put, charity, although the term "charity" is sometimes avoided, partly because of its connotations of being optional, and partly because of its patronizing overtones.

It seems easy to argue that there is a humanitarian duty to help those who are at least in the most critical situation. Indeed, even libertarians and nationalists will often accept that there can be extensive duties to help; but their view is that these duties are voluntary in the sense that doing the right thing should be a matter for individual conscience, and not external enforcement. On this view, if some people choose not to give charitable aid they can be criticized for their selfishness, but not for breaching anyone's rights.

A somewhat different suggestion is to use the idea of "imperfect duties": we each have a (non-voluntary) moral duty to help others through charity, but as I cannot possibly help everyone who is in severe need, it is implausible to say that any particular person who is a stranger to me has a right to my help. Still, I do wrong if I never help anyone, even if I do not breach anyone's rights. So, for example, some donor countries have decided to concentrate their efforts on a small number of recipient countries, on the ground that they can do more good by concentrating resources rather than spreading them thinly. Suppose that the people they are help are no worse off than others they neglect. If their duty is a humanitarian one, the country can argue that it is doing its duty by helping people in one country but it is not violating the rights of anyone else. However, if the global health duty is a duty of justice, then this defense does not seem to be available. Unless special factors apply, arguably wealthy donor countries could be criticized for neglecting the many people who are in as bad a situation as those who they help, and who it would seem also have a right to their assistance.

Duties of humanitarian assistance typically address to the unmet needs of others, rather than their rights. For this reason the moral obligations they generate are typically thought to be weaker than claims of rights. But perhaps more importantly, as Brian Barry points out, typically humanitarian intervention is carried out against an implicit recognition of some sort of status hierarchy, in which the strong or wealthy help the weak or poor, to achieve some particular goal for the weaker party that would be much more difficult without the assistance of the stronger. Relations of justice differ in at least two ways. First, there need not be any implicit status hierarchy between the parties, for the claim is based on rights, not exceptional needs. Second, claims of justice are not usually based on the achievement of some goal, but rather often concern the question of who has rights of control in a particular circumstance.[41] Meeting a claim of justice is more like the repayment of a debt than an attempt to improve the well-being of the badly off. In the case of a debt, all that matters is the particular transaction in hand; it does not matter whether one party is rich or poor, or has behaved well or badly in other situations, or is owed money by others, or how it will spend the payment once received. All of this is irrelevant, and the parties are considered equals in the eyes of the law. In this sense, justice has a narrow focus. Humanitarianism is different not only in the implicit status distinction between donor and recipient but also in paying attention to the broader context. Even if a party is in need, whether it is right for another party to attempt to meet that need will depend on a host of contextual factors, including the history of how the need arose, whether the recipient will spend the money wisely, and so on.

For these reasons, many people will seek to find an alternative basis for the global health duty, as humanitarianism seems to reinforce a pattern of status inequality. But still, with the world as it is, few would dispute that there is a duty of humanitarian assistance to take on the global health duty, provided that it really is possible to do some good.[42] As we have seen, however, the content of a humanitarian global health duty will differ from that arising under a justice-based view. Primarily it is determined by questions of appropriate goals and how to meet those goals, rather than the rights of individuals. I will return to this point at the end of this paper. Next, though, we should turn to the question of whether, in fact, developed nations ought to take on the global health duty out of enlightened self-interest.

SELF-INTEREST AND GLOBAL HEALTH

There are a number of ways of making the argument that taking on the global heath duty is in the interests of developed nations. Here are some. First, certain virulent infectious diseases are cosmopolitan by nature; they do not respect political boundaries. Consequently, to avoid worldwide pandemics it is necessary to address infectious diseases throughout the world,[43] and indeed, given the movement of peoples, it can be cost-effective for one country to vaccinate people in another.[44] Second, addressing health needs in general may also be helpful in creating a peaceful and secure world order.[45] Third, it may create beneficial strategic partnerships. Fourth, it may improve the possibilities of trade and therefore lead to economic advances. Fifth, and less nobly, it can offer an opportunity to governments to introduce what are, in effect, trade subsidies for their own industries without breaking free trade agreements. These last arguments, of course, are less likely to be made explicitly.

Equally, though, there are going to be limits to self-interest arguments. Indeed, it could be argued that self-interest is better advanced by spending more on defense, and cultivating only the more affluent and powerful. Furthermore, tough border controls could be much more efficient as a protective measure against infectious diseases, especially for relatively small and isolated countries. Those who aim to persuade others that taking on the global health duty is in the self-interest of developed nations[46] rarely consider the possibility that it might be even more in the self-interest of the developed nations not to do so, in terms of what else they could do with the same resources. We should also note that spending resources on tackling non-communicable disease will be harder to fit into this picture.

However, there is a different type of argument from self-interest, and here I mean the self-interest of individuals, rather than nations, to accept the global health duty. Of course there is, in some cases, a narrow, cynical, argument. Academics, whether medical researchers or political philosophers, have careers to advance, and working on global health is as good a way of doing it as any other; people might enjoy the adventure of working abroad, and so on. But there is a further, deeper and more interesting argument, according to which part of the good life for one person is to be able to do good for others, and that adding value to the lives of others adds value to mine. This observation accords with an Aristotelian ethical tradition where acting in an ethically admirable way need not always require self-sacrifice. Rather, it would be a matter of directing one's attention and energy into certain types of projects rather than others, for the sake of both oneself and those one helps. In some ways, then, it is problematic to classify this as an argument from self-interest, as it really is a way of attempting to undermine the distinction between self-regarding and other-regarding behavior.[47]

To some degree such a perspective has become submerged in individualistic Western thought, but is more present in communitarian Eastern and African thought.[48] The African concept of *ubuntu*—explained by Archbishop Desmond Tutu as the idea that "a person is a person through another person."[49]—perhaps captures much of this idea. An alternative presentation of a similar Bantu thought is stated by Magobe Ramose as "to be human is to affirm one's humanity by recognizing the humanity of others."[50] Such notions as these must surely implicitly or explicitly be moving many of the people who accept the global health duty: that to do so is personally fulfilling precisely because it is a way of responding to the humanity of others. It is also worth reminding ourselves of a point often missed: that the debate about global justice should require us to attend not only to the needs and rights of people in the developing world, but also to their personal and philosophical insights.[51]

THE NATURE OF THE GLOBAL HEALTH DUTY

I have been talking throughout of the "global health duty," but what has become clear, I hope, is that James Orbinski was right to suggest that "language matters."[52] By this I take it he means that different ways of conceptualizing the global health duty will give rise to different understandings of the duty. Indeed, we have already seen that there are many dimensions of possible difference, in terms of agency, content, scope, stringency, enforcement, and what we might call international social structure. Let us take these in turn.

Agency

The question of agency is, in fact, two-fold: who has the duty, and what patterns of action does it require? In respect of the first, we can see that if the duty is humanitarian, then it would fall on whoever is best placed to meet it; if it is one of self-interest, it would fall only on those who would benefit; if it is one of corrective justice, then it falls on whoever— be it individuals, states, or corporations—has committed the wrongs in question. Cosmopolitan justice appears to make the duty fall on everyone, whereas the international law human rights position presents a more complex account. In the first instance the duty falls on governments to satisfy the rights of their citizens. There is then a second-level duty on other citizens and political actors within that country, as well as members of the international community, to exert pressure on the government to do its duty. Finally, it is arguable that if all this fails, there is then a justice-based duty on all citizens of the world to satisfy the right, if they are able. At this point the international rights perspective coincides with the cosmopolitan duty, but it is a matter of last, rather than first, resort.

Content

Questions of agency concern who should act. Questions of content, we can say, are questions of what kind of actions are called for. Much discussion within the area of distributive justice and health has focused either on increasing flows of money[53] or on health care, in terms of access to medical advice, vaccinations, medicines, clinics, health workers, and hospitals.[54] Within a humanitarian framework this is perfectly natural, as humanitarianism, above all, appears to require us to give special attention to emergency situations, for which specific, targeted, interventions are often required.

However, it is increasingly understood that good health care is but one of the determinants of health.[55-57] I cannot go into details here, but for an example, consider the following, which appears in the World Health Organization (WHO) World Health Report for 2000, in the context of attempting to defend the importance of health systems in the face of skeptical arguments:

> [T]he generation and utilization of knowledge—that is, scientific and technical progress—explained almost half of the reduction in mortality between 1960 and 1990 in a sample of 115 low and middle income countries, while income growth explained less than 20% and increases in the educational level of adult females less than 40%.[58]

If broadly correct, these are remarkable figures, and do perhaps less than intended to provide reassurance about the importance of a health system, narrowly conceived, especially as the figures do not separate medical from other scientific and technical advance. If close to 40% of health gain in low- and middle-income countries is the result of adult female education, this seems to give us a powerful clue to future spending priorities, if our aim is not merely to respond to emergencies but to provide significant long-term improvements in health.

Of course there is no reason why a humanitarian must restrict his or her attention to issues of medical care and services. Humanitarians have the option to concentrate only on emergencies, or to broaden their concern. This is not the case, though, for those whose arguments provide a reason for caring about health, as distinct from health care. Such arguments—for example, those based on human rights—provide no basis for restricting attention to medical services, but rather need to consider the underlying social and material conditions that make people ill or well, and allow them to sustain their health.

We should note, though, a possible difference between cosmopolitan and the international law human rights views. In its pure form the cosmopolitan view provides that the duties one has to all people in the world are the same as the duties one owes to the people in one's own country. For people in the developed world, this appears to entail that there are duties to contribute to very expensive, specialized health services throughout the world. By contrast, the international law human rights view concentrates on circumstances in which human rights are violated, and it is arguable that many medical services currently offered in the developed world go beyond the threshold of what would be required to meet the human right to health. In that sense, the international law human rights view is less demanding than the cosmopolitan view. On the other hand it is much more specific in its demands, as it will focus on particular ways in which governments have failed to meet human rights, for example in the failure to provide appropriate therapies for the treatment of HIV/ AIDS in developing countries.[59]

Scope

The question of scope is here understood as the question of to whom the duty is owed. And once more, the answer will vary depending on the moral foundations of the duty. In the case of corrective justice, as we noted, the duty is restricted to those who have been wronged. In the case of humanitarian assistance, the duty is, strictly speaking, not owed to anyone. On this view one can act wrongly by not contributing to humanitarian aims, yet one wrongs no person. The self-interest view has yet another answer to the question of scope. Insofar as one can talk of duties (which in the self-interest context essentially amount to prudence), those duties extend only to people or countries that will generate a return of some sort. On the cosmopolitan view, the duties are owed to everyone, and this is also true on the international law human rights view. Where they differ, as we have seen, is in terms of issues of agency and content.

Stringency

The issue of stringency can be dealt with reasonably quickly, as we have introduced the issue before. Earlier we noted that a failure to further one's own self-interest is foolish but not morally problematic (there is a complication about a government's moral duty to act in the interests of its own citizens, but we can leave this to one side here). If one fails to act on a humanitarian duty, one has acted wrongly, but has not wronged any person in particular. However, failing to act on a duty of justice is to wrong a particular person, and generally to have breached his or her rights. In this sense, it seems, the duty is more stringent, at least in the sense that a higher level of moral disapproval will generally be thought to be justified.

Enforcement

In the case of most of the sources of duty we have discussed, there is no enforcement mechanism beyond individual conscience and perhaps social pressure. However, there are two important exceptions. The first is that in some cases corrective justice can, at least in principle, be pursued in international law. Examples might include the recruitment of essential health workers, in breach of international instruments regulating the practices,[60] and the behavior of commercial companies in making misleading claims about their products, leading consumers into behavior that is detrimental to their health. However, this avenue appears very limited. The second exception is, of course, the human right to health. As yet, case law in this area is also limited, but the approach holds the promise that the threat of pursuing the minister of health through international courts could provide a powerful spur to action.[61]

International Social Structure

We noted earlier an argument from Brian Barry that humanitarian concern generally reinforces a pattern of superior and inferior, whereas justice treats parties as equals.[62] The idea, presumably, is that humanitarian aid has the air of "noblesse oblige," in which the more fortunate feel responsible for alleviating the situation of those in an inferior position, while justice has no such connotations. Of course, one can imagine exceptions; one country might give another humanitarian aid this year, while next year the situation is reversed. Equally, those pursuing claims of justice may often perceive themselves as in an inferior position. Nevertheless, as a generality Barry's observation seems quite right; few will want to exist on charity, but they will have no qualms about making claims of justice. Charity can be humiliating in a way in which justice rarely is. Hence, we might suggest, humanitarianism will tend to reinforce the existing social standing of nations, whereas arguments from justice, at least in some cases, will make some steps towards challenging existing relations.

CONCLUSION

In conclusion, I think we have amply demonstrated Orbinski's claim that language matters. It may also be clear that I see a good deal of promise for the international law account of the human right to health, but developing the reasons for this will have to wait for another occasion.[63]

NOTES

1. Benatar S, Daar A, and Singer PA. (2003). Global health: the rationale for mutual caring. *International Affairs* 79:133–134.
2. Department for International Development, UK. (2009). *Eliminating World Poverty: Building Our Common Future.* London: DFID, pp. 15–19.
3. Nagel T. (2005). The problem of global justice. *Philosophy and Public Affairs* 33:113–147.
4. Pogge T. (2002). *World Poverty and Human Rights.* Cambridge: Cambridge University Press.
5. Sreenivasan G. (2002). International justice and health: a proposal. *Ethics and International Affairs* 16:81–90.
6. Barry B. (1982). Humanity and justice in global perspective. *Nomos* XXIV:219–252.
7. See also Sreenivasan, 2002.
8. Barry, 1982.
9. Nagel, 2005.
10. Orbinski J. (2008). *An Imperfect Offering.* London: Rider, p. 341.
11. Nozick R. (1974). *Anarchy, State and Utopia.* Oxford: Blackwells.
12. Steiner H. (2005). Territorial justice and global redistribution. In Brock G and Brighouse H, eds. *The Political Philosophy of Cosmopolitanism.* Cambridge: Cambridge University Press, 28–38.
13. Caney S. (2005). *Justice Beyond Borders.* Oxford: Oxford University Press.
14. For further discussion see Miller D. (2007). *National Responsibility and Global Justice.* Oxford: Oxford University Press, and Brock G. (2009). *Global Justice.* Oxford: Oxford University Press.
15. Many of those who accept cosmopolitanism block this inference by also arguing for a form of pluralism in which we have some cosmopolitan duties to all people on earth, but also special duties of justice to our co-nationals (Brock [2009]). Intuitively, this is a much more plausible position than radical cosmopolitanism, but for simplicity of presentation I will concentrate here primarily on the radical view, which does not add the further pluralist element.
16. Singer P. (1972). Famine, affluence and morality. *Public Affairs* 1, pp. 231–232.
17. Sreenivasan, 2002, p. 83.
18. Pogge, 2002, pp. 13–20.
19. For discussion see Thompson J. (2002). *Taking Responsibility for the Past.* Cambridge: Polity, and Miller (2007).
20. Benatar et al., 2003, pp. 111–114.
21. Hammonds R and Ooms G. (2004). World Bank policies and the obligations of its members to respect and fulfill the right to health. *Health and Human Rights* 8:27–60.
22. Kapur D and McHale J. (2005). *Give Us Your Best and Brightest: The Global Hunt for Talent and Its Impact on the Developing World.* Washington: Center for Global Development.
23. Brock G. (2009). Health in developing countries and our global responsibilities. In Dawson A, ed. *The Philosophy of Public Health.* Farnham: Ashgate, pp. 73–83.
24. Unless, of course, the previous injustice has been so deep and pervasive and has done so much damage that the only, or best, way of making up for it now is to accept a comprehensive global health duty.

25. Pogge T. (2005). Human rights and global health: a research programme. *Metaphilosophy* 36:182–209.

26. To be clear, I am not suggesting that cosmopolitans must reject doctrines of human rights; to the contrary, many cosmopolitans will argue that a concern for human rights either follows from, or is consistent with, cosmopolitan foundations. The point, rather, is that it is possible to accept the doctrine of human rights from a variety of moral or political standpoints, of which the cosmopolitan position is merely one.

27. See, for example, the Dec. 13, 2008, edition of *The Lancet*, which includes, among other articles, Backman G, Hunt P, et al. Health systems and the right to health: an assessment of 194 countries," which is an extensive audit of the degree to which 194 countries could be seen to be recognizing the right to health.

28. See also Clapham A and Robinson M, eds. (2009). *Realizing the Right to Health.* Zurich: Rüffer and Rub.

29. Freeman M. (2002). *Human Rights.* Cambridge: Polity, p. 5.

30. For discussion of these arguments see O'Neill O. (2005). The dark side of human rights. *International Affairs* 81:427–439.

31. Griffin J. (2008). *On Human Rights.* Oxford: Oxford University Press.

32. Raz J. (2010). Human rights without foundations. In Tasioulas J and Besson S, eds. *The Philosophy of International Law.* Oxford: Oxford University Press.

33. For example, Griffin (2008) and Ignatieff M. (2001). *Human Rights as Politics and Idolatry.* Princeton: Princeton University Press.

34. Maritain J. (1951). *Man and the State.* Chicago: Chicago University Press, p. 77.

35. Maritain, 1951, p. 76.

36. Rawls J. (1993/1996). *Political Liberalism.* New York: Columbia University Press, pp. 135–172.

37. Griffin J. (2001). Discrepancies between the best philosophical account of human rights and the international law of human rights. *Proceedings of the Aristotelian Society,* 101(1), p. 28.

38. Adams K. (2008). Zimbabwe's health system is in a state of collapse. *British Medical Journal* 337a:2637.

39. Truscott R. (2008). Zimbabwe faces a 'health disaster' as hospitals close. *British Medical Journal* 337a:2710.

40. Before leaving the topic of justice, we also have to acknowledge the possibility that some arguments from justice travel in the reverse direction; that is, arguments that a government's diversion of money to overseas projects is a breach of the trust it owes its taxpayers. Can taxpayers legitimately complain if the government spends money on global health, especially if it leaves even some urgent health needs of its population unaddressed? Again, I raise this possibility not to advocate it but simply to put it on the table.

41. Barry, 1982, pp. 219–252.

42. For recent discussion on whether aid is able to meet its goals, see, for example, Collier P. (2007). *The Bottom Billion.* Oxford: Oxford University Press; Dowden R. (2008). *Africa: Altered States, Ordinary Miracles.* London: Portobello Books; Easterley W. (2006). *The White Man's Burden.* Oxford: Oxford University Press; Moyo D. (2009). *Dead Aid.* London: Allen Lane; and Riddell R. (2007). *Does Foreign Aid Really Work?* Oxford: Oxford University Press.

43. DFID, 2009, p. 15.

44. Benatar et al., 2003, p. 133.

45. DFID, 2009, 16.

46. For example, Benatar et al., 2003.

47. Wolff J. (2002). Contractualism and the virtues. *Critical Review of International Social and Political Philosophy* 5:120–132.

48. Benatar et al., 2003, p. 128.

49. Wilkinson J. (2003). South African women and the ties that bind. In Coetzee PH and Roux APJ, eds. *The African Philosophy Reader.* London: Routledge, p. 356.
50. Ramose MB. (2003). Globalization and *ubuntu.* In Coetzee PH and Roux APJ, eds. *The African Philosophy Reader.* London: Routledge.
51. Cf. Widdows H. (2007). Is global ethics neo-colonialism? An investigation of the issue in the context of bioethics. *Bioethics* 21:305–315.
52. Orbinsky, 2008, p. 19.
53. Sreenivasan, 2002.
54. Pogge, 2005.
55. Wilkinson R. (1996). *Unhealthy Societies: The Afflictions of Inequality.* London: Routledge.
56. World Health Organization (2008). *Closing the Gap in a Generation, Report of the Commission of the Social Determinants of Health.* Geneva: WHO. http://www.who.int/social_determinants/final_report/en/index.html
57. Marmot M. (2004). *Status Syndrome: How Your Social Standing Directly Affects Your Health and Life Expectancy.* London: Bloomsbury.
58. WHO (2000). *World Health Report 2000, Health Systems: Improving Performance.* Geneva: WHO, p. 9. http://www.who.int/whr/2000/en/index.html
59. Heywood M. (2009). South Africa's treatment action campaign: combining law and social mobilization to realize the right to health. *Journal of Human Rights Practice* 1:14–36.
60. Brock G. (2009). Health in developing countries and our global responsibilities. In Dawson A, ed. *The Philosophy of Public Health.* Farnham: Ashgate.
61. I thank Gorik Ooms for powerfully making this point to me.
62. Barry, (1982).
63. This paper was first presented as a talk to the UCL Institute of Global Health, and then to audiences in Exeter and Madrid, and I'm very grateful to the audiences for their comments. I am also extremely grateful to Sarah Richmond, Gillian Brock and James Wilson, and especially to Joseph Millum for extensive written comments on earlier drafts.

REFERENCES

Adams, Kate. (2008). Zimbabwe's health system is in a state of collapse. *British Medical Journal* 337a:2637

Backman, Gunilla, Hunt, Paul, *et al.* (2008). Health systems and the right to health: an assessment of 194 countries. *Lancet* 372:2047–2085

Barry, Brian. (1982). Humanity and justice in global perspective. *Nomos* XXIV:219–252

Benatar, S., Daar, A., and Singer, P. A. (2003). Global health: the rationale for mutual caring. *International Affairs* 79:107–138

Brock, Gillian. (2009a). *Global Justice.* Oxford: Oxford University Press

Brock, Gillian. (2009b). Health in developing countries and our global responsibilities. In Angus Dawson, ed. *The Philosophy of Public Health.* Farnham: Ashgate, pp. 73–83

Caney, Simon. (2005). *Justice Beyond Borders.* Oxford: Oxford University Press

Clapham, Andrew, and Robinson, Mary, eds. (2009). *Realizing the Right to Health.* Zurich: Rüffer & Rub

Collier, Paul. (2007). *The Bottom Billion.* Oxford: Oxford University Press

Department for International Development, UK. (2009). *Eliminating World Poverty: Building Our Common Future.* London: DFID

Dowden, Richard. (2008). *Africa: Altered States, Ordinary Miracles.* London: Portobello Books

Easterley, William. (2006). *The White Man's Burden.* Oxford: Oxford University Press

Freeman, Michael. (2002). *Human Rights*. Cambridge: Polity

Global Health Watch. (2008). *Alternative World Health Report 2*. London: Zed Books

Griffin, James. (2001). Discrepancies between the best philosophical account of human rights and the international law of human rights. *Proceedings of the Aristotelian Society* 101(1):28

Griffin, James. (2008). *On Human Rights*. Oxford: Oxford University Press

Hammonds, Rachel, and Gorik Ooms. (2004). World Bank policies and the obligations of its members to respect, protect and fulfill the right to health. *Health and Human Rights* 8:27–60

Heywood, Mark. (2009). South Africa's treatment action campaign: combining law and social mobilization to realize the right to health. *Journal of Human Rights Practice* 1:14–36

Ignatieff, Michael. (2001). *Human Rights as Politics and Idolatry*. Princeton: Princeton University Press

Kapur, Devesh, and McHale, John. (2005). *Give Us Your Best and Brightest: The Global Hunt for Talent and Its Impact on the Developing World*. Washington: Center for Global Development

Maritain, Jacques. (1951). *Man and the State*. Chicago: University of Chicago Press

Marmot, M. (2004). *Status Syndrome: How Your Social Standing Directly Affects Your Health and Life Expectancy*. London: Bloomsbury

Miller, David. (2007). *National Responsibility and Global Justice*. Oxford: Oxford University Press

Moyo, Dambisa. (2009). *Dead Aid*. London: Allen Lane

Nagel, Thomas. (2005). The problem of global justice. *Philosophy and Public Affairs* 33:113–147

Nozick, Robert. (1974). *Anarchy, State, and Utopia*. Oxford: Blackwells

O'Neill, Onora. (2005). The dark side of human rights. *International Affairs* 81:427–439

Orbinski, James. (2008). *An Imperfect Offering*. London: Rider

Pogge, Thomas. (2002). *World Poverty and Human Rights*. Cambridge: Cambridge University Press

Pogge, Thomas. (2005). Human rights and global health: a research programme. *Metaphilosophy* 36:182–209

Ramose, Mogobe B. (2003). Globalization and *ubuntu*. In P. H. Coetzee and A. P. J. Roux, eds. *The African Philosophy Reader*. London: Routledge

Rawls, John. (1993/1996). *Political Liberalism*. New York: Columbia University Press

Raz, Joseph. (2010). Human rights without foundations. In J. Tasioulas and S. Besson, eds. *The Philosophy of International Law*. Oxford: Oxford University Press

Riddell, Roger. (2007). *Does Foreign Aid Really Work?* Oxford: Oxford University Press

Singer, Peter. (1972). Famine, affluence and morality. *Philosophy and Public Affairs* 1: 229–243

Sreenivasan, Gopal. (2002). International justice and health: a proposal. *Ethics and International Affairs* 16:81–90

Steiner, H. (2005). Territorial justice and global redistribution. In G. Brock and H. Brighouse, eds. *The Political Philosophy of Cosmopolitanism*. Cambridge: Cambridge University Press, pp. 28–38

Thompson, Janna. (2002). Taking responsibility for the past: reparations and historical injustice. Cambridge: Polity.

Truscott, Ryan. (2008). Zimbabwe faces a 'health disaster' as hospitals close. *British Medical Journal* 337a:2710

Widdows, Heather/ (2007). Is global ethics neo-colonialism? An investigation of the issue in the context of bioethics. *Bioethics* 21:305–315

Wilkinson, Jennifer. (2003). South African women and the ties that bind. In P. H. Coetzee and A. P. J. Roux, eds. *The African Philosophy Reader*. London: Routledge

Wilkinson. R. (1996). *Unhealthy Societies: The Afflictions of Inequality.* London: Routledge

Wolff, Jonathan. (2002). Contractualism and the virtues. *Critical Review of International Social and Political Philosophy* 5:120–132

World Health Organisation. (2000). *World Health Report 2000, Health Systems: Improving Performance* http://www.who.int/whr/2000/en/index.html

World Health Organisation. (2008). *Closing the Gap in a Generation, Report of the Commission of the Social Determinants of Health* http://www.who.int/social_determinants/final_report/en/index.html

CHAPTER 5

Justice in the Diffusion of Innovation[*]

ALLEN BUCHANAN, TONY COLE, AND
ROBERT O. KEOHANE

Contemporary theorists of distributive justice do not make the mistake of thinking that the problem of justice is that of fairly dividing a fixed stock of goods. They acknowledge that what is available to distribute changes as our productive capacities develop; that what is produced and how much is produced are subject, within constraints, to choices that human beings make; and that these choices should be guided by principles of justice. To that extent, their views are at least *consistent* with a remarkable fact about modern society: the prominence of innovation in our lives, especially in the form of new technologies developed through the application of scientific knowledge. Yet the significance of innovation for justice—the opportunities for promoting justice that it creates, and the risks of injustice that it poses—has not been adequately appreciated by theorists of justice.[1]

In the first section of this article, we explain why a theory of justice must take the fact of innovation seriously and focus attention on one important problem of justice in innovation: the fact that when powerful innovations do not diffuse widely, but are available only to some, this creates opportunities for domination and exclusion. In the second section, we advance a proposal for a new international institution designed to ameliorate this problem. In the third section, we strengthen the case for our proposal by comparing it both to the status quo and to a prominent

* This chapter is reprinted from an article in Journal of Political Philosophy

proposal for international institutional change advanced by Thomas Pogge. The fourth section explains how our proposal could be integrated into existing international law. Our aim is not to provide a full-blown theory of justice in innovation or a detailed blueprint for its institutional embodiment. Instead, it is to bring innovation to center stage in thinking about justice, to demonstrate that serious efforts to achieve justice in innovation will require institutional innovation, and to stimulate deeper consideration of the issues we address by articulating a concrete institutional proposal.

TOWARD A THEORY OF JUSTICE IN INNOVATION

The Need for an Account of Justice in Innovation

Innovation is significant from the standpoint of justice because it can have either positive or negative effects on justice. Depending on what is created and to whom it becomes available, innovation can worsen existing injustices or create new injustices, or it can lessen existing injustices. Justice in innovation is not restricted to the just distribution of *existing* beneficial innovations for two reasons. First, as the much-discussed case of essential medicines makes clear, the fact that vitally important innovations are *not* occurring can be a concern of justice. Because of lack of market demand in developing countries, medicines that could save the lives of millions of people in these countries, at relatively low cost, may not be developed. If justice implies a human right to health care (even of a rather limited sort), this situation is not merely unfortunate but unjust. Second, if restricted access to important innovations resulted in unjust inequalities of political power or in other forms of wrongful domination, this may contribute to injustices of other sorts.

One final, less obvious connection between justice and innovation is worth considering. In extreme cases the effect of limited access to the innovation would be a concern of justice if those who lacked access were excluded from participation in the most important forms of cooperative interaction. To understand this possibility, consider the much-discussed possibilities of biomedical enhancements of normal human capacities, using an analogy with disability rights. The Americans with Disabilities Act requires that "reasonable accommodation" be made to the special needs of persons with disabilities. For example, in public buildings, such as courthouses, curb breaks and ramps must be provided so that persons in wheelchairs can have access. Suppose that the cumulative effect of a number of biomedical enhancements, including significant enhancements in cognitive capacities and capacities for communication and coordination, was to enable those who had them to interact with other "enhanced cooperators" in a new, more complex, and extremely productive kind of economic cooperation. If most people became "enhanced cooperators" but some did not, the unenhanced might be unable to participate, or might be able to participate only in a minimally competent way, in the most important forms of cooperation in their society. They would in effect be the newly disabled. If the exclusion of people with physical disabilities from important sites of interaction is a matter of justice—a question of their *rights*, not just a matter of charity—then exclusion due to lack of access to powerful innovations would seem to be a matter of justice as well.

Given these possibilities, it is clear that taking the fact of innovation seriously in theorizing about justice requires not only including the products of innovation as subject to principles of just distribution, but also efforts to influence which innovations occur. Such efforts may be needed both to prevent innovations that would worsen existing injustices or create new injustices, and to encourage innovations that would lessen existing injustices. Accordingly, we can define "justice in innovation" as the conformity of both the distribution of the fruits of the processes of innovation, and of the character of the innovation process itself, to the requirements of justice. Justice in innovation may require a proactive stance: that is, it may be necessary to shape the innovation process in the name of justice, either to try to avoid the production of justice-degrading new technologies or to harness the innovation process for the purpose of promoting justice.

How Innovation Can Promote Justice

To the extent that thinking about justice has focused on innovation, concern about the negative impact of innovation on justice has been prevalent. Some observers have worried that if biomedical enhancements of normal human capacities become available to some but not all, this will worsen existing injustices. For example, genetic enhancements are likely to be affordable, at least at first, only to those who already benefit from injustices in the distribution of social goods.[2] Similarly, "the digital divide"—the fact that some people lack access to computers—can itself contribute to political and social inequality and may also exacerbate existing injustices in the distribution of other goods, including wealth.

Less attention has been given to the potential of innovation for promoting justice. To correct this imbalance, we offer the following examples of technological innovations that may have significant justice-*promoting* effects. In each case the innovation in question could be seen as promoting justice by reducing unjust advantages that some people enjoy or by empowering individuals so that they can better exercise their rights.

(1)   Some cognitive enhancement drugs are most efficacious for the less bright; to the extent that existing social arrangements unfairly disadvantage those with lower intelligence or lower intelligence results in part from socioeconomic injustices, making such drugs available to the latter could be justice-promoting.[3] Such pharmaceutical cognitive enhancements might prove more cost-effective than some educational interventions.

(2)   Cheap calculators help "level the playing field" for those who are mathematically challenged, thus reducing injustices that may arise from the ways in which society rewards those with superior math skills, or penalizes those who lack them.

(3)   Medical innovations can remove disabilities that interfere with opportunities individuals ought to have as a matter of justice or that prevent them from exercising their rights.

(4)   Cell phones allow cheap, rapid coordination of economic and political activities; this can help people to lift themselves out of poverty and enable them to exercise their rights of political participation more effectively.

(5)   Internet access to medical information reduces knowledge asymmetries between physicians and patients, and this in turn can reduce the risk that patients' rights will be violated.

(6)   Cell phone cameras provide checks on police behavior, thus helping to reduce violations of civil and political rights or at least facilitate remedial action when they occur.

Disagreement and Uncertainty About Justice

Each of the preceding six innovations appears to reduce certain *inequalities*—but not all inequalities are *injustices*. To know which inequalities are unjust, and hence whether particular innovations are affecting justice positively or negatively, one needs an account of justice. Theorizing about justice is notoriously afflicted, however, with both disagreement and uncertainty. There is disagreement between consequentialists and deontologists, between proponents of "positive" rights and libertarians, between egalitarians, prioritarians, and sufficientarians, and among egalitarians as to what the "currency" of egalitarian justice is (well-being, opportunity for well-being, or resources). In addition, there is uncertainty as to how to move from a given theory's abstract, highest-level principles to lower-level principles with clearer implications for policies and institutions. For example, even if one assumes one knows what the proper principles of distributive justice are for what Rawls calls the basic structure of society, it is not clear which principles of justice should guide particular policies or decisions about rationing scarce medical resources.[4] Given that there is no indication that this disagreement and uncertainty is likely to be resolved in the foreseeable future, how should thinking about justice in innovation proceed? How *can* it proceed in a principled way?

A Provisional Starting Point: The Injustice of Extreme Deprivation

Most theories of justice converge on the belief that what might be called extreme deprivation is presumptively unjust, at least when it is undeserved and unchosen. People suffer extreme deprivation when they lack adequate food and shelter and safe drinking water, when they are afflicted with serious preventable diseases, and when their physical security is seriously compromised by the threat of violence, as in the case of civilians in war zones.

We proceed on the assumption that whatever else it should be concerned with, a theory of justice in innovation should treat extreme deprivation as a matter of concern, in two ways. It should provide guidance both (a) for reducing the risk that innovations will produce or exacerbate extreme deprivations and (b) for helping to ensure that the power to innovate will be harnessed to help ameliorate existing extreme deprivations. The strategy is to consider policies regarding innovation that address the concern about extreme deprivation, without waiting for a resolution of the disagreement and uncertainty that characterize current theorizing about justice. Surely there is enough agreement that some harms should be included to allow us begin to grapple with the problem of justice in innovation.

Exclusion and Domination

To focus *only* on extreme deprivation, however, is too restrictive, for reasons already indicated: it overlooks the fact that innovation can be a concern of justice when limited access to innovations results in unjust exclusion or domination. The analogy with disability rights shows that unjust exclusion can occur without severe deprivation and that these are distinct injustices. Even if a person with mobility limitations is not impoverished and leads an otherwise comfortable life, he or she may rightly complain of injustice if he or she is barred from access to public buildings.[5]

Political inequalities can also be unjust even when they do not result in extreme deprivation. The fact that women in the United States lacked the right to vote until 1920 was an injustice, apart from whatever contribution it made to the extreme deprivation that some women suffered. Similarly, if some innovation in electronic communications conferred advantages in influencing national political processes, in ways that are incompatible with the commitment to broad effective political participation embodied in democratic institutions, this would be an injustice, even if those who were disadvantaged suffered no extreme deprivation.

Some inequalities in political power are inevitable even in the most democratic societies; and some inequalities in political power are not unjust, including those that result from special excellence in the qualities of political leadership. But under modern conditions, in which the state wields such great power over our lives, inequalities in political power have the potential to exacerbate existing injustices and undermine justice where it exists.[6] So political inequality is a proper *concern* of justice even if the people involved are integrated into the society and political inequality is not in itself unjust.[7] Both instrumental considerations grounded in the strategic nature of political inequalities and views according to which political equality is valuable in itself converge on the conclusion that political inequalities are a proper concern of justice, independently of their propensity to create or sustain extreme deprivation. For brevity, we will use the phrase "basic political and economic inequalities," not to refer to just any unequal distribution but only to (a) seriously unjust inequalities in political power and (b) lack of access to important sites and forms of social cooperation that is of comparable consequence to the exclusion suffered by persons with disabilities in societies that do not take disability rights seriously. Our suggestion is that an account of justice in innovation should not be limited to a concern about extreme deprivation, but should also address the potential impact of innovation on "basic political and economic inequality" understood in this way.

It could be argued that the impact of innovation on extreme deprivations is a higher priority, from the standpoint of justice, than the impact on basic economic and political inequality. Whether or not that is so may depend upon the resolution of deep disputes in the theory of justice—in particular whether some form of prioritarianism is the correct view. We have already explained why we think it is appropriate to avoid attempts to resolve such disputes before embarking on an attempt to develop a principled practical response to the issues of justice in innovation.

From the standpoint of many persons in developing countries, the main concern about innovation is its potential impact on extreme deprivation, but for most of those who live in developed countries the impact on basic political and economic inequalities may be more pressing. Given that this is so, there are two reasons to include basic economic and political inequality, not just deprivation, in our provisional conception of justice in innovation. First, it is a *legitimate* concern for people generally, regardless of whether they live in developed or developing countries, even if deprivation is the more serious moral concern. Second, practical thinking about justice in innovation must take the problem of political feasibility seriously, and generally that requires engaging the interests of the better off. An approach to justice in innovation that focuses not only on deprivation but also on basic social and political inequalities is more likely to gain the support of those who are critical for its success.

The threads of the argument thus far can now be pulled together. Because of the prominence of innovation in modern life, thinking about justice should take seriously the potential of innovations both to worsen injustices and to ameliorate them. We begin with undeserved and unchosen severe deprivation, but expand this narrow focus to encompass "basic political and economic inequalities"—seriously unjust inequalities in political power and exclusion from the most important sites and forms of productive cooperation.

Types of Institutional Strategies

There are three basic types of institutional strategies for the pursuit of justice in innovation: (1) *prohibition* of innovations that would worsen existing injustices or create new injustices; (2) *creation* of innovations to ameliorate existing injustices; and (3) *diffusion* of innovations in order to avoid injustices that would arise from differential access to them or to promote justice by ameliorating or removing existing unjust disadvantages.

Prohibition

Voluntary abstention from the development and diffusion of valuable innovations would likely fail, due to familiar free-rider and assurance problems, and is also in tension with the scientific ethos of discovery. Regulation (coercively backed prohibitions) to try to stop development and/or diffusion of innovations thought to have unjust inequality-increasing effects is hardly more promising, for at least two reasons. First, the innovation process is by its nature highly unpredictable and the effects of an innovation on justice, whether for good or for ill, may be especially hard to predict. Consequently, a coercively backed prohibition strategy might deprive us of valuable innovations that would turn out to be consistent with the demands of justice or that might even promote justice. Second, if a certain line of research and development is prohibited in one country or regional governance regime (such as the European Union), it is likely to be taken up in less regulated locales, as has happened across a wide range of cases, including gene therapy and human embryonic stem cell research. For a number of reasons, including the lack of regulatory capacity in many countries, an effective, world-wide scheme of regulatory prohibition, while conceivable in principle, is unlikely in the foreseeable future.

Creation

There are many examples of private and government efforts to spur innovations of various sorts, such as research grants, government contracts, and public and private prizes. Few of these efforts are explicitly directed toward issues of justice in innovation. An exception may be the U.S. Orphan Drug Act, which provides research grants and extended patent life for drugs developed to treat serious diseases that afflict small numbers of people. One plausible interpretation of the purpose of this legislation is that it is designed to ameliorate the unfairness of a situation in which the direction of drug research and development is determined by market demand rather than need, to the life-threatening disadvantage of those with rare diseases. Several more recent proposals to ameliorate the "essential medicines" problem, including one by the philosopher Thomas Pogge, which we consider in detail below, can also be seen as efforts to stimulate the creation of drugs for the purpose of promoting justice. We will use the term "the Creation Problem" to refer to obstacles to the creation of innovations designed to address injustices, including unprofitability under standard market conditions.

Diffusion

Limited or slow diffusion of a beneficial innovation can be problematic from the standpoint of justice for either or both of two reasons: once created, innovations do not mitigate problems of inequality unless they are diffused widely to the disadvantaged; and if diffusion is too limited or occurs too slowly it may actually produce new injustices, either by giving unacceptable advantages in political power to those who do have access to them or in excluding those who lack access to them from important sites or forms of economic cooperation. For convenience, we will use the term "the Diffusion Problem" to refer to both of these phenomena.

A wide range of existing programs, projects, and organizations can be seen as exemplifying the strategy of promoting the diffusion of technologies in order to avoid or mitigate injustices due to lack of access or to promote justice by removing existing unjust disadvantages. An illustrative list might include the following:

(1)   Private and government efforts to bridge the "digital divide" by providing subsidized or free computers, high-speed and/or wireless Internet service, etc.

(2)   Private and government programs designed to diffuse more widely the extremely valuable cognitive enhancement technology commonly known as literacy

(3)   Vaccine delivery programs in less-developed countries, where infectious diseases are still a major contributor to childhood mortality

(4)   Donation or reduced pricing of "essential medicines" through arrangements between governments and pharmaceutical companies (in particular, antiretroviral HIV/AIDS medications)

(5)   "Compulsory Licensing," as recognized by the World Trade Organization's (WTO's) Doha Declaration on TRIPS and Public Health, which acknowledges the right of states to grant licenses for producing essential medicines without the permission of intellectual property (IP) rights holders, if certain standards are met

These examples of diffusion policies are not part of an overall strategy formulated in response to the articulation of goals of justice in innovation. Instead, they reflect an uncoordinated, piecemeal approach. In the next section, we outline a systematic proposal for promoting justice in innovation that emphasizes the Diffusion Problem but also does something to address the Creation Problem. The core of this proposal is a new institution—the Global Institute for Justice in Innovation (GIJI). The proposal focuses on one important impediment to diffusion: the monopoly pricing that results from the current IP rights regime.[8] Our proposal is to modify the IP rights regime in a way that preserves its valuable functions while remedying or at least significantly ameliorating its institutional failures.

THE GLOBAL INSTITUTE FOR JUSTICE IN INNOVATION

The GIJI would be an international organization designed to construct and implement a set of rules and policies governing the diffusion of innovations, on the basis of a sound set of principles. It would operate under conditions of accountability, according to rule-governed procedures, and would seek gradually to inculcate norms that specified appropriate behavior with respect to the diffusion of innovations.[9] The GIJI would be created by a multilateral treaty, with permanent staff, and international legal authority to make decisions that would not automatically be incorporated into the domestic law of its member states, but would become enforceable only as a result of political and constitutional processes undertaken by each member state. In this sense, the GIJI would be similar to the WTO, the rules of which are directly effective only on the international level, rather than the European Union, which requires as a condition of membership that certain rules be directly applicable in domestic legal proceedings. Such an arrangement would limit the sovereignty costs of the GIJI.

A subsidiary activity of the GIJI would be to encourage the creation of useful innovations, for example through prizes and grants for justice-promoting innovations and through offering extended patent life for innovations that have a positive impact on justice. But its major efforts would be directed toward the wider and faster diffusion of innovations in order to ameliorate extreme deprivations and reduce their negative impact on basic political and economic inequalities, as defined in the first section above.

The GIJI would actively promote diffusion entrepreneurship—that is, efforts by nongovernmental organizations (NGOs) and others to accelerate the diffusion of justice-promoting innovations. Indeed, the GIJI could give awards or prizes to firms that had consistently exceeded its diffusion standards, thus providing the firms with reputational benefits.[10] Its most important asset, however, would be what we will term a "licensing option," under which the GIJI would authorize compulsory licensing on a country-by-country basis of innovations that are diffusing too slowly. "Too slowly" here means that the innovations are failing to realize their potential for making significant gains in promoting justice or are exacerbating existing injustices, in the form of extreme deprivation or basic political and economic inequalities.

Member governments of the GIJI would enact legislation authorizing the relevant domestic authorities to initiate administrative actions to issue compulsory licenses for IP as authorized by the GIJI. Since this proposal to allow centrally directed compulsory licensing of IP in these cases is, to our knowledge, a new idea, we will focus on it in what follows.

The Licensing Option

Licenses would be granted to firms or other entities selected by the GIJI free of charge or for nominal fees, and would be distributed so as to reduce the price of the innovation to competitive levels. Thus, if the current slow diffusion of the product is due to monopoly pricing, freely distributing the license would accelerate diffusion. Some innovations, however, diffuse slowly because they are of little value. This is why the GIJI would have a licensing *option*. It would act only where there is evidence that the obstacle to diffusion would be removed by authorizing compulsory licenses and creating a competitive market for the innovation in question.

It is important to understand the political implications of the GIJI's authorization option. Without imposing supranational authority over governments, such authorization would render mandatory licensing by a developing country internationally legitimate. In view of the broadly representative nature of the authorizing body, to be discussed in more detail below, it would be hard for companies, in such a situation, to claim unfairness. The GIJI would therefore greatly strengthen the bargaining position of countries that had well-founded claims of insufficient diffusion. At the same time, however, it would protect firms against attempts by opportunistic governments to abuse compulsory licensing by seizing private property. This proposal, therefore, does not try to suppress or avoid politics (a quixotic venture in international relations) but to shape politics in desirable ways.

If the GIJI's threat to authorize mandatory licensing has sufficient credibility, and imposes sufficiently high threat of loss on the firm, exercise of the GIJI licensing option should be a rare event. The threat of mandatory licensing would deter producers from exercising the capacity for monopoly pricing that IP rights confer. Producers would know that they can keep their full IP rights by refraining from monopoly pricing in the case of innovations whose slow diffusion would have a negative impact on justice. Producers would know that they could avoid the negative publicity of being warned about mandatory licensing, and could receive public praise and reward (through the prizes and grants policy) if they act in ways that promote justice. Over time, this array of incentives could help foster the norm of taking justice into account in the innovation process.

Stages of Intervention

Exercise of the licensing option would be a last resort. The GIJI would construct a "watch list" of innovations that warrant scrutiny from the standpoint of inadequate diffusion. Producers of innovations on the watch list would be notified, without public announcement, that they are on it and that if diffusion does not improve, a *publicized* warning of potential liability to mandatory licensing will be issued in due course. If there is no significant improvement or evidence of significant efforts on the part of the producer to bring about improvement, the GIJI would initiate its internal process for authorizing mandatory licensing and announce that it was doing so. Such authorizations would be both (1) time-limited and (2) area-specific. Compulsory licensing would be authorized for a limited time period only, say from one to as much as ten years, depending upon projections as to how long it would take to achieve a significant increase in diffusion, and the time required for the licensee to receive an adequate return on its initial investment. If the Diffusion Problem were limited to certain less-developed countries in which access to the innovation is critical (as is the case with medicines to combat malaria, for example), then the innovator would lose IP rights only with respect to that market. After the GIJI had authorized compulsory licensing, there would be another period in which the firm whose products were under scrutiny could change its policies to promote diffusion, providing another opportunity for compromise before mandatory licensing was imposed.

We have to consider the likelihood that firms and states supporting them might use this opportunity not to adjust their own policies but instead to put pressure on weaker states not to exercise their authority to invoke compulsory licensing. To reduce this risk, several measures would be necessary. There would have to be a clear legislative statement of observable "pressuring" actions that were inappropriate in conjunction with a GIJI process for compulsory licensing, and of the period of time in which they were inappropriate (any time after the GIJI started considering compulsory licensing for a given product in a given country). Inappropriate actions would include any actions that would be reasonably interpreted as a punishment or threat toward a country that utilized a GIJI authorization for compulsory licensing. Such actions would include withdrawing products from a country's market or raising prices/royalties on them except as part of a general policy applying to a set of similar countries, or threatening to do so; or threatening the withdrawal of other forms of international aid, or of support on an unrelated issue in another forum. On a complaint by a state against a company or another state, a GIJI process would be set in motion involving conciliation, an arbitral panel, and the GIJI's appellate tribunal, as necessary.

If a state or company were found responsible for such actions, it would be put on probation. Complaints against companies or states on probation would be put on a fast track, bypassing the conciliation stage and shortening time periods for each stage in the process, while nonetheless remaining within the limits of due process. Lists of states and companies on probation would be published, and penalties for repeat offenses would be steeply increased. Such a process would strongly discourage coercive interference with a state's decision to utilize the GIJI's authorization of compulsory licensing, while not violating due process or mixing judicial with legislative functions.

Given that compulsory licensing would be time-limited and area-specific, and that the option need rarely be exercised for its purpose to be realized, this proposal can be properly characterized as a modification of existing IP rights, not a radical overturning of them.

Compensation

With respect to the crucial question of compensation, we can imagine a continuum, at one extreme of which there would be no compensation. Such a policy would have the advantage of deterring monopolistic practices and would enable the GIJI to operate on a relatively small budget. But there are three decisive objections to a no-compensation policy. First, innovation would be discouraged, especially innovations designed to help poor people in poor countries, since it is precisely these innovations that would be subject to GIJI authorization of compulsory licensing. Second, significant alterations would be necessary to many contemporary international agreements, including TRIPS and numerous bilateral investment treaties, which require that some level of compensation be paid upon the compulsory licensing of a patent. Third, it is virtually unimaginable that such a policy would be endorsed by wealthy countries that are home to the most innovative firms in such fields as pharmaceuticals and electronics, and whose ratification of a treaty for justice in innovation would be essential for the GIJI to have a meaningful impact.

At the opposite extreme of this continuum would be a policy guaranteeing full market value compensation. If credible, such a policy would not significantly discourage innovation and would generate support from powerful firms. However, such a policy would essentially use public funds to pay monopolistic prices to private firms. This would be unpopular with democratic publics, and it would be difficult to raise sufficient sums to finance many such licenses. Furthermore, it would not deter monopolistic behavior—quite the contrary, it might encourage it.

It seems clear that neither zero compensation nor compensation at the full (monopolistic) market value of the innovation is satisfactory. Hence some middle ground will have to be found. A "fair price," representing a substantial but not exorbitant rate of return for the company, would have to be paid. In our view, current theorizing about justice does not ground a unique determination of "fair price" here; instead, there is probably a range of reasonable alternatives. One of the first actions of the GIJI would be to devise a set of procedures through which a fair price would be determined. The trick is to pick a pricing system that creates the right incentives, given the goals the licensing option is designed to promote, and avoids any clear unfairness to any of the parties concerned. Since anything less than paying the monopoly price could somewhat discourage innovation, the GIJI might find that its diffusion strategy would be more effective if combined with subsidies for the creation of promising drugs, compensating for the speculative but sometimes alluring prospect of very large monopolistic profits in the long run.

Compensation would be paid directly by the GIJI, rather than through the traditional approach of the payment of royalties from sales of licensed products, in order to avoid the price increases that would result from royalties designed to pay the "fair price" determined by the GIJI. Such an approach would be consistent with the GIJI's goal of increasing diffusion of innovations, as a lower price would maximize the number of individuals able to afford the innovation in question.

Political Decision-Making by the GIJI

One of the major functions of the GIJI would be to assess the justice implications of the pace at which useful innovations were diffusing to disadvantaged people, either those suffering severe deprivation or those laboring under burdens of basic economic and political inequalities. Carrying out this function would be contentious and large amounts of money could be at stake, so the GIJI's decision-making arrangements need to be carefully designed. We sketch only one possible design here, to suggest the feasibility of our proposal and to promote discussion.

The GIJI would have an administrative unit with analytical competence and the authority to propose exercise of the GIJI's licensing options and other actions. The model here is something between the WTO, the Secretariat of which is relatively small and definitely subordinate to the membership, and the World Bank and the International Monetary Fund, which are operated by much larger administrative organizations that make many decisions with only general supervision from their boards. The executive head of the GIJI could not order licensing of IP rights on his or her own, but could propose licensing to an assembly of the GIJI.

The assembly, which would meet annually, would be composed of representatives of developed and developing countries, NGOs with substantial records of service to disadvantaged people (such as Save the Children and Oxfam), and firms holding patents. Participating NGOs would have to satisfy familiar requirements of transparency, financial integrity, independence from governments and corporate interests, and responsiveness to the preferences and needs of those individuals and groups they claim to represent or on whose behalf they claim to act.

Each of the four constituencies would elect its representatives at the assembly. As in the Montreal Protocol Fund, governments of developed and developing countries would have equal numbers of representatives, elected separately from these constituency groups. One possibility would be an assembly of 32 representatives, consisting of eight industrialized countries, eight developing countries, eight NGOs, and eight innovation-producing firms. It is important that the numbers not be too large; the Montreal Protocol Fund body, with 14 members, has operated much better than the unwieldy universal bodies involved in the Kyoto Protocol and post-Kyoto negotiations.

Decisions by the GIJI assembly to authorize compulsory licensing would require a super-majority for immediate action, coupled with a majority of the votes in three of the four categories of representatives. Demanding immediate action, NGOs and developing countries could not join with one or two industrialized countries to exercise a licensing option; on the contrary, they would have to get a majority of either industrialized countries or firms. There could be a provision for relaxing this requirement after a delay (say, of one year) in order to give IP rights holders and others time to voice disagreement. The idea is to promote deliberation and compromise, but not to give any one group (such as major drug companies supported by the United States) a veto.

Accountability

The basic structure and key procedures of the GIJI would be deliberately designed to promote accountability. The composition of the GIJI assembly would ensure that the organization is accountable, not just to the states that ratify the treaty that creates it—both with developed and developing economies—but also to various publics whose interests are represented by NGOs, and to the community of innovators. Furthermore, accountability would be enhanced by the stipulation that all major organizational actions, including acquisitions and changes in operating rules, are subject to administrative due process.

Proposals to authorize compulsory licensing could be made only under a set of rigorous due process requirements. First, the executive head of the GIJI would have to make a public announcement of intention to propose compulsory licensing of a specific set of IP rights in a specific country for a specified period of time, and the GIJI would have to provide clear means for comments and discussion. This procedure would be similar to the "notice and comment" procedures of U.S. administrative law, which require federal agencies to publish potential rules, allow time and opportunity for interested parties to complain and make suggestions, and require a reasoned response from the agency proposing the rules. After the required period of perhaps 45 or 60 days has elapsed, the GIJI would have to reissue its proposed order for compulsory licensing, at which point it would formally be put on the docket of the assembly. Decisions of the GIJI could be reviewed for conformity to due process standards by an appellate tribunal, roughly modeled on the Appellate Body of the WTO.[11] That is, there would be a public set of procedures that encouraged compromise but provided for rulings by expert panels that could be appealed to the appellate tribunal, composed of judges selected for relatively long terms. The appellate tribunal would hear cases in public and issue public decisions providing reasons, which could serve as precedents to develop a body of GIJI law.

Funding

The GIJI's funding would come chiefly from member states, on a sliding scale, according to ability to pay. On the model of the World Bank subscription system,[12] countries would commit funds as necessary in large amounts—funds that would be essential to ensure that IP rights holders subject to compulsory licensing received fair compensation.[13] As noted above, the GIJI would pay compensation directly, subsidizing diffusion of the innovation. Having funds readily available would enhance the credibility of the GIJI's warnings that it was intending to order compulsory licensing, and contribute to its deterrence of monopolistic pricing,

Is the Proposal a Morally Unacceptable Modification
of Existing IP Rights?

This proposal does not assume that innovators are morally responsible for injustices that result from inadequate diffusion of their products. On the contrary, we do not believe that innovators have any special moral obligation to promote justice through the diffusion of their products.

The GIJI's ability to order compulsory licensing only assumes that the moral rights innovators have regarding their creations do not preclude the very limited form of interference with existing legal IP rights that properly exercised compulsory licensing entails. At any rate, our proposal is directed toward those who view the existing IP rights regime as roughly within the bounds of the reasonable and the morally acceptable, not toward radical natural rights views that ascribe extremely broad, indefensible "natural" moral rights to innovators. Moreover, our argument is *comparative:* given a reasonable construal of the existing IP rights regime as an instrument designed to serve a plurality of widely held values, our proposed modification of it does a better job of balancing those values. It ameliorates a very troubling side effect of monopoly pricing without an unacceptable decrease in incentives for innovation.

Is the Proposal Politically Realistic?

One could expect the GIJI to be greeted with at least cautious enthusiasm by developing countries and NGOs. Of course, their bargaining strategies will temper their public support, since they will be working for more favorable terms; but in fact they have much to gain and little to lose from the proposal. The proposal will not be as attractive to powerful developed states and the innovation-creating firms based in these states. If the GIJI is to work, it will require the support of these states, including especially the United States and Japan, and the European Union; and at least acceptance by major firms—which might itself be a necessary condition for support by powerful states. Without making unrealistic assumptions of altruism, what incentives would powerful states have to help create and to sustain the GIJI?

Before focusing on the positive incentives, it is important to note that the GIJI does not threaten the constitutional sovereignty of states: that is, their legal supremacy and independence.[14] States would retain their ability to make final decisions on issues of importance to them. All member states would retain the ability to determine for themselves how much control to deliver to the GIJI, and would also retain the right to decide whether to take up any authorizations they received. The GIJI's rulings would not have direct effect within domestic jurisdictions, and could not override domestic laws. Moreover, states could withdraw from the organization, with due notice.

Like the WTO, the GIJI would constitute an *exercise* of sovereignty by states. Members of the GIJI would be publicly committed not to thwart the purposes and actions of the organization—for instance, by threatening retaliation against the GIJI for ordering compulsory licensing of IP rights owned by their own firms, if these acquisitions were judged by the appellate body to have been carried out in conformity with its rules and procedures. Like all international legal agreements the GIJI would limit the legal freedom of action of states, but it would not affect their constitutional sovereignty: their fundamental right to make decisions for themselves.

There are four major positive reasons for developed countries and their firms to support the GIJI. The first and most general is that more rapid diffusion of innovations would accelerate economic development worldwide—a long-term goal of developed countries, as it is in their interests to enhance both prosperity and the chances for a peaceful and more democratic world order. Wide diffusion of innovations would create conditions facilitating the creation of more innovations in more diverse ways, some of which would almost certainly rebound to the advantage of people in developed countries.

Since appeals to general interests are often not persuasive to firms or governments, we rely more heavily on three more self-interested reasons to support the proposal. The most concrete of these three reasons is that the GIJI's role in evaluating patents for potential compulsory licensing could reduce the potential arbitrariness of current compulsory licensing procedures. Decisions at the GIJI would be reached within a system in which both developed countries and IP rights producers themselves are active participants. Developing countries would retain the power to order compulsory licenses without sanction by the GIJI. However, any decision to order a compulsory license that either had previously been rejected by the GIJI or was never submitted to the GIJI would be difficult to defend in the public arena, and arguably inconsistent with claims that it was being pursued for the public good.

The third and fourth reasons are both reputational. The GIJI would provide significant reputational advantages to IP rights holders involved in disputes about alleged monopolistic pricing that harms disadvantaged people. At present, these disputes take place in an open public sphere, in which interest groups with the best sound-bites and the media play a large role. Major drug companies were quite bruised, for example, by the campaigns against them at the beginning of the millennium with respect to pricing of AIDS drugs—campaigns that often portrayed the companies as rapacious profit-seekers unconcerned about the welfare of poor AIDS sufferers in Africa. The GIJI would give the companies and their supporters a forum for their own defense: if a GIJI that was regarded as legitimate by attentive world publics ruled in favor of the company, this would provide compelling support for its reputation.

The fourth reason concerns the reputations of countries rather than firms. By supporting the GIJI the developed countries would be making a powerful symbolic statement at relatively low cost to themselves. The reputation of the rich countries for being willing to help poor ones has been badly damaged by their reneging on promises in the Uruguay Round of trade negotiations (1987–1994) to reduce trade barriers to agricultural products. While the various agricultural lobbies in rich countries may make fulfillment of those pledges impossible, moving ahead with a GIJI could demonstrate good faith. There is no denying that the GIJI would be a "hard sell" for drug companies and other patent holders whose business plans count on monopolistic returns on successful innovations to compensate them for huge upfront investments, many of which yield no commercial products. However, the ability of the GIJI to authorize licensing on a national basis, rather than globally, would mean that patent holders would retain their IP rights in countries in which diffusion was indeed adequate, these being the countries in which current revenue from the innovation in question would predominantly come. Public pressure and attention to the problem of innovation diffusion, in industrialized democracies, would be essential for this proposal to gain sufficient traction to be politically feasible. But in the end, this is a modest proposal that would not fundamentally disrupt the activities of innovation-creating companies, and that might induce them to devise ways to accelerate diffusion of their innovations in ways that rebounded to their long-term benefit.

THE COMPARATIVE MERITS OF THE GIJI

A plausible case for institutional innovation must be comparative in two ways: it must show that the proposed institution is superior both to current efforts to solve the problem it addresses and to the best-developed rival proposal currently on the table. The current response to this problem is the provision in international legal agreements for compulsory licensing of essential medicines through domestic legal systems.

The best-developed proposal for an alternative system is that of Thomas Pogge for a new drug patent system that would be responsive to the global disease burden.

Compulsory Licensing as Currently Employed

Existing compulsory licensing does not fare well in comparison with our proposal for a GIJI for several reasons. First, although existing compulsory licensing is supposed to be accompanied by compensation, there are no provisions for ensuring that states actually render fair compensation—or indeed any compensation at all. While failure to do so might technically give rise to the possibility of a claim at the WTO, this will happen only if the IP rights holder's home state is willing to publicly insist upon payment for the company in question. However, political concerns mean that it is highly unlikely such a claim will be brought where the state undertaking compulsory licensing was one of the poorer developing countries attempting to ensure the availability to its citizens of an essential medicine or other important innovation. Moreover, even if a claim were brought, the dispute resolution process would be formally controlled by the IP rights holder's home state, with the IP rights holder itself having only indirect influence over how the dispute is argued, or whether a given settlement offer should be accepted. In contrast, the GIJI would provide fair compensation without the need of intervention from the IP rights holder's home state, and all dispute resolution processes undertaken at the GIJI would be engaged in directly by the IP rights holder itself. As noted above, these processes operate in accordance with due process, including the possibility of appeal with review by the appellate body.

Furthermore, existing compulsory licensing is unilateral, at the discretion of a single state, with no accountability mechanism, whereas a decision by the GIJI to authorize compulsory licensing would occur through the operation of a multilateral institution, with credible provisions for accountability not only to states, but to other stakeholders as well. Lack of accountability might seem to advantage weak states that are most likely to need to exercise the option of compulsory licensing. However, weak states are subject to powerful pressures from strong states (where most IP rights are held) to not exercise this option. Since GIJI authorization of licensing is multilateral, with robust accountability, it would provide opportunities for weaker states to benefit from initiatives with respect to diffusion without having to resort to politically risky efforts to invoke compulsory licensing on their own authority. Multilateralism provides some protection for weak states that act as part of, or on behalf of, a larger group.

Pogge's "Patent 2" Proposal

In several influential papers, Thomas Pogge has offered an institutional proposal designed to address both aspects of the "essential medicines" problem: the lack of access to life-saving drugs that millions of people suffer because of monopoly pricing under the current IP rights system, and the failure to develop drugs that would be of great benefit to millions of people. Both of these deficiencies derive from the lack of market demand resulting from poverty. Pogge proposes to leave intact the existing IP rights system (what he calls the Patent 1 option) but to create an alternative: innovators could opt for Patent 2, which requires them to make public all information about their innovation and forgo all regular IP rights, but which makes them eligible to be rewarded by disbursements from an international fund in proportion to the positive impact of their innovation on the global burden of disease.[15]

Pogge has made a major contribution by emphasizing the moral importance of the issue of the diffusion of life-saving drugs and by putting forward an ingenious proposal that is responsive to firms' interests and the many incentive problems that arise in this area. We see his proposal as a very valuable prod to discussion, rather than as an attempt to provide the "last word," and in that spirit we offer criticism and defend our alternative in comparison to his Patent 2 proposal.

Our proposal for a GIJI is in one sense much broader than Pogge's Patent 2 proposal, which is limited to one kind of innovation, namely patentable drugs, and is designed to address only one aspect of justice in innovation, namely the problem of extreme deprivation. In contrast, the GIJI takes into account the relevance of innovation to justice more generally and identifies legitimate interests in justice—namely, the concern about basic economic and political inequality—beyond the preoccupation with extreme deprivation. This difference, however, is not critical. In principle, Patent 2 could be instituted as part of a broader effort on innovation, with adaptations of Pogge's ideas to other types of innovation that might affect basic economic and political inequality more than extreme deprivation.

A key feature of Pogge's Patent 2 proposal is that its exercise is entirely voluntary. This voluntariness may seem to be an advantage of Pogge's scheme over ours: no potentially intrusive institution would be created under Patent 2, and opposition by firms to a legal regime providing for it would presumably be muted by the voluntary adherence provision. Drug companies could decide, case by case, whether to invoke Patent 1 or Patent 2 protections. However, the voluntary nature of Patent 2 is a double-edged sword, since firms might never invoke the Patent 2 option. Never invoked,

Patent 2 would be like unfinished monuments in the desert: testimonies to failed ambition. The big question about Patent 2, therefore, is whether firms will invoke it.

Whether they will do so depends upon how credible the promise of reward is. For the promise to be sufficiently credible to induce drug producers to forgo the known benefits of the Patent 1 option, two things must be true. First, drug producers must have confidence that the promised funds will be available, perhaps many years in the future. We call this the *funding assumption*. Second, the firms must have confidence that the procedure for identifying the disease burden reduced by a drug, and therefore the Patent 2 rewards due to drug companies, will be reliable and fair. Call this the *reliability assumption*. In our view, both of these assumptions are so problematic as to threaten the credibility and therefore feasibility of Pogge's proposal.

The problem with the funding assumption is that it is inconsistent with what is known about the trustworthiness of international funding pledges. The most notorious of these pledges is that made in UN General Assembly Resolution 2626 (1970), adopted without a vote. This resolution pledged that "Each economically advanced country will progressively increase its official development assistance to the developing countries and will exert its best efforts to reach a minimum net amount of 0.7 per cent of its gross national product ... by the middle of the Decade."[16] Thirty-seven years later, only five small European countries exceeded that target, and U.S. aid stood at 0.16% of gross national income.[17] Looking at this track record on the most publicized commitment in the world political economy over the past 40 years, it is very unlikely that any drug company would rely on any promises about long-term funding for the Patent 2 scheme. Providing public funds to drug companies is unlikely to be politically popular: competing demands will always seem more urgent and desirable.

The reliability assumption is also problematic because of the difficulty of obtaining agreement, even among experts, on reliable measurements of the impact of a particular drug on the global burden of disease. One source of difficulty is the fact that in many cases the decline in the incidence of a particular disease will be the result of a number of factors, including the decline of other diseases, in cases where comorbidity is prominent, and the cumulative effect, over many years, of a combination of medical, environmental, and cultural changes. Furthermore, the assessment authority would have an institutional incentive to understate the value of a patented innovation, in order to reduce the payment that it needs to make. Given the complex causation and the inherent negative bias of the assessment institution, it seems unlikely that drug producers will forgo the well-trod path to profits in exchange for an unpredictable outcome in a very problematic process for determining who gets rewarded and how large the reward is.

We conclude that although Patent 2 does not require any restriction of existing IP rights and thus might be thought to be superior to the GIJI on this count, this advantage will be nullified if drug producers have insufficient incentives to take up the Patent 2 option in the first place. *The central problem with Pogge's proposal is that neither the funding assumption nor the reliability assumption is credible.* Due to the weakness of these essential assumptions, Patent 2 is very unlikely to be implemented in its current form.

In contrast, both the funding and reliability requirements of the GIJI's policies are much less demanding. The resources required by the GIJI on an annual basis are moderate, to cover administrative costs. On the other hand, in the event of compulsory licensing, the GIJI would indeed have to pay substantial compensation from the fund provided by the contingent state commitments described in the earlier section entitled "Funding." That is, the GIJI does require a "war chest" of contingent resources provided by states, but the existence of the IMF and World Bank demonstrates that such contingent commitments are feasible. Moreover, while the compensation paid should be fair, it need not equal or even approximate the income IP rights holders would generate from monopoly pricing, and the deterrent effect of authorized compulsory licensing, and provisions for consultation, should ensure that the provisions are rarely exercised. Consequently, the level of funding required for the GIJI will be significantly lower than that required for Patent 2. Although the GIJI does include a provision for prizes and grants, its main focus is not on rewarding those who create products that the market would not otherwise produce, but rather on making sure that what does get produced becomes widely available, rapidly enough, to promote justice or at least not to worsen injustices.

It is important to understand why the poor track record on international financial commitments is a debilitating problem for Patent 2 but not for the GIJI. For the GIJI's threat of internationally authorized compulsory licensing to provide an effective incentive for innovators to lower prices or take other measures to accelerate the diffusion of their products, it only needs to make credible claims of a much more limited sort than those required under the Patent 2 scheme. Innovators must believe that they are at risk of authorized compulsory licensing if their product is diffusing so slowly that it is likely to be perceived by the GIJI to be contributing significantly to existing injustices or to be failing to make a significant contribution to ameliorating a significant injustice. Generally speaking, the risk that one may completely lose one's monopoly for a highly valued innovation would seem to concentrate the mind more effectively than the speculation that one could be rewarded—and continue to be rewarded—decades from now, if states hold fast to their pledges to create and sustain a reward fund.

Pogge might reply that it counts heavily in favor of his proposal that it includes a powerful creation strategy as well as a diffusion strategy, while the GIJI focuses only on the latter. In other words, he might argue that the GIJI's inclusion of provision for prizes and grants is an insufficient response to the fact that drugs that would have a large positive impact on justice are not being created due to the inadequate incentives provided by the existing IP rights system. This is a fair point in the sense that institutional proposals that would stimulate the creation of new innovations would certainly be welcome; but if, as we have argued, Pogge's proposal would be a dead letter, it does not count in favor of his scheme. In any event, no institution can reasonably be expected to do everything. It is true that the distinctive thrust of the GIJI, the licensing option and free licensing authority, is directed toward the Diffusion Problem, not the Creation Problem. Revisions of Pogge's Patent 2 proposal that made it institutionally more credible for helping to solve the Creation Problem would be welcome. Indeed, if our proposal for a GIJI effectively resolved the Diffusion Problem, the task of designing an institution to solve the Creation Problem might be easier to fulfill. Its designers would no longer have to address both problems simultaneously with one instrument—a job that in institutional design, as in economics, is often difficult or impossible.

The problem of inadequate diffusion of innovations is of sufficient importance to warrant consideration in its own right, independently of the problem of essential medicines. Many innovations that could have an important impact on justice are *not* like antimalarial drugs: access to them will be beneficial not just to those in less-developed countries, but to virtually everyone, and the problem they present for justice is not that they are unlikely to be produced by the market. Consider, for example, biomedical technologies that extend years of vigorous life or that augment the immune system, or drugs that enhance important cognitive skills. The problem here is not that there is insufficient market demand to stimulate research and development; rather, it is the risk that these valuable innovations will not be available except to the better off or that they will not become available to most people quickly enough to avoid significant injustices.

THE STATUS OF THE GIJI UNDER INTERNATIONAL LAW

While there are a number of international agreements that would potentially affect the actions of the GIJI, attention here will be paid only to
the two most important types: (1) the WTO's TRIPS agreement and
(2) investment treaties.

TRIPS

As a mandatory agreement for all WTO Members, TRIPS has a far-
reaching global impact, and thus the degree to which the GIJI and its
actions would conform to the requirement of TRIPS is extremely important. The following discussion will explain why a WTO member state that
grants a compulsory license as a result of a decision by the GIJI would not
be in violation of its obligations under TRIPS.

While questions have previously been raised regarding the acceptability of compulsory licensing under TRIPS, its acceptability as well as the
freedom of states to decide the reason for compulsory licenses being
granted was explicitly confirmed in the 2001 Doha Declaration on the
TRIPS Agreement and Public Health. Moreover, while TRIPS does specify some reasons for which compulsory licenses might be granted under
domestic law, these are not stated to be exclusive. Consequently, so long
as compulsory licensing under the GIJI operates in a manner consistent
with the constraints on compulsory licensing enunciated in TRIPS, no
WTO liability would attach to any appropriate action taken in accordance
with a GIJI compulsory licensing decision.

Compulsory licensing restrictions under TRIPS are predominantly
found in Article 31, which lists 12 procedural standards that must be met
in order for any grant of a compulsory license to be TRIPS-consistent. For
the purposes of discussion of the GIJI, four are particularly important,
and thus will be discussed here.

Under Article 31(a), decisions to grant a compulsory license must be
made on an individualized basis. That is, licenses cannot be granted for all
products of a particular type, such as "all pharmaceuticals." Rather, each
individual product must be considered for compulsory licensing on its
own merits. As the GIJI process specifically involves evaluation of innovations on an individualized basis, this provision clearly presents no
obstacle to the GIJI.

Under Article 31(b), an attempt must be made prior to compulsory
licensing to obtain authorization from the patent holder to license the
patent on reasonable commercial terms and conditions. Exceptions exist

to this rule, including where a national emergency or other urgent circumstance exists. However, while in some circumstances the GIJI may indeed need to rely upon this "national emergency" exception, it will usually not be necessary. The GIJI is institutionally designed to ensure that direct discussions with patent holders occur for a reasonable time prior to any decision to order a compulsory license. Consequently, unless a "national emergency" makes a rapid compulsory licensing order necessary, the requirements of Article 31(b) will be met—and if a "national emergency" has occurred, Article 31(b) will not be applicable.

Under Article 31(h) the patent holder must receive "adequate remuneration" to compensate it for any losses due to the compulsory license. While there is no clear agreement regarding the meaning of "adequate" as used in this provision, the goal of the GIJI to pay "fair" compensation, at a rate higher than the 2% to 5% royalty rate conventional in compulsory licensing, would seem to ensure that the compensation paid by the GIJI will indeed be more than "adequate."

Under Articles 31(i) and (j), the compulsory licensing decision must be subject to review by an authority superior to the body making the original decision. While appeals may not be available within the domestic legal system in which compulsory licensing was ordered, this requirement is clearly met by the incorporation within the GIJI of an appellate tribunal.

The GIJI, then, is designed in a manner that would make enactment of its compulsory licensing decisions consistent with the TRIPS obligations of the GIJI member state concerned. Ideally, to remove any doubt, this would be reflected through the enactment of a special amendment to TRIPS clarifying that no grant of a compulsory license taken in accordance with a GIJI decision could give rise to a claim for violation of WTO obligations. However, even if such an amendment were not able to be passed at the WTO, an additional protection exists for developing countries enacting GIJI-ordered compulsory licenses, in the form of the dispute settlement system of the WTO.

As TRIPS is a WTO text, any claim that a state was in violation of its TRIPS obligations in enacting a GIJI-ordered compulsory license would have to be resolved through state-to-state arbitration, rather than through individual patent holders directly bringing a claim against the GIJI member state in question. However, members of the GIJI will find it politically enormously difficult to justify bringing a WTO case against a state that has merely implemented a GIJI decision, when the complaining state itself had previously agreed by becoming a member of the GIJI that the GIJI's procedures were fair. Non-members of the GIJI would, of course, face no such obstacle. However, as already argued, there is no reason to believe that any compulsory licensing decision made by the GIJI would result in WTO liability even were a claim brought.

Investment Treaties

An enormous number of investment treaties now exist around the world, and a great number of them explicitly include reference to IP rights as constituting a form of "investment." Consequently, it is possible that the granting of a compulsory license by a developing country, in accordance with a decision by the GIJI, would give rise to a claim for compensation under an investment treaty.

While investment treaties all contain a variety of grounds on which an investor can claim compensation from a state, those based on the manner in which an investment has been treated, such as "fair and equitable" treatment, would be very unlikely to serve as the basis for a claim for any action taken in accordance with a GIJI decision, due to the procedural safeguards included in the design of the GIJI. In addition, the traditional claim for "expropriation," made when a state takes the property of a foreign investor, could not be made with respect to compulsory licensing done in accordance with a GIJI decision, as the investor would retain the patent in question, but would merely be required to allow others to produce licensed versions of the product in question.

Compulsory licensing could, however, give rise to a claim of "indirect expropriation," which occurs when a state regulates an investment in a manner that leaves formal ownership of the investment with the foreign investor, but effectively takes away the benefits of the investment. While arguments would be available to any GIJI member state forced to defend such a claim, the unresolved nature of contemporary international investment law regarding indirect expropriation means that it is impossible to be certain that a compensable indirect expropriation would not be found.

Moreover, the structure of investment treaty dispute resolution means that the patent holder would have the right itself to institute arbitration in order to secure compensation for its alleged losses. Thus, unlike at the WTO, developed states would not be able simply to reject claims for compensation by their investors who have allegedly suffered losses as a result of a GIJI decision.

Nonetheless, while investors control their own claims under an investment treaty, it is important to remember that the treaty is nonetheless between the two states, with the investor itself having no direct role in its implementation or interpretation. As a result, the risk of claims being raised under investment treaties as a result of a GIJI compulsory licensing decision could be significantly reduced merely by requiring states joining the GIJI to sign a declaration that no compulsory licensing granted in accordance with a GIJI decision would give rise to a claim under any investment treaty to which it was a party.[18] Alternatively, even greater protection could be gained if individual agreements were signed by GIJI member states that were parties to investment treaties stating that compulsory licensing granted in accordance with a GIJI decision would not give rise to a claim under the specific treaty in question.

This would, of course, not be a complete solution, as claims could still be made under investment treaties involving non-GIJI member states. Moreover, investment arbitration tribunals have recognized the right of investors to qualify as an "investor" under a specific treaty merely by undertaking the formalities of incorporation in a state party to the treaty, so long as the treaty itself permitted this. Consequently, the risk of an investment treaty claim could not be entirely eliminated by such agreements.

Nonetheless, the possibility of a compensation order being made by an investment arbitration tribunal against a developing country due to a GIJI decision could be adequately addressed by having such claims paid by the GIJI itself where the state acted in accordance with GIJI rules and instructions. In this way the burden would be spread among all GIJI members, thus minimizing the financial burden on any individual state.

CONCLUSION

One of the morally unacceptable features of the contemporary world is that innovations that would be of immense value to severely deprived people, and that would ameliorate unjust economic and political inequalities, are not widely available even though the marginal costs of providing them are low. One source of this problem is the patent system, which stimulates innovation by giving monopoly rights to patent holders. Monopoly pricing by patent holders combines with lack of resources by those who need the innovations the most to generate deprivation and inequality. The *diffusion of innovation* is blocked by the features of dominant institutions.

Since this is an institutional problem, we propose an institutional solution: a Global Institute for Justice in Innovation. This institute would offer prizes and other incentives for innovation, but its major task would be to promote the diffusion of existing justice-affecting innovations through a multi-step process. Quiet encouragement of more rapid diffusion could be followed, when unsuccessful, with public "naming and shaming" of firms that restricted access to their products through monopoly pricing or other means. But the institute would also have a standing compulsory licensing option for IP rights whose owners were not sufficiently promoting diffusion to disadvantaged people. If informal measures did not succeed, the GIJI could authorize states to issue compulsory licenses for innovations that were not diffusing at a sufficiently rapid rate. Such proposals would have to be accepted by supermajority vote of an assembly in which developed countries, developing countries, NGOs, and firms holding IP rights would be equally represented. Fair compensation, according to previously publicized procedures and guidelines, would be paid by the GIJI, drawing from funding by its member states. Finally, in applying any authorized measures, the GIJI would be subject to procedures of global administrative law, including oversight by an independent appellate tribunal.

We anticipate that many of these elaborate procedures would not need to be invoked. We expect that the mere threat of compulsory licensing would accomplish a great deal, without its frequent exercise. Much good would therefore be done, at low cost and without incursions on state sovereignty or frequent use of coercion.

Our proposal is indebted to the pioneering work of Thomas Pogge, who has emphasized the importance of access to new drugs—a major part of the innovation diffusion problem—and has made an institutional proposal of his own. In our view, however, his Patent 2 proposal has institutional flaws inherent in the lack of credibility of long-term promises by states fairly to assess the value of drugs in relieving the global disease burden, and on that basis to provide an adequate flow of royalties to drug firms. We deem both Pogge's *reliability assumption* and his *funding assumption* highly problematic. We argue that our proposed GIJI would be more credible and more effective, and would accomplish enormous good at very low cost. It would not solve the problem of creating new innovations—for which a revised version of Pogge's proposal might be valuable—but it could go a long way toward solving the Diffusion Problem, which currently is the source of so much unnecessary misery and unjust inequality in the world.

NOTES

1. An important, though as we shall argue, partial exception is the work of Thomas Pogge in his Patent 2 proposal, which we consider in some detail later in this article. © 2009 The Authors. Journal compilation © 2009 Blackwell Publishing Ltd, 9600 Garsington Road, Oxford OX4 2DQ, UK and 350 Main Street, Maiden, MA 02148, USA.

2. See, for example, Francis Fukuyama (2002). *Our Post-human Future*. New York: Farrar, Straus and Giroux, pp. 9–10.

3. Anders Sandberg and Nick Bostrom. (2006). Converging cognitive enhancements. *Annals of the New York Academy of Science* 1093:201–227; and Nick Bostrom. (2008). Smart policy: cognitive enhancement in the public interest. In Leo Zonneveld, Huub Dijstcblowem, and Danielle Ringoir, eds. *Reshaping the Human Condition: Exploring Human Enhancement*. The Hague/London: Rarthenau Institute and British Embassy, Science & Innovation Network, and Parliamentary Office of Science and Technology, pp. 29–36.

4. Allen E. Buchanan. (1984). The right to a decent minimum of health care. *Philosophy & Public Affairs*, 13:55–78; Normal Daniels. (2001). Justice, health, and healthcare. *American Journal of Bioethics* 1:2–16; and Dan W. Brock. (2003). Ethical issues in the use of cost-effectiveness analysis for the prioritization of health care resources. In T. Edejer et al., eds. *Making Choices in Health: WHO Guide to Cost-Effectiveness Analysis*. Geneva: World Health Organization, pp. 289–312.

5. The Americans with Disabilities Act and the international Convention on the Rights of Persons with Disabilities both recognize this point.

6. Michael Walzer. (1983). *Spheres of Justice; A Defense of Pluralism and Equality*. New York: Basic Books, Chapter 1.

7. According to some democratic theorists, what might be called basic political equality—having secure standing as an equal participant in the most fundamental political processes in one's society—is itself *a* requirement of justice, because it is required for a proper public recognition of the equality of citizens. On this view, inequalities in political power that are incompatible with or that tend to undermine this fundamental equality are unjust, independently of their tendency to produce other bad effects, including extreme deprivations, violations of particular civil and political rights, or distributive unfairness. See Thomas Christiano. (2008). *The Constitution of Equality: Democratic Authority and Its Limits*. Oxford: Oxford University Press.

8. It should be noted that when "monopoly pricing" is referred to in the current article, this includes the partial or complete refusal to sell in a given market by an IP rights holder, as this refusal is based upon an inability to receive the monopoly prices insisted upon.

9. On institutions see Stephen D. Krasner, ed. (1983). *International Regimes*. Cambridge, MA: MIT Press, introduction; and Robert O. Keohane. (1989). *International Institutions and State Power*. Boulder, CO: Westview. On accountability see Ruth Grant and Robert O. Keohane. (2005). Accountability and abuses of power in world politics. *American Political Science Review*, 99:29–43.

10. Geoffrey Brennan and Philip Pettit. (2004). *The Economy of Esteem*. Oxford: Oxford University Press.

11. The combination of notice and comment procedures with judicial review of due process is a feature of administrative law, as developed in the United States since the Administrative Procedures Act of 1946, and now spreading to international organizations. Sec Benedict Kingsbury, Nico Krisch, and Richard Stewart. (2005). The emergence of global administrative law. *Law and Contemporary Problems* 68:15–62.

12. Article 5 of the Articles of Agreement of the International Bank for Reconstruction and Development, the largest unit in the World Bank Group, provides that 20% of the subscription of each member is subject to call when needed for ordinary obligations of the Bank, and 80% is basically held in reserve to guarantee loans issued by the Bank. For the Articles of Agreement, see http://siteresources.worldbank.org/EXTABOUTUS/Resources/ibrd-articlesofagreement.pdf.

13. This should include the possibility of extra payment necessary to secure the cooperation of IP rights holders in cases in which the rights holder possessed non-public information essential for the manufacture of the licensed product.

14. Hedley Bull. (1977). *The Anarchical Society.* New York: Columbia University Press, p. 8.

15. Thomas W. Pogge. (2005). Human rights and global health: a research program. *Metaphilosophy* 36:182–209.

16. Michael A. Clemens and Todd J. Moss. (2005, Sept.). *Origins and Relevance of the International Aid Target.* Working Paper Number 68, Center for Global Development, p. 8.

17. Anup Shah. US and foreign aid assistance. http://www.globalissues.org/article/35/iis-and-foreign-aid-assistance, last updated April 13, 2009.

18. Naturally, the language used to describe the declaration is only intended to convey the substance of the declaration, and the precise wording of the document itself would need to be different.

PART TWO

The Relationship Between Ideal and Non-ideal Theory

CHAPTER 6

ᴄᴠᴏ

[handwritten: categorisation classification]

Non-ideal Theory: A Taxonomy with Illustration

GOPAL SREENIVASAN

Derek Parfit concludes his seminal analysis of egalitarianism with the rueful observation, "Taxonomy is unexciting, but it needs to be done."[1] No doubt every taxonomer harbors secret hopes that his taxonomy will still prove useful, despite being tedious, though few succeed on anything like Parfit's scale. In a field as neglected as that of non-ideal theory, one might think that greater optimism in this regard could survive without a protective guard of caution. But it is difficult to know, since tedium can be such a challenge to surmount.

Any analysis of the category of the "non-ideal theory of justice" should begin with John Rawls' division of non-ideal theory into two branches, *partial compliance theory* and *transitional theory*.[2] My central aims here are to explain and illustrate this division, as well as to complicate it, defend it, and finally to extend it. That is to say, I shall embrace taxonomy with a vengeance.

After setting out (and also complicating) Rawls' division of non-ideal theory in the first section, I proceed in the second section to defend his division against those who neglect its transitional justice branch. In the third section, I argue that we should extend the category of non-ideal theory beyond the terms of Rawls' division, and I illustrate my extension in the domain of international distributive justice. I close by examining the relations between my extension of non-ideal theory and its two existing Rawlsian branches, in the context of that illustration.

RAWLS' DIVISION OF NON-IDEAL THEORY

On Rawls' account of justice, ideal theory describes a well-ordered institutional arrangement: institutions are well ordered when they are both
just and known to be just; and when individuals both accept and comply
fully with the requirements these institutions impose on them.[3] This suggests two rather different ways in which circumstances may fail to be ideal.
On the one hand, background institutions may not be just; on the other
hand, individuals may not fully comply with the standing requirements
placed on them. For each kind of defective case, there is a corresponding
branch of non-ideal theory.[4]

To prescribe for the case in which individuals do not fully comply with
the requirements of justice, there is non-ideal theory as *partial compliance*
theory. Partial compliance theory embraces several different kinds of
question. Let me give three examples. To begin with, partial compliance
theory specifies what happens to an individual's obligations when others
fail to do their fair share within some distributive scheme.[5] Next, it
includes the theory of punishment and restitution, since a good deal of the
criminal law is plausibly regarded as articulating requirements of justice.
Finally, not all of the (legal) requirements that agents actually face in the
real world are themselves fully just. Hence, partial compliance theory also
includes the question of civil disobedience: of when justice *permits* (or
even requires) *noncompliance* with standing requirements that are less
than fully just.[6]

To prescribe for the case in which background institutions are not
just, there is non-ideal theory as *transitional* theory. Transitional theory
specifies the obligations that individuals have to bring just institutions
into existence. There are also two ways in which background institutions
may fail to be just: they may be unjust or they may not exist at all. An
individual may therefore be obligated to do his or her part either to reform
existing institutions or to introduce just ones from scratch (e.g., in the
state of nature).[7]

Of course, the distinction between these branches of non-ideal theory
is merely an analytic convenience. While the case in which background
institutions fail to be just *and* individuals fail to comply fully with the
requirements of justice is thereby made to appear logically "special,"
everyone knows that in our actual circumstances it is all too ordinary.
This means that partial compliance theory and transitional theory will
often apply together. Consider, for example, two of the central questions
raised by the aftermath of regime change—whether and how to punish
wrongdoers from the old regime, and whether and how to compensate
victims of that regime. Answers to these questions combine partial

compliance theory (theory of punishment and restitution) with transitional theory (reform of existing, but previously unjust institutions).

Among other things, reflection on the aftermath of regime change illustrates how jointly applying partial compliance theory and transitional theory can involve more than mere addition. To accommodate its questions properly requires that we mildly complicate Rawls' account of non-ideal theory in two respects, once for each branch. On the first branch, we should recognize explicitly that *institutions*, and not simply individuals, can be in partial compliance.[8] It follows that we should also allow for the possibility that punishment and restitution may be called for in response to institutional injustice (and not only to individual injustice). Historical examples include reparations for slavery or genocide and prohibitions on an aggressor nation's maintaining an army (e.g., article 9 of Japan's constitution).

On the second branch, we should explicitly distinguish two moments in the reform of previously unjust institutions. On the one hand, there is a *forward-looking* moment, in which just institutions—or at least, more just institutions—are introduced. (I take it that this is the moment one naturally associates with an obligation to reform existing institutions.) However, the introduction of just institutions may not suffice to inaugurate a steady state of ideal justice. This is not simply because the newly introduced institutions are themselves likely to be only imperfectly just. Rather, even if the new institutions were perfectly just (in themselves), achieving a steady state of ideal justice might still require some sort of collective reckoning with what went before—require it, that is, in addition to introducing the new institutions. Whether some such reckoning is in fact required is doubtless something best decided on a case-by-case basis. But in the absence of some guarantee that the answer will always be negative, transitional theory has at least to provide for the possibility that punishment or restitution for previous institutional injustice(s) will be required. It therefore needs a second, *backward-looking* moment, in which to ask whether that possibility is realized in the case at hand.[9]

IN DEFENSE OF TRANSITIONAL JUSTICE

As I mentioned at the outset, the enterprise of non-ideal theory has been sadly neglected, at least in philosophy.[10] Such attention as non-ideal theory has received has mostly been directed to the theory of punishment or of civil disobedience.[11] That is, it has mostly been directed to topics that are subjects in their own right, in addition to falling within the scope of non-ideal theory.

The major exception to this generalization is Liam Murphy's book, *Moral Demands in Nonideal Theory*. Murphy is concerned with the first question among the three examples I gave earlier of questions addressed by partial compliance theory. He asks "what a given person is required to do in circumstances where at least some others are not doing what they are required to do."[12] To illustrate Murphy's question, let us follow him in supposing that the ideal requirements of justice are given by (roughly) utilitarianism: individual agents are required to maximize everyone's total well-being, to the best of their individual ability. Under circumstances of partial compliance (i.e., usually), some other agents will shirk their responsibilities under this principle. The predictable consequence is that total well-being remains at a suboptimal level. However, if a given individual *can* relieve some of the shortfall, by contributing still more to total well-being (i.e., by sacrificing more), then utilitarianism will require (even) more of that agent under partial compliance than it does under full compliance. Murphy claims, very plausibly, that this is not fair.

Consider an example. In 2002, there were more than 43 million people without health insurance in the United States, and the cost of funding primary care physician services for this group was about US $9.68 billion annually.[13] Suppose that ideal theory (somehow) singled a million people out to foot this bill at an annual cost of $9,680 each. If only half of the designated million contributors actually paid up, it would seem unfair to respond by asking the half who did pay to pay another $9,680 each. Moreover, it would be unfair *even if* they could "afford it," in the sense that losing an additional $9,680 would not disqualify them from the requirement to contribute, as defined by whatever criteria of justice singled them out in the first place.

Partial compliance theory should therefore insulate responsible agents from the unfairness of having to pick up the slack caused by the noncompliance of others. To this end, Murphy introduces a "compliance condition," which holds (roughly) that the costs to an agent of complying with the requirements of beneficence should not be higher under partial compliance than they are under full compliance.[14] While the details he adds in interpreting this condition need not concern us, the general idea is certainly plausible.[15]

Unlike Rawls, however, Murphy seems to identify non-ideal theory *wholly* with its partial compliance branch.[16] Now it would be natural to suppose that the issue of whether non-ideal theory also has a transitional justice branch depends on a recent dispute about the role of institutions within a theory of justice.[17] Roughly, the dispute concerns whether some fundamental principles of justice apply to institutions alone—and, therefore, to individuals only mediately via institutions—or whether instead fundamental principles of justice always apply simply and directly to individuals. Since Murphy and Rawls take different sides in this dispute, one might think that Murphy's narrow view of non-ideal theory flows from his skepticism about institutions as the primary subject of justice.[18] However, while entirely plausible, this conjecture is mistaken.

One way to see that it is mistaken is to restrict our attention to the relations in which *individuals* stand to fundamental requirements of justice that apply directly to them (if need be, we can add the assumption that there are some). Suppose that, by and large, individuals do not fully satisfy these requirements. Two quite distinct questions can be raised against this background. The first question takes the violations themselves as given, and asks what kinds of responses to them are warranted (e.g., how should violators be treated or how should the requirements for non-violators be adjusted, if at all?). By contrast, the second question focuses precisely on the violations themselves, and asks how they can be permissibly eliminated (or at least reduced). I take it that the first question operates on what is recognizably the territory of partial compliance theory. But, while perhaps less easily recognized (given Rawls' formulations), the territory on which the second question operates is effectively that of transitional theory. For what it asks is how justice can be (more) fully achieved, beginning from circumstances where justice is at best incompletely achieved.

A different way to see the mistake in the plausible conjecture is to distinguish two relations in which *institutions* may stand to the requirements of justice. The first relation is the topic of the dispute between Murphy and Rawls. Let us call it *institutions as subject.* The second may be introduced by considering the extent to which institutions facilitate the implementation of the requirements that justice imposes on individuals. In particular, consider whether certain institutions are effectively compulsory—that is, whether they must exist or otherwise be created—in order to implement justice fully. If they are effectively compulsory, there is a distinct relation in which institutions may stand to the requirements of justice. Let us call it *institutions as (compulsory) handmaiden.* To illustrate, recall the administration of justice in Locke. According to Locke, the law of nature applies directly to individuals in the state of nature, who are also authorized to enforce it. However, the pre-political enforcement of this law (i.e., of justice) is defective: it suffers from three famous "inconveniences." Fully to remedy these inconveniences requires individuals to introduce political institutions (i.e., to leave the state of nature).

Suppose we agree with Murphy that institutions are not the primary subject of justice. This still leaves open the possibility that they are its (compulsory) handmaiden. It is implausible to deny institutions this weaker role. But then justice actually requires certain institutions to be introduced, at least for its full implementation. Hence, non-ideal theory has to assign *transitional* obligations that refer explicitly to institutions (as in Rawls' original formulation), even if only in the service of achieving full compliance with requirements of justice that apply directly to individuals.[19]

Moreover, on the plausible assumption that introducing such supporting institutions has certain start-up costs (in addition to the cost of continuing maintenance), non-ideal theory will have to assign some one-time transitional costs. Taking account of these costs contradicts the generalization of Murphy's compliance condition that results from explicitly identifying non-ideal theory with partial compliance theory: the costs to an agent of complying with the requirements of beneficence should not be higher under non-ideal theory than they are under ideal theory. The contradiction follows from the fact that costs under ideal theory do not include transitional costs. Indeed, in the special case of the administration of justice, this point goes through without assuming that there are one-time start-up costs, since under full compliance there would be no need for the relevant institutions at all (and, hence, not for their *maintenance* costs either).

In fact, this argument can be made from still weaker premises. We do not need to rely on the assumption that some institutions are effectively compulsory (and not simply instrumentally useful) for the full implementation of justice. Suppose, instead, that institutions make no necessary contribution to justice, not even an instrumentally necessary contribution. Ideal justice, then, can be fully achieved without any institutions at all.[20] Now imagine that, despite being superfluous in this sense, institutions nevertheless exist, simply as a matter of fact. Finally, suppose that these existent institutions are unjust. None of these premises violates any of Murphy's strictures, since he allows that fundamental principles of justice can apply to institutions. He insists only that such principles must be the *same* principles that apply to individuals.[21]

The brute existence of institutional injustice suffices to raise questions corresponding both to transitional theory's forward-looking moment (how to pursue institutional reform) and to its backward-looking moment (whether to seek punishment and restitution for institutional injustice). But the backward-looking question alone suffices to raise the prospect of one-time transitional costs that cannot be dismissed from the standpoint of justice.[22] Consider a society with an institution of slavery, for example. Even if the institutional reforms that justice requires of this society are purely negative (simple abolition), and even if we assume that abolition can be implemented for free, the prospect of transitional costs remains firmly in place, since reparations may be due for slavery. If they are, then non-ideal theory still needs a principle of (transitional) justice to assign those one-time costs,[23] and the previous generalization of Murphy's compliance condition will fail again, since the costs it counts under ideal theory do not include the costs of reparation either.

I conclude that even skeptics about institutions as the primary subject of justice should broaden their conception of non-ideal theory. Specifically, they should recognize a transitional justice branch, in addition to the partial compliance branch.

BEYOND RAWLS

To illustrate the transitional justice branch with a philosophical example, we have to turn to Rawls' late work, *The Law of Peoples*.[24] The only example of an obligation to transfer resources that Rawls accepts in the international case belongs to this second branch of non-ideal theory. His "duty to assist burdened societies" is explicitly an obligation of transitional justice, since its aim is to assist "burdened societies" to *become* "well-ordered."[25] Moreover, from Rawls' point of view, this has a crucial consequence— namely, that the duty toward a given society *expires* once that society has been become well-ordered. One of his main objections to principles of global distributive justice is that they lack a "target and a cut-off point."[26] In other words, the objectionable principles are proposed in ideal theory, and so entail permanent obligations. We shall return to the subject of international distributive justice.

But I should now like to suggest a more radical expansion of the category of "non-ideal theory," one that goes beyond either Murphy or Rawls. To begin with, we should notice an assumption they both share—namely, that ideal theory is *prior to* non-ideal theory. On their conceptions, non-ideal theory proceeds by reference to the content of an ideal theory of justice, and thereby presupposes it. Rawls is explicit on this point:

> Non-ideal theory asks how this long-term goal [*of ideal state*] might be achieved, or worked toward, usually in gradual steps. It looks for policies and courses of action that are morally permissible and politically possible as well as likely to be effective. So conceived, non-ideal theory presupposes that ideal theory is already on hand. For until the ideal is identified, at least in outline—and that is all we should expect—non-ideal theory lacks an objective, an aim, by reference to which its queries can be answered.[27]

The rough idea is that before we can take any steps forward, we need to know where we are supposed to end up. Otherwise, we cannot know whether any given step is a step in the right direction.

This priority assumption operates on both the partial compliance branch and the transitional justice branch of non-ideal theory, as Murphy and Rawls understand them. Under partial compliance, we need to know what the ideal principle of justice is—in Murphy's case, the principle of beneficence—and what fair shares it assigns, in order to know how the "compliance condition" operates. Otherwise, we will be unable to specify the limits it sets on individual sacrifice. Similarly, in transitional justice, we need to know what the ideal institutions are—in Rawls' *Law of Peoples*, some description of a "well-ordered society"—in order to know what agents are obligated to introduce (and how). Otherwise, we will be unable to specify the "cut-off" point on the duty of assistance.

Of course, I do not deny that non-ideal theory *can* work like this. But I want to suggest that it need not. More strongly, there exists a kind of non-ideal theory for which the priority assumption fails. On this conception, non-ideal theory functions as an anticipation of ideal theory. Its prescriptions anticipate the ideal requirements of justice rather than presupposing them. To do so, non-ideal theory has to make assumptions about the minimum requirements that *any plausible and complete* ideal theory of justice will include. In this vein, it can define targets for practical action *before* a complete ideal has been worked out, even in outline. Furthermore, if our assumptions about the minimum demands of justice are defensible, we can be confident that steps toward these targets are steps in the right direction.

Let us call this third conception "non-ideal theory as *anticipatory theory*."[28] A comparison with supervaluationism about vague predicates may be instructive.[29] What the two theories have in common is that, in each case, the subset of what all the disagreeing contenders agree upon is counted as correct. With supervaluationism, the contenders are "precisifications" of some vague predicate. With an anticipatory theory of justice, the contenders are plausible and complete specifications of the requirements of ideal justice. Both approaches reach a core of agreement by circumventing existing disagreements instead of resolving them.[30]

I take it that non-ideal theory so conceived is coherent and distinctive. What remains to be seen is whether it has any significant instantiations. Elsewhere I have proposed an anticipatory theory of *international distributive* justice: it holds that any plausible and complete ideal theory of international distributive justice will minimally include an obligation on the richest nations to transfer 1% of their gross domestic product (GDP) to the poorest nations. For concreteness, I specify this as an obligation incumbent on the "Group of Seven" (G7) nations of the Organization for Economic Co-operation and Development (OECD).[31] For 2007, this amounted to an obligation to transfer some US$280 billion. By contrast, in 2007, official development assistance from the G7 was 0.23% of GDP, or $69.446 billion.[32] So even a 1% transfer would clearly be a step of some kind. The question is whether we can know that it would be a step in the right direction.

To establish the 1% proposal as an instance of anticipatory theory, we would have to show that it can be secured without having to resolve various debates in ideal theory about international distributive justice.[33] The aim would be to demonstrate that a 1% transfer is philosophically, and not simply intuitively, secure as a step towards justice between nations and to do so before an ideal theory is settled or in hand.

A good start on this demonstration can be made by observing that a 1% obligation belongs to the core of agreement already shared among a significant coalition of rival moral theories and positions. For instance, it can be endorsed by utilitarians,[34] by global egalitarians and prioritarians of various kinds,[35] and by decent humanitarians[36]—not to mention by many decent, ordinary people.[37] Each of these groups can endorse the proposal for its own reasons.

Unlike supervaluationism, however, anticipatory theory does not simply rest content with the core of such agreement as it happens to find ready-made. It also aims to *expand* the scope of existing agreement among plausible and complete ideal theories. There are at least two strategies by which anticipatory theory can expand this scope. To begin with, it can seek to demonstrate that certain disagreements in ideal theory do not need to be resolved for the purposes of non-ideal theory. In the next section, we shall see how an application of this strategy suffices to bring Rawls and his followers into the 1% coalition.

In addition, anticipatory theory can seek to demonstrate that other disagreements or objections are actually resolvable on the basis of premises that are not controversial in ideal theory. In effect, this strategy aims to make latent agreement manifest. Elsewhere I argue that applications of this second strategy suffice to resolve disagreements about the magnitude of the proposed obligation and about how to spend the money (see note 24). Both arguments refer to a more specific version of the proposal, which divides 1% of the G7's GDP into quarters, with one quarter allocated to cover existing development commitments and the remaining three quarters spread over three fundamental determinants of health in developing countries—health care and public health; education (especially for girls and women); and basic nutrition and income support.

For example, what the argument about magnitude seeks to show is that, so targeted, the G7's transfer would very plausibly yield a disproportionate "bang for the buck" in terms of individual well-being. It thereby offers to explain, on empirical grounds, how an obligation can be both light enough in its burden (on the G7) to avoid being "too demanding" and yet also bountiful enough in its effects (on the well-being of the globally worst off) to be worthy of the status of "minimum obligation."

INTERSECTIONS

Altogether I have distinguished three different kinds of non-ideal theory: partial compliance theory, transitional theory, and anticipatory theory. Naturally, there may be still others. But as these three are all consistent, one might pursue non-ideal theory along any one of these branches or along any combination of them. Earlier I illustrated how partial compliance theory might combine with transitional theory. Let me close by illustrating how anticipatory theory might combine first with transitional theory, and then with partial compliance theory.[38]

Recall the disagreement between Rawls and his cosmopolitan critics about whether obligations to transfer resources between nations are permanent obligations of ideal theory or temporary obligations of transitional non-ideal theory. I myself think the 1% obligation is plausibly regarded as a part of ideal theory, and so as a permanent obligation. But for the purposes of anticipatory non-ideal theory, it makes no sense to insist on this. Hence, we may begin by regarding it as a transitional obligation. To this end, it suffices to add a suitable cut-off point to the 1% obligation. In keeping with the more specific version of the proposal mentioned above, we can say that the obligation cuts off when no country (better still: or Indian state or Chinese province) has an average life expectancy of 10 years or more below the global average.[39] When the final cut-off point has been reached, the question of whether the G7's obligation to transfer 1% of GDP annually to the poorest nations is a permanent obligation or not will acquire practical purchase. But until then, non-ideal theory can safely ignore the question, and Rawls and his followers can participate fully in the 1% coalition.

Under full compliance by the G7, $210 billion (0.75% of their GDP) would be transferred annually to improve the fundamental determinants of health in poor countries. If this transfer were directed to the world's poorest quintile (1.316 billion people), it would fund a per capita package of almost $160. Unfortunately, in the real world, not a single G7 nation spends (anything like) 1% of GDP on official development assistance—in 2007, France was closest, at 0.39%.[40] However, Murphy's basic observation in partial compliance theory retains its force here, even though its beneficiary must be identified hypothetically: it would be unfair to require a "fully compliant G7 nation" to transfer more than 1% of GDP, so as to pick up some of the slack caused by the rest of the G7's partial compliance.[41]

To insulate "full compliers" against this unfairness, we should therefore regard 1% of GDP as a fixed ceiling on each G7 nation's obligation under partial compliance by other G7 nations. That is to say, the non-ideal theory of justice would require nothing further of any G7 nation that managed to raise its official developmental assistance to 1% of GDP annually. In anticipatory non-ideal theory, 1% constitutes a G7 nation's "fair share" of improvements to the fundamental determinants of health in poor countries.

Now, since anticipatory theory centrally proceeds by avoiding unnecessary disagreements, we should actually distinguish our conclusion here—that 1% is fixed as the ceiling on a G7 nation's obligation under partial compliance—from the *rationale* that fair distributive shares are fixed under partial compliance. While I myself find the rationale very plausible, it is not wholly uncontroversial (cf. note 15). However, there is no need to rely on it either. For in anticipatory theory, the ceiling on a G7 nation's obligation winds up being fixed at 1% anyhow, as a side effect of avoiding controversies about the magnitude of obligations of international distributive justice under *full* compliance (i.e., of escaping the objection of being "too demanding"). Thus, full compliers with obligations of anticipatory theory are insulated against the unfairness that Murphy identifies, but without presupposing that partial compliance theory requires them to be so insulated.

Finally, we should consider separately whether the *floor* is fixed under obligations under partial compliance, as well as the ceiling. The floor is fixed under a G7 nation's obligation under partial compliance when that nation remains responsible for no less than its "full" obligation, even when no other G7 nation complies. Even if the ceiling is fixed on a G7 nation's obligation, it does not follow that the floor is fixed under that obligation as well. Nevertheless, 1% of GDP has the further advantage of being small enough that no G7 nation can plausibly claim that solitary compliance will put it at any serious relative disadvantage within its peer group (i.e., the G7). Hence, noncompliance by the rest of the G7 does not relieve a given G7 nation of its obligation to transfer the full 1% of GDP.

The tenability of this solitary compliance scenario helps to answer an important practical objection to the 1% proposal—namely, that there is no obligation to line the pockets of the corrupt. For concreteness, let me explain how it helps with reference to a specific G7 country. I shall take Canada as my example, since it is my own country. In 2007, Canada's official development assistance was 0.28% of GDP, and 0.72% of Canadian GDP (1% minus existing official development assistance) was $9.068 billion.[42] Under full compliance (by the G7), Canada would be responsible for transferring that $9.068 billion at a rate of $160 per capita. In other words, in anticipatory non-ideal theory, Canada's fixed share of improvements to the fundamental determinants of health among the world's poorest quintile would cover a population of 56.68 million people.

Since partial compliance by the rest of the G7 does not relieve Canada of its obligation to transfer a full 1% of GDP, Canada should simply go it alone, if need be, and transfer $9.068 billion annually at the full compliance rate of $160 per capita. To do so, Canada would have to choose a mix of jurisdictions where life expectancy is 10 years or more below the global average (our cut-off point, recall), up to a total population of 56.68 million people. Transferring its 1% on this basis would enable Canada to cover the same fixed share of the world's poorest quintile as it would cover under full compliance by the G7—neither more nor less. Yet, in that case, corruption is irrelevant to Canada's action unless it is so prevalent that insufficient non-corrupt (and badly off) jurisdictions exist for Canada to reach its fixed share of 56.68 million people effectively.

To illustrate the case for the irrelevance of corruption to Canada's action, let us use the "control of corruption" dimension of the worldwide governance indicators (WGI) reported by Kaufman, Kraay, and Mastruzzi as our measure of corruption.[43] Scores for the six dimensions of the WGI are reported annually, both on a scale of −2.5 to 2.5 (with zero as the mean of a normal distribution) and as a percentile rank.[44] For consistency with our previous calculations, I use the control of corruption scores for 2007; and for simplicity, I focus on sub-Saharan Africa.

Excluding Mauritius and Namibia (where life expectancy is less than 10 years below the global average), and counting only the part of the population living below the $1.25-a-day poverty line calculated by the World Bank (conserving purchasing power parity [PPP]), Canada could still reach its fixed share of 56.68 million people if it limited its choice of jurisdictions to the top nine sub-Saharan African countries in the WGI control-of-corruption ranking.[45] Six of those countries rank in the top half of the WGI's worldwide ranking—Botswana,[46] South Africa, Ghana, Rwanda, Madagascar, and Lesotho. The other three are very close to the 50th percentile—Burkina Faso and Mali (46th percentile) and Benin (43rd).

As long as countries *near the middle* of the WGI's worldwide control-of-corruption ranking are reasonably counted as "non-corrupt," Canada can reach its fixed share of the world's poorest quintile by simply avoiding corrupt jurisdictions altogether. Alternatively, as long as resources transferred to countries near the middle of this ranking are reasonably counted as spent "effectively," Canada can still reach its fixed share effectively by limiting its transfers to the nine indicated countries.[47] But then corruption is no impediment to Canada's action on its 1% obligation. At worst, it is an impediment to later full compliers, which is not Canada's problem.[48]

[handwritten annotation: 'who will be left in corrupt countries']

NOTES

1. Parfit, D. (1991). *Equality or Priority?* Lindley Lecture, University of Kansas, p. 34.

2. Rawls, J. (1999). *A Theory of Justice*, revised edition. Cambridge, MA: Harvard University Press. As we shall see in the second section, this claim is actually controversial, though nevertheless correct. Let me also serve notice that "transitional justice," as it emerges on the second branch of Rawls' division, means something quite different from what it means in the political science literature burgeoning under the same name. (It has nothing in particular to do with the "aftermath of regime change.") However, it is not merely a coincidence that the same expression is used in these two contexts. For some examination of what might be learned from each topic about the other, see Sreenivasan, G. (2012). What is non-ideal theory? In Williams, M., and J. Elster (eds.), *Transitional Justice*. New York: New York University Press. The present chapter abridges that one.

3. Rawls, *Theory of Justice*, 1999, Sections 2, 39, 69.

4. For completeness, I should note that Rawls also adds that ideal theory "works out the principles that characterize a well-ordered society *under favorable circumstances*" (*Theory of Justice*, 1999, 216; emphasis added). This suggests a further branch of non-ideal theory, to prescribe for the case in which circumstances are *un*favorable. However, I shall not pay much attention to this idea myself. While the circumstances addressed by non-ideal theory will obviously be less favorable than those addressed by ideal theory, it is not clear to me how much more weight this way of drawing the contrast will bear. On the one hand, given Rawls' stipulation of the "circumstances of justice" (*Theory of Justice*, 1999, Section 22), there is a limit both to how "favorable" the circumstances assumed by ideal theory can be and to how "unfavorable" those assumed by non-ideal theory can be. On the other hand, insofar as non-ideal theory is still *theory*, rather than policy or administration (say), its assumptions and precepts will inevitably remain simplified and idealized to some extent. I am indebted here to an unpublished paper by Alex Tuckness, which emphasizes the quoted passage.

5. Murphy, L. (2000). *Moral Demands in Nonideal Theory*. New York: Oxford University Press. I expand on this example further below.

6. Rawls, *Theory of Justice*, 1999, Sections 53, 55–59.

7. Rawls, *Theory of Justice*, 1999, 99, 293–294.

8. Strictly speaking, this complication is better understood as a clarification, since Rawls himself is well aware of the point (*Theory of Justice*, 1999, 215–216).

9. This is not exactly the same as the question interjected by the first complication, for here the question operates with a wider scope than its counterpart has in partial compliance theory. Under partial compliance, the concern is limited to how far institutional injustice *merits* punishment or restitution (or both), either in general or in a particular case. By contrast, in this backward-looking moment, the case for punishment or restitution can

extend beyond the (de)merits of an institutional injustice, to include various instrumental requirements of achieving a steady state of ideal justice.

10. For a thoughtful account of how considerations of partial compliance and of unfavorable circumstance might factor in moral reflection on the problems of migration, see Carens, J. (1996). Realistic and idealistic approaches to the ethics of migration. *International Migration Review* 30:156–170. Carens frames his analysis in terms of a contrast between "idealistic" and "realistic" approaches to morality, instead of a contrast between "ideal" and "non-ideal" theory (for that matter, he does not use the expressions "partial compliance" or "unfavorable circumstances" either). It seems to me that Carens' contrast is actually better suited than Rawls' to capturing the theoretical relevance of variation in how "favorable" actual circumstances are (cf. note 4). Among other reasons, Carens' contrast saliently applies to the enterprise of *ideal theorizing*, in Rawls' sense, taken all by itself. Indeed, Rawls' own ideal theory aims to strike some kind of balance between the terms of Carens' contrast, as Rawls' aspiration to a "realistic utopia" makes clear. See Rawls, J. (1999). *The Law of Peoples*. Cambridge, MA: Harvard University Press, Section 1.

11. We could add just war theory to this list, which Rawls includes under partial compliance theory (*Theory of Justice*, 1999, 8; see also *Law of Peoples*, 1999, 91–105).

12. Murphy, 2000, 5.

13. Astor, A., M. Danis, and G. Sreenivasan (2003). Providing free care to the uninsured: how much should physicians give? *Annals of Internal Medicine* 139(9):W78.

14. Murphy, 2000, 77.

15. For some skepticism, see Arneson, R. (2004). Moral limits on the demands of beneficence? In Chatterjee, D. (ed.), *The Ethics of Assistance: Morality and the Distant Needy*. New York: Cambridge University Press, pp. 33–58. Paul Hurley is more sympathetic (Hurley, P. [2003]. Fairness and beneficence. *Ethics* 113:841–864). I accept that, as baldly formulated in the text, the compliance condition is subject to counter-examples. For example, if two children are drowning in the proverbial pond, and one can save both of them at little cost or risk to oneself, then one is obliged to save both; and this remains the case *even if* there is another bystander, equally capable of saving the children, who refuses to help. But I am simply taking it for granted that the compliance condition can be refined to accommodate such cases.

16. Murphy, 2000, 5, 135. Murphy is not alone in neglecting the transitional justice branch of non-ideal theory. For example, see Feinberg, J. (1973). Duty and obligation in the non-ideal world. *Journal of Philosophy* 70:263–275; Nielsen, K. (1985). Ideal and non-ideal theory: how should we approach questions of global justice?" *International Journal of Applied Philosophy* 2:33–41; and Phillips, M. (1985). Reflections on the transition from ideal to non-ideal theory. *Noûs* 19:551–570. But my defense of Rawls' division of non-ideal theory concentrates on Murphy, since in his case a plausible explanation can be conjectured for the neglect. George Sher also treats "non-ideal theory" and "partial compliance theory" as interchangeable, but then he goes on to include transitional questions in his illustrative list of questions falling within their ambit. See Sher, G. (1997). *Approximate Justice: Studies in Non-Ideal Theory*. Lanham, MD: Rowman and Littlefield, 1–2.

17. See Murphy, L. (1998). Institutions and the demands of justice. *Philosophy and Public Affairs* 27:251–291; and Cohen, G. A. (1997). Where the action is: on the site of distributive justice. *Philosophy and Public Affairs* 26:3–30; and Cohen, G. A. (2008). *Rescuing Justice and Equality*. Cambridge, MA: Harvard University Press. Murphy and Cohen oppose Rawls here. See, e.g., Rawls, J. (1993). *Political Liberalism*. New York: Columbia University Press, lecture 7. Other contributions to the debate include Pogge, T. (2000). On the site of distributive justice: reflections on Cohen and Murphy. *Philosophy and Public Affairs* 29:137–169; and Scheffler, S. (2006). Is the basic structure basic?

In Sypnowich, C. (ed.), *The Egalitarian Conscience*. Oxford: Oxford University Press, pp. 102–129.

18. Murphy himself makes no such claim.

19. Both of the diagnoses I have offered treat "transitional" obligations as requiring individuals to do something to *improve* compliance with principles of justice that apply directly to individuals. The second diagnosis takes the "something" to involve introducing or reforming facilitating institutions. It would be reasonable to ask, in relation to the first diagnosis, what distinctive steps individuals might take to improve compliance by others (i.e., steps that are not tantamount to introducing or reforming institutions). But profitably to discuss this question requires a tolerably clear account of what counts as an "institution." I have offered two diagnoses partly in order to avoid that thicket.

20. Of course, in one sense, this is an extremely strong assumption. However, since it is all grist to Murphy's mill, it makes for a dialectically weak premise.

21. Murphy, 1998, 252–253.

22. This prospect cannot be dismissed as long as (we may safely assume that) punishment or restitution *is* sometimes warranted for institutional injustice.

23. The relevant principle most clearly counts as a principle of *transitional* justice if the ground for paying reparations is not (or not only) a matter of an intrinsically warranted response to slavery, but rather (or also) a matter of what is instrumentally required to maintain abolition as the new steady state. Compare note 9.

24. This section excerpts from Sreenivasan, G. (2002). International justice and health: a proposal. *Ethics and International Affairs* 16:81–90; and Sreenivasan, G. (2007). Health and justice in our non-ideal world. *Politics, Philosophy, and Economics* 6:218–236.

25. Rawls, *Law of Peoples,* 1999, 106, 111, 118.

26. Rawls, *Law of Peoples,* 1999, 115–19. Rawls has in mind Beitz, C. (1999). *Political Theory and International Relations,* revised edition. Princeton: Princeton University Press; and Pogge, T. (1994). An egalitarian law of peoples. *Philosophy and Public Affairs* 23:195–224.

27. Rawls, *Law of Peoples,* 1999, 89–90. Cf. *Theory of Justice,* 1999, 8 and 216; and Beitz, 1999, 170–171.

28. If it is compulsory to model conceptions of non-ideal theory as "theoretical responses" to some set or other of "non-ideal circumstances," then anticipatory theory can be modeled as responding to the *absence of a (settled) ideal theory of justice.*

29. Fine, K. (1975). Vagueness, truth and logic. *Synthese* 30:265–300.

30. Alternative comparisons might be to Sunstein, C. (1995). Incompletely theorized agreements. *Harvard Law Review* 108:1733–1772; or to Rawls, 1993, lecture 4. But since their subjects are closer to home, those comparisons may also distract and mislead. For example, there is no commitment in anticipatory theory to Rawls' idea of "public reason" or its attendant strictures.

31. The G7 are Canada, France, Germany, Italy, Japan, the United Kingdom, and the United States.

32. Figures from OECD. (2008). *OECD in Figures.* Paris: OECD, 13 and 61.

33. For an overview, see Caney, S. (2005). *Justice Beyond Borders.* Oxford: Oxford University Press, chapter 4.

34. For example, Singer, P. (2002). *One World.* New Haven, CT: Yale University Press, 192.

35. For example, Pogge, T. (2002). *World Poverty and Human Rights.* Oxford: Blackwell, chapter 8.

36. For example, Sachs, J. (2005). *The End of Poverty.* New York: Penguin Press, chapter 15; and Bono.

37. Singer and Pogge explicitly endorse a 1% minimum; Sachs and Bono endorse the United Nation's Pearson target of 0.7%.

38. This section excerpts from Sreenivasan, G. (2008). Global health and non-ideal justice. In Singer, P. and A. Viens (eds.), *The Cambridge Textbook of Bioethics*. Cambridge: Cambridge University Press, 369–375.

39. Of course, progress towards this goal will also raise the global average. So let us say that the obligation "finally cuts off" when the global average is 10 years below the top national average and the condition given in the text is satisfied.

40. OECD, 2008, 60.

41. Anticipatory theory leaves open the possibility that, in the ideal theory of justice, rich nations will be obligated to transfer *more than* 1% of GDP to poor nations. But for its non-ideal purpose of setting interim targets for practical action, anticipatory theory ignores this possibility and simply concentrates on the minimum requirements of ideal justice.

42. OECD, 2008, 60, 12.

43. Kaufmann, D., A. Kraay, and M. Mastruzzi (2009). *Governance Matters VIII: Aggregate and Individual Governance Indicators, 1996–2008*. World Bank Policy Working Paper No. 4978. Available at http://ssrn.com/abstract=1424591. According to the United Nations Development Programme, the WGI are "the most widely quoted and used governance indicator source in media, academia and among international organizations." UNDP. (2009). *Governance Indicators: A User's Guide,* 2nd edition, 56. Available at http://www. undp.org/oslocentre/flagship/democratic_governance_assessments.html. Of course, any particular choice of measure will be controversial. The WGI are used here simply for illustration. For some criticism, see Razafindrakoto, M., and F. Roubaud (2006). *Are International Databases on Corruption Reliable? A Comparison of Expert Opinion Surveys and Household Surveys in Sub-Saharan Africa.* DIAL Document de travail DT/2006–17. Available at http://dial.prd.fr/dial_publications/PDF/Doc_travail/2006–17_english.pdf; and for further discussion, see Kaufman, D., and A. Kraay (2008). Governance indicators: where are we? where should we be going? *World Bank Research Observer,* 23, 1–30.

44. The data are available and also mapped online at www.govindicators.org.

45. I took 2007 life expectancy data from World Health Organization. (2009). *World Health Statistics 2009.* Geneva: WHO. In 2007, life expectancy in Mauritius was 73 years and in Namibia it was 59 years. Worldwide it was 68 years. I took both population and poverty data for 2007 from UNDP. (2009). *Human Development Report 2009.* New York: Oxford University Press.

46. Botswana actually ranks in the top quartile, but its total population is only 1.9 million.

47. To illustrate one of the many trade-offs that arise in non-ideal theory: Canada could further limit its choice to South Africa, Ghana, Rwanda, and Madagascar (all clearly above the 50th percentile worldwide for control of corruption), and still reach its target of 56.68 million people, if it counted the population below the $2-a-day (PPP) poverty line instead. Madagascar, the lowest-ranked in this group, was in the 56th percentile.

48. For helpful comments, I am very grateful to Melissa Williams and Joe Millum.

REFERENCES

Arneson, R. (2004). Moral limits on the demands of beneficence? In D. Chatterjee (ed.), *The Ethics of Assistance: Morality and the Distant Needy.* New York: Cambridge University Press, pp. 33–58.

Astor, A., M. Danis, and G. Sreenivasan (2003). Providing free care to the uninsured: how much should physicians give? *Annals of Internal Medicine* 139(9):W78.

Beitz, C. (1999). *Political Theory and International Relations.* Revised edition. Princeton: Princeton University Press.

Caney, S. (2005). *Justice Beyond Borders*. Oxford: Oxford University Press.

Carens, J. (1996). Realistic and idealistic approaches to the ethics of migration. *International Migration Review* 30:156–170.

Cohen, G. A. (1997). Where the action is: on the site of distributive justice. *Philosophy and Public Affairs* 26:3–30.

——(2008). *Rescuing Justice and Equality*. Cambridge, MA: Harvard University Press.

Feinberg, J. (1973). Duty and obligation in the non-ideal world. *Journal of Philosophy* 70:263–275.

Fine, K. (1975). Vagueness, truth and logic. *Synthese* 30:265–300.

Hurley, P. (2003). Fairness and beneficence. *Ethics* 113:841–864.

Kaufmann, D., A. Kraay, and M. Mastruzzi (2009). *Governance Matters VIII: Aggregate and Individual Governance Indicators, 1996–2008*. World Bank Policy Research Working Paper No. 4978. Available at: http://ssrn.com/abstract=1424591

Kaufman, D., and A. Kraay (2008). Governance indicators: where are we? where should we be going? *World Bank Research Observer* 23:1–30.

Murphy, L. (1998). Institutions and the demands of justice. *Philosophy and Public Affairs* 27:251–291.

——. (2000). *Moral Demands in Nonideal Theory*. New York: Oxford University Press.

Nielsen, K. (1985). Ideal and non-ideal theory: how should we approach questions of global justice? *International Journal of Applied Philosophy* 2:33–41.

OECD. (2008). *OECD in Figures*. Paris: Organisation for Economic Co-operation and Development.

Parfit, D. (1991). *Equality or Priority?* Lindley Lecture, University of Kansas.

Phillips, M. (1985). Reflections on the transition from ideal to non-ideal theory. *Noûs* 19:551–570.

Pogge, T. (1994). An egalitarian law of peoples. *Philosophy and Public Affairs* 23:195–224.

——. (2000). On the site of distributive justice: reflections on Cohen and Murphy. *Philosophy and Public Affairs* 29:137–169.

——. (2002). *World Poverty and Human Rights*. Oxford: Blackwell.

Rawls, J. (1971). *A Theory of Justice*. Revised edition, 1999. Cambridge, MA: Harvard University Press.

——. (1993). *Political Liberalism*. New York: Columbia University Press.

——. (1999). *The Law of Peoples*. Cambridge, MA: Harvard University Press.

Razafindrakoto, M., and F. Roubaud (2006). *Are International Databases on Corruption Reliable? A Comparison of Expert Opinion Surveys and Household Surveys in Sub-Saharan Africa*. DIAL Document de travail DT/2006–17. Available at http://www.dial.prd.fr/dial_publications/PDF/Doc_travail/2006-17_english.pdf

Sachs, J. (2005). *The End of Poverty*. New York: Penguin Press.

Scheffler, S. (2006). Is the basic structure basic? In C. Sypnowich (ed.), *The Egalitarian Conscience*. Oxford: Oxford University Press, pp. 102–129.

Sher, G. (1997). *Approximate Justice: Studies in Non-Ideal Theory*. Lanham, MD: Rowman and Littlefield.

Singer, P. (2002). *One World*. New Haven: Yale University Press.

Sreenivasan, G. (2002). International justice and health: a proposal. *Ethics and International Affairs* 16:81–90.

——. (2007). Health and justice in our non-ideal world. *Politics, Philosophy, and Economics* 6:218–236.

——. (2008). Global health and non-ideal justice. In P. Singer and A. Viens (eds.), *Cambridge Textbook of Bioethics*. Cambridge: Cambridge University Press, pp. 369–375.

——. (2012). What is non-ideal theory? In M. Williams and J. Elster (eds.), *Transitional Justice*. NOMOS LI New York: New York University Press.

Sunstein, C. (1995). Incompletely theorized agreements. *Harvard Law Review* 108: 1733–1772.

Tuckness, A. (unpublished). *Nonideal Theory and Justice.*

UNDP. (2009a). *Human Development Report 2009.* New York: Oxford University Press.

——. (2009b). *Governance Indicators: A User's Guide.* 2nd edition. Available at http://www. undp.org/oslocentre/flagship/democratic_governance_assessments.html

WHO. (2009). *World Health Statistics 2009.* Geneva: World Health Organisation.

CHAPTER 7

൚

The Bioethics of Second-Best

ROBERT E. GOODIN

The real world is messy in all sorts of ways. Those who try to model the world, whether as scientists trying to explain it or as moralists trying to change it, attempt to abstract from that messiness. They seek models that are simpler, cleaner, more transparent than the reality that those models are attempting to mirror.[1]

Newtonian physics envisages balls colliding on a frictionless plane. Of course, when Fast Eddie shoots pool on a real table, he had better remember about friction. But starting with an idealized Newtonian model and then factoring in friction serves as a pretty good guide.

Something analogous is generally supposed to be true when it comes to moral philosophy. In terminology owing to Rawls (capturing an idea that is much older), moral philosophers distinguish between "ideal theory" and "non-ideal theory."[2] They accept the need to make adaptations to ideal theory when applying it to the real world, of course, adjusting for the ways in which the actual differs from the ideal.[3] But it is generally assumed that making those sorts of adjustments will be no more problematic for moral philosophers than it is for Fast Eddie hunched over the green felt.

Starting with the ideal may indeed be ideal, in all sorts of ways.[4] But in one important respect it might be seriously in error. It is wrong simply to assume—as moral philosophers typically do, without much elaboration or discussion—that when it comes to applying their moral theories, starting from the ideal and tacking back to the real will be relatively straightforward. They dismiss all that with a cavalier "mutatis mutandis" ("changing those things than need to be changed"). As this chapter will show, it may not be that easy.

From basic economic theory, we know that the second-best state of affairs might be very different indeed from the first-best. Compensating variations in several dimensions might be—and typically are—required to make up for shortfalls in others. Where that is so, the right thing to do in some non-ideal world cannot be simply and straightforwardly read off ideal theory's prescription of what to do under ideal circumstances.

After briefly cataloging various different respects in which the real world might deviate from the world as presupposed in ideal theorizing, I shall introduce the General Theory of Second-Best and explain the trouble it makes for reading real-world prescriptions directly off ideal-theory pronouncements. The upshot of that discussion is that adjusting our (ideal) theory to the (real, messy) world may be genuinely problematic. Alternatively, we might try instead to adjust our (real, messy) world so that it better fits the conditions presupposed by ideal theory. If we succeeded completely in that, then the prescriptions of ideal theory would be directly applicable, without the sort of adjustments that the Theory of Second-Best makes so fraught; but unless we can completely instantiate all the conditions presupposed by ideal theory, we might still fall afoul of the Theory of Second-Best. Yet another response to the problems posed by the General Theory of Second-Best would be to abandon the quest for excessively idealized moral theories, and instead to theorize the world in the vicinity of where we actually find ourselves. Depending upon the empirical facts of the matter, any of those strategies might work. And, again depending on empirical facts of the matter, any of them might be usefully supplemented by a pair of partial solutions, decomposing problems into their component parts or making policy choices that are relatively robust across changing circumstances. In the end, however, there is no sure solution to the problems posed by the Theory of Second-Best—only a suite of more or less imperfect alternatives.

WHY SECOND-BEST?

The need for "second-best" solutions arises when, and because, "first-best" solutions are unavailable. Moral philosophers mark this as a contrast between "ideal" and "non-ideal theory," and it is that moral usage with which this chapter is concerned. (There may be various other non-moralized senses in which a state of affairs can be non-ideal. My focus here, however, is on deviations from the ideal that are problematic in a more principled way.)[5]

At a minimum, we can say that a state of affairs is morally non-ideal if resources are inadequate to meet morally obligatory tasks.[6] This shortfall might occur in any (or several or all) of the following dimensions:

1. Material resources might not be adequate. Morally, let us suppose that we ought to prevent all preventable deaths. But given available technology, we simply cannot manufacture enough vaccine in time. Or if health dollars are strictly limited, we might decide to settle for a second-best procedure that is "much cheaper and almost as good" for treating one condition, in order to free up health dollars to treat some other condition.[7] Or a poor country that relies on foreign donors with short time horizons might opt for a narrowly targeted health intervention that shows quick results rather than investing in more comprehensive primary health care infrastructure: the latter would be better, but the foreign funds would not be available for it.[8]

2. Ideational resources might not be adequate. We might have inadequate or even erroneous understanding, either normatively (of True Morality) or empirically (of how the world works). We might not know how to make a vaccine. Or we might not understand why morally we should make and distribute it widely, even if we could.

3. Institutional resources might not be adequate. Morally, let us suppose that we ought to harvest organs from willing donors immediately upon "death" and transplant them promptly into the most appropriate recipients. But in the absence of institutional arrangements for wide-scale tissue-matching and priority-setting, we often end up giving the organ to someone who, while appropriate, is almost certainly not the *most* appropriate.[9]

4. Motivational resources might not be adequate.[10] Morally, we ought to allocate scarce medical resources to the people who need them most badly. But we just cannot bring ourselves to give the last available dose of vaccine to some stranger who is at greater risk rather than giving it to our own child, or to send scarce vaccine to people who are at greater risk abroad rather than allocating it to our fellow countryfolk.

5. Resources for coordination might not be adequate. Suppose that morally we ought to prevent the spread of life-threatening diseases. But even though the requisite motivational, material, ideational, and institutional resources are all in place for doing so, we might still be thwarted by a failure to coordinate our well-meaning efforts. The correct sequencing of pharmacological, psychological, and social interventions is crucial, let us suppose; different people necessarily have to be responsible for each, and coordinating their concerted action is beyond our powers.[11]

The *source* of the shortfalls that make a situation less than ideal might be in our own resources or in the resources of others. Thus, for example, we may be unable to inoculate everyone in time because of a lack of material resources to manufacture enough vaccine on our part, or on the part of others (e.g., foreign manufacturers). Or the reason we are unable to get the vaccine to those most in need might have to do with a failure of other people's motivations rather than our own. (We declined to give the last dose of vaccine to our own child only to watch someone else appropriate it for her own child who was not particularly needy, either.) And so on.

In terms of *responses*, it is an open question what morally we ought to do when we are in situations where resources are inadequate in any of these ways to meet morally obligatory tasks. One alternative would be to take the world as we find it and do the best we can in those (admittedly, non-ideal) circumstances. Another alternative would be to try to transform the situation—to make the non-ideal circumstances more nearly ideal, wherever we can. Both strategies risk falling systematically afoul of the General Theory of Second-Best, as I shall go on to argue in the later section "Strategies for Bridging the Real–Ideal Gap."

SECOND-BEST MIGHT BE COMPLETELY DIFFERENT

Whether or not we should "settle for less than the best" often provokes heated disputes—and rightly so. (I shall say more of that shortly.) What is involved in settling for second-best is, in contrast, typically taken to be relatively straightforward.[12] Wrongly so, I now want to argue.

Second-best solutions can, and often do, display peculiar features that pose particular challenges of institutional design and policy choice. While these are nowise unique to bioethics, neither is bioethics in any way immune to them.

The General Theory of the Second-Best

Long ago, economists Richard Lipsey and Kelvin Lancaster proved the General Theory of Second-Best. Put into what passes among economists as plain English: "The general theorem of the Second-Best states that if one of the conditions [characterizing the optimal outcome] cannot be fulfilled a Second-Best . . . situation is achieved only by departing from all other optimum conditions."[13]

Lipsey and Lancaster's formal proof is tied to specifics of the standard economic set-up (well-ordered preferences, general equilibrium, and so on). And their very strong conclusion ("only by") derives from some of the very particular stipulations that economists conventionally make concerning preference functions.

There is, however, a weaker—and more genuinely general—version of the Theory of Second-Best that is independent of any such assumptions. In that weaker and more general form, which will be my focus in this chapter, the Theory of Second-Best says this: If the first-best state of affairs cannot be obtained, the second-best state of affairs is not necessarily identical to the first-best in any respect. Whereas Lipsey and Lancaster's stronger version would say "necessarily not," the weaker and more general version that I shall be discussing says merely "not necessarily."

The phenomenon is a familiar one across broad swathes of life, once you come to think about it. Here is one homely example. Suppose that the car that I want as my first-best car has three attributes: it is a (1) new (2) silver (3) Rolls-Royce. But suppose that there is no such car available at the moment, and for some reason I really must acquire a car immediately. Hence I have to settle for second-best. The Theory of Second-Best cautions me that my second-best car will not necessarily be one that displays more rather than fewer of the same attributes as my first-best car. Thus, for example, the second-best from my point of view would probably be a (1) week-old (2) black (3) Jaguar rather than a (1) new (2) silver (3) Toyota, if those were the only two cars on offer. That is true, even though the Jaguar displays none of the same features as my first-best car, and even though the Toyota displays two out of the three. That is precisely the point of the Theory of Second-Best.

Avishai Margalit offers another example, this one drawn from the teachings of the Church of Rome:

> The Catholic Church believes that being a nun is the ideal life. It is the life of perfection for women. The Catholic Church also believes that the sacrifice entailed in giving up sexuality and motherhood is such that most women cannot attain the ideal of becoming nuns. The Second-Best for a woman is not to become a nun with a lax attitude toward the prohibition of sexuality, but instead to become a mother.[14]

Similar examples pervade the public sphere as well. Imagine a health-promotion campaign. Were we promoting a healthy lifestyle, what characteristics would that have? The first-best lifestyle would (let us imagine) include at least the following attributes: (1) no tobacco, (2) little alcohol, and (3) regular exercise. However, nobody would seriously think that the second-best lifestyle would be one that displayed perfectly two out of those same three attributes without any regard to the other. The second-best lifestyle would definitely not be one in which you neither smoke nor drink but never get any exercise at all, either.

The same phenomenon recurs when implementing social ideals more broadly. The best society, let us imagine, is one that is both (1) free and (2) equal. But suppose the only way to maintain perfect equality is to interfere with people's freedom in some respect (e.g., with their freedom to bequeath large sums of money to their heirs). While the first-best society is, *ex hypothesi*, one that maximizes freedom and maximizes equality, the second-best society is probably not one that maximizes one of those values completely, regardless of the cost to the other. Instead, the second-best society is probably one that scores pretty highly on both freedom and equality without literally maximizing either (e.g., imposing confiscatory taxes on bequests, but only above a certain size).

That thought is sometimes expressed in terms of "value trade-offs."[15] Of course, there may be some values you refuse ever to trade off for any others. Maybe some values stand in a strict hierarchical relation to others, such that any difference (no matter how small) on the top-ranked value trumps for you any difference (no matter how large) on the lower-ranked value. Much more typically, however, you would probably be prepared to give up a little bit of value-attainment in one dimension in exchange for a certain amount in the other.[16]

When the choice situation forces us to trade off one value for another, we have to decide which matters more (and by how much) in that situation. In such cases, worries can arise concerning the commensurability of values. Can they really be compared, in ways that would allow us coherently to trade more of one for less of the other?

Sometimes, however, choosing among options does not require any invidious comparisons across values. All the comparisons can sometimes be done within the same value. Cass Sunstein offers various examples of that sort in his discussion of "health–health trade-offs" that arise in risk regulation. For example, "Regulations designed to control the spread of AIDS and hepatitis among health care providers may increase the costs of health care, and thus make health care less widely available, and thus cost lives. . . . A ban on carcinogens in food additives may lead consumers to use noncarcinogenic products that carry greater risks in terms of diseases other than cancer."[17]

Note that what you regard as second-best will always depend on the interaction between your evaluative standards (preferences, values) and your options (the feasible set over which you can effectively choose). But note well: It is not as if you change your standards when confronted with new options. The same standards apply; they simply apply differently over a different feasible set.

Note too that the Theory of Second-Best applies only to multiple-attribute decision problems. If there is only one criterion (or if there are multiple criteria that are hierarchically ranked in a lexical order, such that only one is in play at any given time), then it is necessarily the case that what is second-best will be whatever is as similar as possible to what is first-best on that lone or lexically prior criterion.

There are two factors driving the Second-Best phenomenon. One is "suboptimization." That is the error of optimizing on only a subset of all the dimensions that are actually important to you. In so doing, you get the right result with respect to that subset of dimensions to which you are paying attention—but the wrong result with respect to the other dimensions that you are ignoring in the process. That is how you end up choosing the new silver Toyota—by fixating on two criteria that matter to you but ignoring the third (that you want a luxury car, not merely a new black one). Typically, the right thing to do all-things-considered differs from the right thing to do only-some-things-considered.

The second factor driving the Second-Best phenomenon is "interaction" across those dimensions. So, for example, education interacts with health which interacts with employment: the more education people have, the better able they are to make healthy lifestyle choices and the better able they are to take advantage of employment opportunities; and the healthier people are, the better able they are to hold down a job. That is why policymakers need to consider the entire suite of education–health–employment policies all together in a holistic manner, rather than just attending to them separately.[18]

STRATEGIES FOR BRIDGING THE IDEAL–REAL GAP

With the Theory of Second-Best firmly in view, let me now return to show what trouble it can make for moral philosophers trying to apply their "ideal theories" to the real world.

Make the Theory Fit the World

Standard practice, as I have said, is for moral philosophers to develop their theories of what should be done in "ideal" conditions, abstracting from various messy features found in the real world. When they come to apply those theories to the real world, they (like Fast Eddie) then simply make such adjustments to ideal theory's prescriptions as are required in light of those non-ideal facts about the real world.

Sometimes those are modest tweaks; other times they are major bolt-ons. For an example of the latter, notice that in the world of ideal theory, no one would ever break the law and no government would enact an unjust law. Both things sometimes happen in the real world, however. To accommodate that fact, Rawls needed to bolt a theory of corrective justice and a theory of civil disobedience onto his theory of justice when applying it to the non-ideal real world.[19]

Even when whole new branches have to be added to ideal theory in applying it to the real world, however, writers like Rawls tend to presume that the basic structure set by ideal theory remains unchanged. Adjustments will inevitably be required at the margins, and extra bits will have to be added on the edges. But writers like Rawls assume that the great bulk of ideal theory's prescriptions will remain the same when applied to the real world.[20]

Furthermore, however far in the background, this assumption is not just some incidental oversight: it goes to the heart of the methodology. If systematic, thoroughgoing revisions to ideal theory's prescriptions would be regularly required in applying it to any real-world situation, then there would be no point in starting with ideal theory. If tacking back to the real world requires us to rethink everything afresh, then it is not at all clear what ideal theorizing has bought us.

The Theory of Second-Best, however, warns that rethinking everything afresh may be precisely what is required. There is no reason to think that the second-best (i.e., non-ideal) world is necessarily identical to the first-best (i.e., ideal) world in nearly every respect. There is no reason necessarily to think that most (or indeed any, much less most) of ideal theory's prescriptions will carry over unchanged (or even just minimally changed) in a world that is less than ideal in any respect.

So the Theory of Second-Best stands in the way of any assumption that what is second-best can necessarily be read easily and straightforwardly off ideal theory's description of what is first-best. Adapting ideal theory to a non-ideal world might require systematic root-and-branch changes to ideal theory's prescriptions.

Or it might not. All that my weaker and more general version of the Theory of Second-Best says is that the second-best does not necessarily bear any close resemblance to the first-best, not that it necessarily does not. Whether it does or not depends just on boring empirical facts of the matter—facts to do largely with interactions across the evaluative domains and the centrality of any given evaluative criterion to the situation under consideration. This is what creates possibilities for *is it really?*

more or less successfully "decomposing problems" in ways I shall discuss shortly.

The Theory of Second-Best in the form I am deploying it is thus better understood as cautionary rather than condemnatory. It does not say that the standard approach of first producing an ideal theory, and then assuming that the prescriptions for the real world are pretty much the same, is necessarily wrong, always and everywhere. There might be some circumstances in which that would work fine. All the Theory of Second-Best, in its weaker and more general form, says is that we must not assume that that is always the case. We must proceed with caution, and we must be prepared for that procedure sometimes to fail badly.

Make the Real World More Nearly Ideal

The standard strategy is to start from ideal theory and adjust it to fit the real world. The Theory of Second-Best tells us that there may be real problems in doing so—that really major changes in ideal theory's prescriptions might be systematically required. Might those problems be avoided by taking the opposite tack, adjusting the real world to fit the presuppositions of ideal theory?

There are many reasons, quite apart from the difficulties posed by problems of identifying the Second-Best, for wanting to make the real world more ideal. Other things being equal, it would be better if we were not lacking in the resources required to discharge our moral responsibilities. And all the resources listed at the outset—material, ideational, institutional, motivational, and coordination—can, in principle, be increased.

Just how hard or costly it will be to increase our moral resources is an open question, and one that is probably best answered differently for different classes of cases. It may be easier to effect changes in your own institutions or your own motivations than those of others, for example. That might suggest, in turn, that we ought to strive primarily to correct our own failings while by and large taking those of others as given. Or for another example, it may be easier to change institutions than motivations: that was Rousseau's thought in "taking men as they are and laws as they might be."[21]

Conceptually, it is important to appreciate that feasibility should always be understood *dynamically*.[22] Writers like Bentham rightly bemoan the way in which "the plea of impossibility offers itself at every step, in justification of injustice in all its forms."[23] It is the job of leaders, as Max Weber says at the end of his essay on "Politics as a Vocation," to make possible tomorrow what is not possible today.[24] True, as moral philosophers are fond of saying, "ought implies can." But that does not excuse us from doing what we ought to do, merely because we cannot do it just at the moment. If we can get ourselves into a position to do what we ought to do (in time still to do it), then that's what we ought to do, *ceteris paribus* (all other things being equal).

Note that *ceteris paribus* clause well, however. It is not necessarily the case that the best thing to do, all things considered, is to transform a non-ideal situation into one that is morally ideal. It is not, anyway, if the resources that would be expended to bring about that transformation could be used in morally more desirable ways. In some non-ideal situations, you can do almost as well as you could in the absolutely ideal one. Then transforming the situation from non-ideal to ideal may well cost more than morally it is worth.

Nor is it the case that making real-world circumstances more like those presupposed by our ideal theory will necessarily make it easier to surmise from ideal theory what we should do in the real world—particularly if we do not succeed in instantiating *all* of the conditions presupposed by ideal theory. Only if we did could we be sure that the prescriptions of ideal theory are precisely right for us in the world that we actually occupy. If any one of the conditions presupposed by ideal theory is missing, then the Theory of Second-Best warns that we might (not necessarily will, but might) need to make systematic alterations right across the board in the prescriptions of ideal theory.[25] Again, if ideal theory's prescriptions were themselves decomposable, and if we could make real-world circumstances completely ideal in every respect relevant to any given prescription, then in that very special case we could know with confidence what to do, judging just from ideal theory alone. But that is contingent, a special case rather than a standard solution.

Theorize the Vicinity of the Actual

If the Theory of Second-Best stands in the way both of fitting ideal theory to the real world and of fitting the real world to ideal theory, then perhaps we ought abandon ideal theory altogether. That is to say, instead of abstracting from all the messiness of the real world and assuming circumstances that are much more ideal than the actual circumstances in which we find ourselves, maybe we should build our moral theories from the start around assumptions that are broadly true of the world we are actually in.[26]

Again, there is much to be said for this strategy, quite apart from any assistance it might provide in avoiding problems posed by the Theory of

Second-Best. Perhaps the main thing to be said for theorizing the vicinity of the actual is that we test our moral theories against our moral intuitions. Those moral intuitions, in turn, have been formed around the standard sorts of cases commonly faced in ordinary life. If we try to theorize a world too far from our lived experience, we have nothing reliable to test our theories against.[27]

While the strategy of theorizing in the vicinity of the actual is thus very tempting, there are also obvious problems with it. For one, it is messy—complicated, confusing—to try to model everything at once, without abstracting anything away. For another, theorizing that is tightly bound to existing circumstances may not give us much guidance as to the direction in which we should attempt to change those circumstances for the better, morally. For yet another, theorizing in this mode may be highly unstable, prescribing very different things as circumstances change, perhaps even only slightly (and maybe rightly so, in line with the Theory of Second-Best.)

Beyond all that, however, would this strategy really solve the problems that the Theory of Second-Best poses for moral theorizing? It would, if our moral theory were crafted to the *exact* circumstances actually obtaining in the real world. Then our moral theory's prescriptions would be precisely right for the circumstances in which we apply it. But the Theory of Second-Best warns us that, if the actual circumstances are different in any respect whatsoever from those presupposed by our moral theory, then that theory's prescriptions might need to be systematically altered. And requiring that moral theorizing be perfectly tailored to actual circumstances, when circumstances are constantly changing, would have us constantly retheorizing—leaving us with little time for acting. If for whatever reason circumstances change faster than our moral theories can, this strategy will offer us no secure protection against the problems posed by the Theory of Second-Best.

One comforting thought (which may or may not actually be true) might be this. Perhaps as a brute empirical fact of the matter, the prescriptions of a moral theory would not change very much with small changes in circumstances. Then a moral theory concocted for one set of circumstances would be pretty much valid anywhere in the near neighborhood of those circumstances. If that is true as a matter of empirical fact, then a moral theory framed around our actual circumstances will provide pretty good guidance anywhere in the vicinity of our currently actual circumstances. (And by the same token, a moral theory framed around ideal assumptions will provide pretty good guidance anywhere in close vicinity to the circumstances presupposed by that ideal theory.)[28] If that is how things empirically turn out, then well and good. But again, that is a purely contingent empirical matter, and it could of course turn out otherwise.

A PARTIAL SOLUTION: DECOMPOSING PROBLEMS

The three grand solutions for coping with the problems posed by the Theory of Second-Best all thus fail, or anyway may well fail. Let us now turn to a pair of more partial solutions that might work in a more limited way. Both turn out also to be vulnerable to threats from the Theory of Second-Best, each in its own way.

Public policymakers commonly try to decompose complex problems into their component parts, assigning responsibility for solving each component to separate actors. That might be a particular sector of society (the family unit), a particular government department (of health or education or employment), or a particular country (the one in which the victim happens to live). In so doing, policymakers are hoping that the best decision within each part, taken one at a time, will still be best when aggregated be best overall.

The basic idea obviously long predates him, but it is to Herbert Simon that we owe the modern formal concept of a "(nearly) decomposable system." Such a system is in his terms characterized by a high degree of modularity. That is to say, interactions occur (almost) wholly within modules, with (almost) none occurring across modules.[29]

Some problems lend themselves to that sort of solution. Others do not. It all just depends on whether we can carve up the problem into (largely) self-contained parts, and assign each part to a (largely) self-contained module. If no matter how we carve up the problem, different parts will interact too much with other parts, then trying to divide up responsibility for the problem among (largely) self-contained units will not yield a good solution.

The Theory of Second-Best however warns us that problems are not necessarily decomposable in the way that this approach requires. If something goes wrong in one dimension, then adjustments may well be required in several other dimensions. Assigning responsibility for each dimension to different agents acting largely in isolation from one another precludes (or anyway hinders) precisely such across-the-board adjustments.

Phrasing the point in economic terms, Lipsey and Lancaster write: "It should be obvious . . . that the principles of the general theory of Second-Best show the futility of 'piecemeal welfare economics.' To apply to only a small part of an economy the welfare rules which would lead to a Paretian optimum if they were applied everywhere, may move the economy away from, not toward, a Second-Best optimum position."[30]

Of course, we wish it were otherwise. Nearly decomposable systems would be quite convenient. They would allow us to take advantage of specialization and of the division of labor. They would admit of easy repair, replacing one malfunctioning module with another without any interruption to any other part of the system. Nearly decomposable systems evolve and adapt more quickly to changing circumstances, and hence enjoy an evolutionary advantage.[31] So we may well wish that we were able to solve all our problems in that way.

But wishing does not make it so. Decomposable solutions presuppose decomposable problems. And where the second-best is radically different in many dimensions from the first-best, devolving responsibility for different dimensions to different non-interacting agents clearly risks folly.

This general principle has implications for public policy and social problem-solving across a wide range. I shall illustrate it here by reference to bioethical issues relating to health care, at three levels: individual physicians, institutional designers, and policymakers.

Role-Differentiated Individual Responsibilities

In addition to the general duties and responsibilities that each of us has merely as a moral agent, most of us also have "special responsibilities" toward certain other people and for certain other actions and outcomes. The role-differentiated responsibilities of health care providers are a case in point.

The key thing to notice, in the context of the present discussion, is that the patient–provider relationship is a bilateral relationship. Each patient is assigned one or more health care providers, who have a special responsibility for that patient's health in a way that they do not for other people who are not their patients. In the words of the oath that the World Medical Association's Declaration of Geneva asks physicians to swear: "The health of my patient will be my first consideration." Or as a philosophical lawyer (who also served as President Ronald Reagan's Solicitor General) elaborates: "Doctors . . . owe a duty of loyalty to their clients, a loyalty which . . . requires taking the medical . . . interests of that client more seriously than the interests of others in similar or greater need, more seriously, indeed, than formulas of either efficiency . . . or fairness . . . would require or even permit."[32]

This strategy of assigning specific individuals responsibility for the health care needs of specific other individuals is clearly a strategy of "decomposition" of the sort just discussed. It involves decomposing the problem of caring for the health of people in general into a problem of one particular person's taking care of particular patients, one at a time.

That would be a good solution, if the problem were itself decomposable. But insofar as there are interactions among the health (and hence health care needs) of different people, such a modular approach is not well suited to coping with them. And we know that this is true: people catch infectious diseases from other people, and the best way to prevent one patient from becoming infected is often by preventing others around him or her from becoming infected. That is what "public health" is all about—and public health programs are, of course, the antithesis of a modular one-person-at-a-time approach to health care.

Not only are there important interactions among patients on the "disease" side of the equation; so too are there important interactions among patients on the "treatment" side. Some of these interactions are positive. Economies of scale can make the unit cost of treating any given patient's condition a decreasing function of the number of other patients afflicted with the same condition.

Some of the more important interactions are negative, however. Insofar as medical resources are scarce relative to the need for them (and they virtually always are), one person's treatment comes to some greater or lesser extent at the cost of some other person's not being treated. The modular approach assigns health care providers responsibility for zealous advocacy of their patients' medical interests—even if at some cost to the greater needs of others who happen not to be their patients. That can easily lead to misallocation of scarce medical resources, unless (contrary to fact, everywhere in the known world) everyone in need of health care had an equally effective advocate whose zeal were strictly proportional to the patients' medical needs.

Sector-Specific Institutional Responsibilities

An analogous failure due to modularity occurs at the level of inter-sectoral institutional design. The tasks of government are divided along functional lines, and responsibility for different functions is assigned to different departments.

Again, that modular solution would work well were the problems to be solved decomposable in form. But typically they are not: population health is powerfully affected not only by what happens in ministries of health but also by policies laid down by other ministries (of housing, of employment, of environment—not to mention of war). Each agency is left concentrating on its own particular remit, which affects performance under others' remits, but in ways that are not its job to take into account.

A classic case of suboptimization ensues, as each maximizes its own dimension of responsibility without reference to the cost on dimensions that are the responsibilities of others. The Theory of Second-Best teaches us that we should generally look at choices in a holistic way, assessing options along all the relevant dimensions at once. By "departmentalizing" policy choice, the modular approach to institutional design does the opposite of that. Each department is assigned some dimension (or some small set of dimensions) to be its own particular responsibility, and it assesses policies within its portfolio under that (those) dimensions.

Were the spillovers across departmental remits either rare or only on high-profile problems, there might be some hope of resolving them through special interdepartmental committees (or, in extremis, at the cabinet table itself). But there is every reason to think that the spillovers are ubiquitous, and that coordination is poor. Certainly when it comes to regulating health risks, "coordination . . . in modern government . . . has not been pursued in any systematic way."[33]

This has grievous effects on public health in a great many ways. For just one example, reflect upon what is now known about the "social determinants of health."[34] Spending just a little money on good public housing can save a lot of money in terms of keeping people out of the hospital. If both those items appeared on the same organization's ledger, the expenditure for housing would be more than balanced by the money it saved. But those items appear on different departments' ledgers. In consequence, the health costs and benefits of policy in other domains are not fully internalized by the departments with line responsibility for the actions in question. And attempts in Canada, for example, to overcome that by giving regional authorities responsibility for a wide range of health and community services, and reallocating resources within that portfolio, did not meet with any conspicuous success.[35]

Another aspect of the institutional carve-up of responsibilities is of course between countries: pollution, transmission of infectious diseases, consequences of trade and labor market policies of one country on the health of citizens in other countries, and so on.[36] There are clearly health-related spillovers across national borders. However much cross-border cooperation there may be in mounting coordinated responses to them, it is undeniably the case that dividing responsibility among different jurisdictions leads to a certain amount of suboptimization, as each inevitably prioritizes the health needs of its own citizens over those of others.

One Issue at a Time

Herbert Simon was fond of saying that the human mind is essentially a serial processor: it tends to focus attention on one thing at a time.[37] When treating patients, physicians would never dream of attending to just "one condition at a time," ignoring the way a patient's various conditions and treatments might interact. Yet when making public policy on health, we tend to do precisely that.

We are familiar with one-thing-at-a-time thinking on health through the various "wars" serially waged. Just think of the March of Dimes against polio, the War on Cancer, and so on.[38] Such one-thing-at-a-time thinking is reflected in public health schemes of "vertical" service delivery in developing countries (such as National Immunization Days) that target specific interventions in ways not fully integrated into the rest of the health system.[39] One-thing-at-a-time thinking is seen, too, in the way that even complex omnibus legislation, like the 2010 U.S. health care reform bill, is almost invariably discussed in terms of just one lightning-rod feature at a time—rotating serially, in that case, between "death panels," abortion, and the "public option."

More generally, time on the legislative calendar is strictly limited. It is conventional wisdom that there is room on the agenda of the UK Parliament for only about 20 major pieces of legislation to be enacted in any given year,[40] so limits of parliamentary scheduling as well as public attention also ensure that we cannot discuss everything all at once.

But if the policy problems are interrelated (and all are, if only in that they all compete for scarce budgetary resources), then we risk falling afoul of the Theory of Second-Best by not considering them all at once— or at least trying to do so. The modular approach of thinking about issues one at a time (or a few at a time) leads to suboptimization, as we choose policy options that are best on the dimensions we are focusing on without regard to their impact on other dimensions we are not.

This occurs, for example, whenever government departments (of health or anything else) are told to work within some fixed budget. They do the best they can with the resources they have been given. But (the Theory of Second-Best teaches us) they might well have done something completely different had they been allocated either more or less resources.

Just as we fall afoul of the Theory of Second-Best when taking the budget allocated to health (or anything else) as given, we do so when-ever taking the institutional allocation of responsibilities as given.

Redrawing the departmental allocation of responsibilities to make the same organization responsible for both health and housing, insofar as those importantly interact, would reduce the institutionally induced suboptimization that comes from different departments focusing on each area in isolation. Redrawing the national boundaries to make all of sub-Saharan Africa a single decision-making unit would reduce institutional barriers to a consolidated HIV/AIDS strategy for the region.

When discussing the "internalization of externalities," political economists often talk about redrawing the boundaries so as to make a single unit responsible for regulating all activities that affect one another within some policy domain. Making a single decision-making body responsible for an entire watershed—the Rhine, for example—would eliminate the risk of some upstream jurisdiction doing something that would make economic sense only if it could count on passing on the costs to downstream jurisdictions. The case for broadening jurisdictional responsibilities from within the Theory of Second-Best reinforces that: doing what is best for the Rhine valley overall is importantly different from doing what is best for the Rhine taken one stretch at a time. Decomposition does not work well given interactive systems of that sort.

Strategies for Overcoming the Errors of Decomposition

When we have erred in trying to decompose problems that are not truly decomposable, and in consequence have run into Second-Best–style problems, there are basically two standard ways of solving the problem. Both involve taking a more holistic approach, but the one takes a holistic approach from the bottom up, while the other takes a holistic approach from the top down.

Both strategies are nicely illustrated in recent public-service reforms in the UK. For an example of the bottom-up approach to getting a more holistic perspective, consider the 1991 innovation whereby GP fundholders became consolidators with responsibility and resources to manage comprehensively all of the medical needs of their patients. All the interacting factors involved in providing good medical care to a patient were thus placed in a single person's hands. For examples of the top-down approach, consider the creation of the U.S. Department of Homeland Security or the UK "joined-up government" initiative that tried to coordinate to the multiply interacting aspects of public policy at the very center of government.[41]

Neither is a wholly satisfactory approach. The bottom-up approach is (or can be if it works well) good for capturing the interactions among policies affecting any given person, but it is no good for capturing interactions among people. And, as I have already said, it crucially supposes that each person has an equally effective advocate. The top-down approach is (or can be if it works well) good for capturing the interactions among policies within a jurisdiction, but it is no good for capturing interactions among jurisdictions.

Each of these standard approaches has merit, in its place. But something more is needed to solve the widest-scale errors of Second-Best arising from trying to decompose problems that are not decomposable. For that, we need to implement something more like the "all-encompassing" approach suggested by the "internalizing externalities" story about the Rhine just told. We need to create some authority that is actually responsible for all of the interacting elements taken together. For problems that importantly interact all across the globe, that would mean a world government with responsibility for those problems.

ANOTHER PARTIAL SOLUTION: ROBUSTNESS AGAINST CHANGING POSSIBILITIES

So far I have been using the term "second-best" in a generic way to refer to anything that is not first-best. But of course the same sorts of feasibility constraints that force us to fall back on second-best solutions can, if the literal second-best is also infeasible, also drive us to third-, fourth-, or fifth-best solutions.

The Theory of Second-Best applies all the way down. Just as the second-best solution might be unlike the first-best in every respect, so too the third-best will often be unlike the first- or second-best in every respect. And so on for the fourth-best, *et seq.*

Now, suppose feasibility constraints change more quickly than policy choices can change. That is a not uncommon phenomenon in public policy in general. Path-dependency and historical lock-in are familiar features.[42] That fact, when combined with the Theory of Second-Best, has important implications for policy choice.

Suppose we have chosen the third-best alternative, because that was the best available at the time we had to choose. But now suppose feasibility constraints ease, and the second-best alternative suddenly becomes an option. If (per the Theory of Second-Best) the second-best alternative is wholly unlike the third-best, and if we are locked into a policy fit only for pursuing that very different third-best alternative, we will be unable to avail ourselves of this new opportunity.

Such reflections give rise to a further prescription for policymaking. Whenever feasibility constraints are likely to change more quickly than policy, the rule ought to be: Choose a policy that is robust against changes in feasibility. Thus, in the example just offered, it might be better to opt for a policy to pursue the fourth-best alternative instead of the third-best (which is also genuinely available), if that policy would be more amenable also to be used for pursuit of the second- or even first-best alternatives should they become available.

Like all maxims of policy, this one should provide only a *pro tanto* (so far as it goes) reason for action. If the fourth-best alternative is very much worse than the third-best, or if it seems very unlikely that the second- or first-best alternatives will become feasible anytime soon, then this maxim might well be overridden. So the robustness strategy, like all the others, is highly sensitive to the empirical facts of the matter, and there is no reason to think it will always prove possible or even desirable. Still, "robustness of policy choice against changing feasibility" is a consideration that ought always be borne in mind in choosing among "sticky" courses of action.

Suppose, for a bioethical example, we are trying to prevent the spread of some infectious disease. The first-best alternative would be for everyone to be vaccinated against the disease. But suppose it is technically infeasible to produce enough vaccine to do that. How then ought we best distribute the limited supply of vaccine?

Imagine that, upon inspection of the interaction patterns among the populations at risk, you notice that they sort themselves into several relatively self-contained communities, with only a few people passing between them. (Think for example of how HIV is spread by long-distance truckers to remote communities in southern Africa.) Given that observation, one strategy would be to try to create herd immunity within at least some of those communities (although only some, since—let us suppose—there is not enough vaccine to create herd immunity within all such communities). Another strategy, aimed more at preventing the spread of the disease from one community to others, would be to inoculate everyone who passes between communities (which would also—let us suppose—pretty fully exhaust supplies of the vaccine).

Obviously, once the vaccine has been injected into one person it cannot be extracted and re-injected into another. So once we have implemented one or other of these policies, we are "locked in." The first condition for applying my "robustness" rule thus applies. The second (quickly changing possibilities) might as well. Suppose for example new transportation possibilities suddenly emerge, and lots of new people start plying the trade routes between the various communities.

Absent robustness considerations, the ranking of policy options would seem to be that: (1) immunizing everyone at risk is first-best (but *ex hypothesi* infeasible); (2) immunizing travelers between communities is second-best, because if successful it promises to protect all communities; (3) immunizing enough people to create herd immunity within some communities is third-best, because it leaves other whole communities unprotected.

That is the way things look, without taking robustness considerations into account. Given robustness considerations, however, and the real risk that there might suddenly be so many travelers that that strategy collapses, it might be better to opt for distributing the vaccine so as to give at least some communities herd immunity.

CONCLUSION

The upshot is a moderately discouraging one. The trouble that the Theory of Second-Best makes for policy choice is not easily avoided. It issues firm cautions against the strategy of bringing the real world more into line with the presuppositions of ideal theory, in much the same way and for much the same reasons that it warns against the more standard strategy of adapting ideal theory to fit the real world. The Theory of Second-Best further warns that theorizing in the vicinity of the actual might be highly unstable; that problems might not be decomposable; and that there may be no policy choice that would be robust against all possible (or even all likely) eventualities. In short, if we are looking for guarantees, there seem to be none.

The Theory of Second-Best might unsettle any of those strategies. Or it might not. On the weaker and more general form I have been focusing on in this chapter, there is no reason to think it will necessarily always do so. It may turn out that there are some realms in which the second-best really is exactly like the first-best in all respects except one. Whether there are, and what they might be, is an empirical matter, not resolvable by philosophical analysis alone. The most the philosopher can do is to warn policymakers to watch out for interactions across evaluative dimensions and to be sensitive to important evaluative dimensions being omitted altogether.

ACKNOWLEDGMENT

Earlier versions of this chapter have been presented to seminars of the Bioethics Department at the Clinical Center of the National Institutes of Health and of the Georgetown-Johns Hopkins Greenwall Fellows. I am grateful to participants in those discussions, to Joe Millum for written comments, and to Marion Danis, Christine Grady, Chiara Lepora, Frank Miller, and Annette Rid for resource materials. This chapter was written during my time as a Visiting Scholar in the Bioethics Department at the NIH. The views expressed are my own and do not represent the opinions or policies of the National Institutes of Health or the Department of Health and Human Services.

NOTES

1. Black M. (1962). *Models and Metaphors.* Ithaca, NY: Cornell University Press.
2. Rawls J. (1971). *A Theory of Justice.* Cambridge, MA: Harvard University Press, p. 245ff; Sreenivasan *infra*; Simmons AJ. (2010). Ideal and non ideal theory. *Philosophy & Public Affairs* 38(1):5–36. For an excellent application to another realm, that of international migration and global justice, see Carens JH. (Spring 1996). Realistic and idealistic approaches to the ethics of migration. *International Migration Review* 30:156–170.
3. Of course, Rawls does not abstract from absolutely all non-ideal aspects of the actual world. He does not abstract from the sad fact of material scarcity for example: if everyone could have as much as they wanted and questions of distributive justice would not arise (Rawls, 1971, sec. 22, pp. 126–130). What grounds there are for abstracting out some non-ideal facts but not others for purposes of moral theory is a large topic, too little discussed. But I simply note it here in passing.
4. Ideal theory, Rawls says (1971, pp. 8–9, 391), is the more "fundamental." We need it to tell us what we should be aiming at through social reform of non-ideal circumstances (Rawls, 1971, pp. 8–9, 245; Rawls J. [1993]. *Political Liberalism.* New York: Columbia University Press, p. 285; Rawls J. [2001]. *Justice as Fairness: A Restatement.* Cambridge, MA: Harvard University Press, p. 13).
5. Whereas Hume and Rawls (1971, pp. 126ff) following him would define the "circumstances of justice" in terms of scarcity of resources relative to *desires*, I shall define what's morally "ideal" or "non-ideal" relative to moral *rights and duties*.
6. I am uncertain how to incorporate imperfect duties and supererogation here. Maybe it morally matters that people have enough resources to perform some of those acts too, and a state of affairs is also morally non-ideal insofar as they do not.
7. The possibility might be more theoretical than real, judging from the scarcity of reports of such "decrementally cost-effective medical innovations" in the published literature. A survey of medical cost-utility analyses published between 2002 and 2007 found that only 9 of the 2,128 interventions described were less expensive by $100,000 or more than the existing standard of care per quality-adjusted life year sacrificed (Nelson AL, Cohen JT, Greenberg D, Kent DM. [2009]. Much cheaper, almost as good: decrementally cost-effective medical intervention. *Annals of Internal Medicine* 151(9):662–667). It is an open question whether this result is due to a genuine paucity of "much cheaper and almost as good" interventions or whether it is merely an artefact of publication practices.
8. Msuya J. (2003). Horizontal and vertical delivery of health services what are the tradeoffs? Background Study no. 26942 for *World Development Report 2004/5.* Washington, DC: World Bank, p. 17. Available at http://www-wds.worldbank.org/external/default/WDSContentServer/IW3P/IB/2003/10/15/000160016_20031015125129/additional/310436360_200502761000211.pdf.

9. Healey K. (2000). Embedded altruism: blood collection regimes in the European Union. *American Journal of Sociology* 105(5):1633–1657; and Healy K. (2006). *Last Best Gifts: Altruism and the Market for Human Blood and Organs.* Chicago: University of Chicago Press. I employ scare quotes, because brain-dead patients are in other important senses not really dead: indeed, dead organs can save no one's life.

10. This is the case most often discussed in the philosophical literature, by Rawls (1971, pp. 8, 254, 351) when he defines "ideal theory" in terms of "strict compliance" and by others who talk of the "demandingness" of morality (Murphy LB. [2000]. *Moral Demands in Non-ideal Theory.* Oxford: Oxford University Press; Mulgan TP. [2001]. *The Demands of Consequentialism.* Oxford: Oxford University Press; cf. Goodin RE. (2009). Demandingness as a virtue. *Journal of Ethics* 13(1):1–13.)

11. Often, but perhaps not always, institutional arrangements are the solution to such problems—in which case the fourth case collapses into the second.

12. Or third-best, if what would be second-best is also unattainable, and so on: those further iterations will become important in my discussion below; but for now let me speak as if "second-best" embraced all those non–first-best possibilities.

13. Lipsey and Lancaster elaborate: "The general theorem for the second best optimum states that if there is introduced into a general equilibrium system a constraint which prevents the attainment of one of the Paretian conditions, the other Paretian conditions, although still attainable, are, in general, no longer desirable. In other words, given that one of the Paretian optimum conditions cannot be fulfilled, then an optimum situation can be achieved only by departing from all the other Paretian conditions. The optimum situation finally attained may be termed a second best optimum because it is achieved subject to a constraint which, by definition, prevents the attainment of a Paretian optimum." (Lipsey RG, Lancaster KJ. [1956]. The general theory of second best. *Review of Economic Studies* 24:11–33).

14. Margalit A. (2010). *On Compromise and Rotten Compromises.* Princeton, NJ: Princeton University Press, p. 116.

15. As in the title of Arthur Okun's 1975 book, *Equality and Efficiency: The Big Tradeoff.* Washington, DC: Brookings Institution.

16. Barry B. (1965). *Political Argument.* London: Routledge & Kegan Paul, pp. 4–8; cf. Rawls, 1971, pp. 34–45.

17. Sunstein CR. (1996). Health–health tradeoffs. *University of Chicago Law Review* 63(4):1535–1536.

18. As Prince Edward Island tried to do, with little success, by giving a single regional authority responsibility for reallocating resources across a broad range of health and community services (Stoddart GL, Eyles JD, Lavis JN, Chaulk PC. [2006]. Reallocating resources across public sectors to improve population health. In Heymann J, Hertzman C, Barer ML, Evans RG, eds. *Healthier Societies: From Analysis to Action.* Oxford: Oxford University Press.

19. Rawls, 1971, pp. 8–9, 245–246, 575.

20. True, Rawls offers what Simmons (2010, p. 24) calls an "integrated ideal." But while the ideal itself is integrated, the procedure Rawls suggests for pursuit of that ideal in the non-ideal world is not. Simmons thinks otherwise. He (and he thinks Rawls) appreciates that "our attacks on particular, especially offensive injustices may be . . . understandably compelling." Yet he (and he thinks Rawls) supposes that "few devotees of 'partial justice' would be able to sustain their single-minded commitments in the face of clear evidence that their efforts were setting back or permanently blocking movement toward overall social justice." Yet a "one injustice at a time" approach is precisely what Rawls (2001, p. 13) seems to recommend, in saying that "ideal theory . . . should . . . help to clarify the goal of reform and to identify which wrongs are more grievous and hence more urgent to correct."

21. Rousseau, J-J. (1762). *The Social Contract*. For discussions of these options with special reference to global justice debates see Valentini L. (2009). On the apparent paradox of ideal theory. *Journal of Political Philosophy* 17(3):332–355; Lawford-Smith H. (2010). Debate: ideal theory—a reply to Valentini. *Journal of Political Philosophy* 18(3): 357–68; and Ypi L. (2010). On the confusion between ideal and non-ideal in recent debates on global justice. *Political Studies* 58(3):536–555.

22. Gilabert P. (2009). The feasibility of basic socioeconomic human rights: a conceptual exploration. *Philosophical Quarterly* 59(237):659–681. Simmons (2010, pp. 24–25) rightly emphasizes this, and supposes Rawls either does or should do similarly within the logic of his own theory.

23. Bentham J. (1827). Rationale of Judicial Evidence. Vol. 7 in *Works* (ed. Bowring J). Edinburgh: W. Tait, p. 285; Goodin RE. (1982). *Political Theory & Public Policy*. Chicago: University of Chicago Press, Chapter 7.

24. Weber M. (1919/2004). Politics as a vocation. In Weber M. *The Vocation Lectures* (ed. Owen D and Strong TB, trans. Livingstone R). Indianapolis: Hackett, p. 9.

25. There is, furthermore, a bootstrapping issue here: if we are in a non-ideal world and we want to make it ideal, how do we know how to do that? We cannot (necessarily, anyway) use ideal theory to guide us how to get to the ideal, because *ex hypothesi* we are not in the ideal state yet; and the Theory of Second-best tells us that what is the right thing to do in the ideal state might be very different from the right thing to do in non-ideal states.

26. For critical discussion of this option see Simmons, 2010, p. 30ff.

27. Goodin, 1982, Chapter 1.

28. Rawls's (1971, p. 351) talk of "a state of near justice" suggests that that might have been what he was thinking, when assuming that his theory of civil disobedience could be bolted onto his theory of justice without altering much else in that larger theory.

29. Simon HA. (1969). *The Sciences of the Artificial*. Cambridge, MA: MIT Press; Simon HA. (2002a). Near decomposability and the speed of evolution. *Industrial and Corporate Change* 11(3):587–599; Simon HA. (2002b). Organizing and coordinating talk and silence in organizations. *Industrial and Corporate Change* 11(3):611–618.

30. Lipsey & Lancaster, 1956, p. 17.

31. Simon, 2002a and 2002b.

32. Fried C. (1978). *Right & Wrong*. Cambridge, MA: Harvard University Press, p. 176.

33. In the words of Sunstein (1996, p. 1555)—now Administrator of the Office of Information and Regulatory Affairs in the Executive Office of the President of the United States, who is responsible for doing just that.

34. World Health Organization. (2008). *Closing the Gap in a Generation: Health Equity through Action on the Social Determinants of Health*. Final Report of the Commission on Social Determinants of Health. Geneva: World Health Organization.

35. Not in the Prince Edward Island experiment, under the 1993 Health and Community Services Act, anyway (Stoddart et al., 2006).

36. For a sophisticated political-philosophical treatment of the latter issue, see Young IM. (2004). Responsibility and global labor justice. *Journal of Political Philosophy* 12(4):365–388.

37. Simon, 2002b.

38. Observing how the War on Terror was displacing the War on Drugs, one character on "The Wire" is led to ask, "Is our heart not big enough for two wars?"

39. Msuya, 2003.

40. Mechelen DV, Rose R. (1986). Law as a resource of public policy. *Parliamentary Affairs* 36(3):297–314.

41. Bogdanor V, ed. (2005). *Joined-Up Government*. Oxford: Oxford University Press.

42. Arthur WB. (1988). Self-reinforcing mechanisms in economics. In Anderson PW, Arrow KJ, Pines D, eds. *The Economy as an Evolving Complex System*. Reading, MA: Addison Wesley; Arthur WB. (1989). Competing technologies & lock-in by historical events. *Economic Journal* 99:116–31; Pierson P. (2000). Increasing returns, path dependence and the study of politics. *American Political Science Review* 94(2):251–268.

REFERENCES

Arthur, W. Brian. (1988). Self-reinforcing mechanisms in economics. In Phillip W. Anderson, Kenneth J. Arrow, & David Pines, eds. *The Economy as an Evolving Complex System*. Reading, MA: Addison-Wesley

Arthur, W. Brian. (1989). Competing technologies & lock-in by historical events. *Economic Journal* 99:116–131

Barry, Brian. (1965). *Political Argument*. London: Routledge & Kegan Paul

Bentham, Jeremy. (1827). Rationale of judicial evidence. Vols. 6–7 in *Works*, ed. J. Bowring. Edinburgh: W. Tait, 1843

Black, Max. (1962). *Models and Metaphors*. Ithaca, NY: Cornell University Press

Bogdanor, Vernon, ed. (2005). *Joined-Up Government*. Oxford: Oxford University Press

Carens, Joseph H. (1996). Realistic and idealistic approaches to the ethics of migration. *International Migration Review* 30(1):156–170

Fried, Charles. (1978). *Right & Wrong*. Cambridge, MA: Harvard University Press

Gilabert, Pablo. (2009). The feasibility of basic socioeconomic human rights: a conceptual exploration. *Philosophical Quarterly* 59(237):659–681

Goodin, Robert E. (1982). *Political Theory & Public Policy*. Chicago: University of Chicago Press

Goodin, Robert E. (2009). Demandingness as a virtue. *Journal of Ethics* 13(1):1–13

Healey, Kieran. (2000). Embedded altruism: blood collection regimes in the European Union. *American Journal of Sociology* 105(5):1633–1657

Healey, Kieran. (2006). *Last Best Gifts: Altruism and the Market for Human Blood and Organs*. Chicago: University of Chicago Press

Lawford-Smith, Holly. (2010). Debate: ideal theory—a reply to Valentini. *Journal of Political Philosophy* 18(3): 357–68

Lipsey, Richard G., and Kelvin J. Lancaster. (1956). The general theory of second best. *Review of Economic Studies* 24:11–33

Margalit, Avishai. (2010). *On Compromise and Rotten Compromises*. Princeton, NJ: Princeton University Press

Mechelen, Denis van, and Richard Rose. (1986). Law as a resource of public policy. *Parliamentary Affairs* 36(3):297–314

Msuya, Joyce. (2003). *Horizontal and Vertical Delivery of Health Services: What Are the Trade Offs?* Background Study 26942 for World Development Report 2004/5. Washington, DC: World Bank. Available at http://www-wds.worldbank.org/external/default/WDSContentServer/IW3P/IB/2003/10/15/000160016_20031015125129/additional/310436360_200502761000211.pdf

Mulgan, Timothy P. (2001). *The Demands of Consequentialism*. Oxford: Oxford University Press

Murphy, Liam B. (2000). *Moral Demands in Non-ideal Theory*. Oxford: Oxford University Press

Nelson, Aaron L., Joshua T. Cohen, Dan Greenberg, and David M. Kent. (2009). Much cheaper, almost as good: decrementally cost-effective medical intervention. *Annals of Internal Medicine* 151(9):662–667

Okun, Arthur. (1975). *Equality and Efficiency: The Big Tradeoff*. Washington, DC: Brookings Institution

Pierson, Paul. (2000). Increasing returns, path dependence and the study of politics. *American Political Science Review* 94 (2):251–268

Rawls, John. (1971). *A Theory of Justice*. Cambridge, MA: Harvard University Press

Rawls, John. (1993). *Political Liberalism*. New York: Columbia University Press

Rawls, John. (1999). *The Law of Peoples*. Cambridge, MA: Harvard University Press

Rawls, John. (2001). *Justice as Fairness: A Restatement*. Cambridge, MA: Harvard University Press

Rousseau, Jean-Jacques. (1762/1973). The social contract. *The Social Contract and Discourses* (trans. G.D.H. Cole. New ed.). London: Everyman/Dent, pp. 164–278

Simmons, A. John. (2010). Ideal and nonideal theory. *Philosophy & Public Affairs* 38(1):5–36

Simon, Herbert A. (1969). *The Sciences of the Artificial*. Cambridge, MA: MIT Press

Simon, Herbert A. (2002a). Near decomposability and the speed of evolution. *Industrial and Corporate Change* 11(3):587–599

Simon, Herbert A. (2002b). Organizing and coordinating talk and silence in organizations. *Industrial and Corporate Change* 11(3):611–618

Stoddart, Greg L., John D. Eyles, John N. Lavis, and Paul C. Chaulk. (2006). Reallocating resources across public sectors to improve population health. In Jody Heymann, Clyde Hertzman, Morris L. Barer, and Robert G. Evans, eds. *Healthier Societies: From Analysis to Action*. Oxford: Oxford University Press, pp. 327–347

Sunstein, Cass R. (1996). Health-health tradeoffs. *University of Chicago Law Review* 63(4):1533–1572

Valentini, Laura. (2009). On the apparent paradox of ideal theory. *Journal of Political Philosophy* 17(3):332–355

Weber, Max. (1919/2004). Politics as a vocation. In Weber, M. *The Vocation Lectures* (ed. David Owen and Tracy B. Strong, trans. Rodney Livingstone). Indianapolis, IN: Hackett, pp. 32–94

WHO CSDH. (2008). *Closing the Gap in a Generation: Health Equity through Action on the Social Determinants of Health*. Final Report of the Commission on Social Determinants of Health. Geneva: World Health Organization

Young, Iris Marion. (2004). Responsibility and global labor justice. *Journal of Political Philosophy* 12(4):365–388

Ypi, Lea. (2010). On the confusion between ideal and non-ideal in recent debates on global justice. *Political Studies* 58(3):536–555.

PART THREE

Non-ideal Theory

CHAPTER 8

Global Justice and the "Standard of Care" Debates

EZEKIEL J. EMANUEL

O n June 5, 1981, the U.S. Centers for Disease Control (CDC) reported on a cluster of cases of a rare pneumonia among young gay men in Los Angeles.[1] This was the first description of what came to be known as HIV/AIDS.[2] In late 1982, the first case of AIDS in Africa was reported as "slim disease."[3] The first treatment victory in the AIDS battle occurred in 1987, when the efficacy of AZT in delaying the development of symptoms led to U.S. Food and Drug Administration (FDA) approval.[4] In 1994, the stunning results of AIDS Clinical Trial Group (ACTG) study 076 were published.[5] That trial randomized pregnant women in the United States to a relatively long and complex course of AZT or placebo. Beginning in the second trimester, the 076 AZT regimen required pregnant women to take five AZT tablets each day and to get intravenous AZT on the day of delivery, and for the infant to get AZT and to be bottle-fed for 6 months. All of this decreased maternal–fetal HIV transmission by 67%, from 25.5% to 8.3%—a tremendous victory.

While the 076 regimen became standard of care in the United States and other developed countries, it could not—or would not, in the foreseeable future—actually be used in developing countries for economic, cultural, logistical, and other reasons. For one thing, few pregnant women in developing countries received antenatal care, so few women could obtain AZT in the second trimester, and few bottle-fed their babies. And yet the problem was substantial. According to the United Nations

Programme for AIDS (UNAIDS), the transmission rate from infected mothers to newborns in developing countries was estimated at between 25% and 33%. In 1997, more than 500,000 children were estimated to be newly infected, and "[t]he overwhelming majority of these children acquired the infection from their mothers before or around the time of birth, or through breast milk."[6] What to do?

In June 1994, the World Health Organization (WHO) held a meeting to discuss strategies for discovering an effective and affordable intervention to reduce maternal–fetal HIV transmission in developing countries. At that meeting, it was concluded that randomized placebo-controlled trials of shorter and simpler regimens of AZT should be conducted.[7] Fifteen trials were designed, approved, and conducted in 11 countries in Southeast Asia, the Caribbean, and Africa. Of these trials, nine were sponsored by the U.S. National Institutes of Health (NIH) and the CDC, now known as the Centers for Disease Control and Prevention; five by other, non-U.S. governmental organizations; and one by UNAIDS.

In 1997, two articles appeared in the *New England Journal of Medicine* criticizing these short-course AZT trials as unethical—one by Peter Lurie and Sidney Wolfe of Ralph Nader's Public Citizen Health Research Group,[8] and one by Marcia Angell, then executive editor of the *New England Journal of Medicine*.[9] These and subsequent critics[10–14] raised many ethical issues that were not always clearly distinguished.[15] A common theme was that the trials' use of placebo controls was unethical. While in the bioethics literature the debate about placebo-controlled trials in developing countries has centered on whether they violate provisions of the Declaration of Helsinki, the issues go much deeper, and the importance—and mesmerizing power—of the debates about these trials is that they implicate questions of global justice. Indeed, the ethical debate about these trials is a practical embodiment of the debate between cosmopolitan and statist views of global justice. Its resolution suggests two alternative positions on global justice that might be called pragmatic cosmopolitanism and beneficent statism. These alternatives embody our moral intuitions better than either the cosmopolitan or statist view and also provide practical ways of fairly addressing problems in the non-ideal world.

CLARIFYING THE ETHICAL ISSUE

Many issues have been raised about the ethics of conducting placebo-controlled trials in developing countries. The most important can be classified into four distinct groups. The first group of issues concerns "reasonable availability"—that is, whether research participants, members of the larger community or countries in which the trial was conducted, or even whole geographic regions, should be guaranteed access to the drug, vaccine, or other intervention, if it is proven effective in the research trial.[16-18] This principle of reasonable availability, articulated first by the Council for International Organizations of Medical Sciences (CIOMS), provides benefits to research participants and their communities only if an intervention is proven effective. It offers nothing to participants in early-phase research studies—which are not designed to demonstrate anything about effectiveness—nor to participants in later-phase research in which the intervention proves ineffective.[19] Reasonable availability implicates questions about exploitation, particularly (but not necessarily) focused on exploitation of the society or community from which participants are recruited by researchers and sponsors from developed countries.[20]

The second group of issues concerns the quality of informed consent among poor, poorly educated, and frequently illiterate people in developing countries with few health care alternatives to research participation. This raises ethical concerns about autonomy and coercion in research.

The third group of issues concerns the physician–patient fiduciary relationship. Critics argue that researchers, like doctors, have role-related obligations to provide the best treatment to their patients. Not to give the best treatment would be to violate the physician's obligation to his or her patients.

The fourth and final set of issues, which seems the focus of the Angell and the Lurie and Wolfe critiques, and of this chapter, is the "standard of care."[21,22] The standard of care issue concerns the nature of both the experimental interventions that participants in research trials should receive and the treatment—if any—provided to those in the control arm of the trial. (The standard of care issue excludes questions about ancillary care—that is, whether there is an ethical requirement to provide medical tests and treatments beyond those necessary to address the research question. For instance, the standard of care debate does not include whether it is necessary for researchers to provide malaria treatments to participants in HIV/AIDS trials of antiretroviral drugs if the malaria treatments are not necessary to answering the research question.)

The critics of the short-course AZT trials made three main arguments against the use of placebos and in favor of using the 076 regimen as the standard of care in the control arms of the trials. First, they claimed that using placebo in developing countries was a "double standard." After the 076 trial, any maternal–fetal HIV transmission trials in developed countries would have to use the 076 regimen as the control arm because it was the standard of care for pregnant HIV-infected women. To use a placebo would be to expose the research participants to serious and permanent harm when a proven effective therapy existed—a clear violation of the ethical norms of research (see below). If using a placebo control is unethical in maternal–fetal HIV transmission studies in developed countries, then the critics contended it must also be unethical to use placebo controls in maternal–fetal HIV transmission studies in developing countries. To treat HIV-infected pregnant women differently, based just on their country of residence, is to treat equals unequally—a double standard.

Second, critics argued that the short-course AZT trials violated the Declaration of Helsinki, specifically paragraph II 3 in the 4th (1989) revision, and what became paragraph 29 in the 5th (2000) revision, which stated:

> In any medical study, every patient—including those of a control group, if any— should be assured of the best proven diagnostic and therapeutic method.[23]

(In the 5th revision enacted in 2000, "best proven" was changed to "best current" and "prophylactic" was added to "diagnostic and therapeutic" methods. The 2000 provision thus read: "The benefits, risks, burdens and effectiveness of a new method should be tested against those of the best current prophylactic, diagnostic, and therapeutic methods.") This provision of Helsinki required that once the 076 regimen was proven effective in 1994, every participant in a subsequent research trial anywhere in the world should be assured of receiving the 076 regimen or a better treatment. The most recent (2008) revision of Helsinki (in paragraph 32) provides a very limited exception. It states that "The benefits, risks, burdens and effectiveness of a new intervention must be tested against those of the best current proven intervention." However, it allows that "The use of placebo, or no treatment, is acceptable in studies where no current proven intervention exists; or where for compelling and scientifically sound methodological reasons the use of placebo is necessary to determine the efficacy or safety of an intervention and the patients who receive placebo or no treatment will not be subject to any risk of serious or irreversible harm."

BEYOND PLACEBOS

The standard of care debate has focused on the use of placebos in HIV/ AIDS research. But this distorts the issue. The issue is not placebos *per se* but the use of any intervention or test that is deemed "less effective" than the worldwide best standard. For instance, in developed countries, response to HIV/AIDS antiviral therapy is measured by diagnostic tests of CD4 levels and viral load. Early in the HIV epidemic, some trials had assessed whether clinical measures of effectiveness, such as weight gain, could be used in developing countries instead of laboratory measures to monitor the effectiveness of antiviral therapy. This change in monitoring tests would be done for either cost or logistical reasons. Such clinical monitoring would clearly be less than the best diagnostic test used in developed countries.

Similarly, after the controversy about short-course AZT there arose questions about trials designed to test whether a single 200-mg dose of the antiretroviral drug nevirapine, during labor, would be effective in reducing maternal–fetal HIV transmission. Approximately 75% of all HIV transmission from mother to child occurs during and after delivery. One perinatal dose of nevirapine could affect this perinatal HIV transmission, but not the 25% of transmission that occurs *in utero* prior to delivery. Thus, prior to the start of the study it was thought nevirapine would necessarily be inferior to the 076 regimen. A trial of nevirapine against the 076 regimen would subject participants to a treatment likely *ex ante* to be inferior to best available standard of care in developed countries.

As these two cases illustrate, both in terms of diagnostic tests and therapeutic interventions, the ethical issue in the standard of care debate is whether anything whose effectiveness is lower—or whose side effects are greater—than the most advanced, high-technology test or intervention available anywhere in the world can be used in a clinical trial. This clearly goes beyond the use of a placebo.

To some degree, critics have conceded the principle that the best need not always be used in a trial in developing countries. Many have accepted that if a research study is done in a developing country on cardiac disease—an increasing problem in many developing countries—it would not be necessary to provide care in a cardiac care unit, a standard practice in developed countries. As Ruth Faden and Nancy Kass wrote in 1998:

> [N]o one suggests requiring that the most expensive and complex of Western tertiary care, for example, renal dialysis or coronary bypass, must be provided in research conducted in the developing world.[24]

Despite this concession, no principled method has been articulated to differentiate life-saving coronary care units or dialysis from expensive or logistically complex life-saving antiretroviral therapy. The basic issue remains: What should the standard of care in a research trial in a developing country be—and how can that standard be justified?

THE ETHICS OF CLINICAL RESEARCH

A commonly accepted framework for evaluating the ethics of research studies suggests that studies must fulfill eight principles: (1) collaborative partnership, (2) social value of the research question, (3) scientific validity of the research design, (4) fair subject selection, (5) favorable risk–benefit ratio, (6) independent review of the research protocol, (7) informed consent, and (8) respect for the research participants.[25] In a randomized controlled trial, determining what treatment the research participants should receive—whether in the control or experimental cohorts—requires addressing three of these principles: social value, scientific design, and consideration of risk–benefit ratio.

First, the research question must contribute social value. It should produce results that will directly lead to interventions to improve human health, or will contribute generalizable knowledge that can guide future research or be combined with further research results to improve human health. In the case of the AZT trials, even critics agreed that it was socially valuable to find an effective way to reduce maternal–fetal HIV transmission in developing countries that was affordable and culturally appropriate and could be practically provided. The long-course 076 AZT regimen was too costly. At about $1,000 per mother in 1995, the cost was far in excess of what developing countries could devote to a single patient. Moreover, many if not most pregnant women in developing countries never presented to health authorities in the second trimester—when treatment begins under the 076 protocol—to begin treatment, or even at the time of delivery to get intravenous AZT. Bottle feeding—to prevent transmission through breast milk—was expensive and practically difficult and posed problems of diarrhea from contaminated water. It also was stigmatizing because it publicly identified HIV-infected women. Finding an intervention that addressed most if not all these issues, especially the problems of cost and those related to mothers engaging the health system late in pregnancy, if at all, was widely viewed as socially valuable.

Second, assuming the research question is socially valuable, then the study must be designed to produce valid scientific data that can be interpretable. Providing uninterpretable data makes the research worthless and a waste of resources and exposes participants to unnecessary risks. Thus, the use of a specific control must be able to provide valid and reliable data for the question that has social value. In the case of short-course AZT, using either historical data on HIV transmission rates or the 076 AZT regimen as the control would not have provided valid, interpretable data. Historical controls are notoriously unreliable, especially in the case of HIV transmission. HIV transmission rates vary substantially from site to site, and have varied over time in the same site. This was well demonstrated in the very placebo-controlled short-course AZT trials being challenged. In the original 076 trial conducted in developed countries, the HIV transmission rate in the placebo arm was 25.5%.[26] In the short-course AZT trial, the maternal–fetal HIV transmission rate in the placebo arms varied from 18.9% in Thailand, to 24.9% in Côte d'Ivoire, to 27.5% in Burkina Faso.[27–29] The transmission rate in the short-course AZT arms also varied, from 9.9% in Thailand, to 15.7% in Côte d'Ivoire, to 18.0% in Burkina Faso. Using as a historical control the transmission rate of 18.9%—which was the transmission rate in the placebo arm of the Thailand trials—would have made the short-course AZT trial appear ineffective in Burkina Faso. Similarly, using the short-course AZT results from Thailand would have made the short-course AZT intervention appear as effective as the 076 AZT regimen.

What about using active controls—that is, giving every participant in the control arm the 076 regimen? This design was advocated by Lurie and Wolfe and by Angell, but it presents ethical and scientific problems. Ethically, despite their endorsement, this approach should not satisfy Angell or Lurie and Wolfe. Indeed, it would conflict with the Declaration of Helsinki and would violate their own principle. Research participants receiving short-course AZT were not going to receive the best proven treatment. They would be given an intervention that researchers believed *ex ante* was inferior to the 076 regimen. Scientifically, a trial of the 076 AZT regimen versus short-course AZT would provide data but would not answer the critical public health questions of whether the short-course regimen reduced maternal–fetal HIV transmission and whether the amount of reduction was worth the cost, given alternative potential uses of the limited resources. At the outset of the trials, the data on the 076 regimen suggested that it generated an HIV transmission rate of 8.3%. What if the short-course AZT arm produced a transmission rate of 18.0%, as it did in Burkina Faso? What would these data mean? Would the short-course AZT have been recognized as having "worked"? Would it be "worth" using? From a methodological—scientific validity—perspective, a placebo arm was critical to knowing whether short-course AZT really reduced HIV transmission compared to no intervention at all.

Even if a placebo control were scientifically necessary, its use would not necessarily be ethical. For that, the study would have to fulfill the risk–benefit principle. This principle requires researchers to ensure that the research participants receive the medical interventions they are entitled to relative to the scientific objective. The reason for this is to ensure that the risk–benefit ratio for the participants is favorable. If participants in the trial do not receive the medical services they are entitled to, then by participating in the trial they are likely to be at risk for additional harms they would otherwise not experience. For instance, if the research study is evaluating a new drug for tuberculosis (TB), then the research participants in the control arm must receive the TB medications they are entitled to. Only by receiving the TB medications they should receive based on standard clinical practices can we be sure their risk–benefit ratio is minimized and they are not exposed to additional risks unnecessarily. If they received placebo instead of standard TB medications, then they would be at increased risk compared to routine clinical care.

In some cases, because of poor local medical services, potential participants already might not be receiving the medical tests and treatments they are entitled to. Should researchers then *not* be required to provide these tests and treatments, because withholding them would not change the risk–benefit ratio and would not expose participants to additional risks? There is a reasonable argument that the researchers are not making the people worse off in this non-ideal circumstance; they are just using the bad level of local care to conduct research. But to justify this conduct of research would be to justify exploiting the injustice of the world.

The risk–benefit principle does not require that participants receive all the medical interventions they are entitled to. It does not require that the research team give them asthma or hypertension treatments, if the research is focusing on cervical cancer treatments. The researchers are not responsible for rectifying all the injustices of the health care system (assuming they fulfill all the other relevant ethical criteria, so that the research is valuable, scientifically sound, etc.). Conversely, if there is a randomized trial of an experimental asthma treatment, and the participants in the trial are identified because they happen not to be receiving the care they are entitled to, such as steroid inhalers, they are likely to suffer serious side effects and risks. The researchers are exploiting an injustice—the bad care the participants are receiving—to conduct the research study. However, this requirement does not mean the research team needs to provide these asthma patients with medications for some other condition, such as TB; ethically, the researchers are required only to provide medical interventions the participants are entitled to that are relevant to the scientific objective of evaluating a new asthma agent.

Importantly, entitlement is a matter of justice, not of what people in fact receive. The entitlement to health care services is determined not by what any citizen of any country actually receives, but by what a just distribution of resources would entitle the people to. Entitlement to health care services is a matter of justice. People may actually be receiving too much or too little compared to what they should, as a matter of justice, receive. Determining what research participants are entitled to links the standard of care debate with the principles of justice.

Consider the Tuskegee syphilis study, in which hundreds of black sharecroppers with syphilis were followed from 1932 until 1972, without being treated with penicillin when it became available. There were many ethical violations in this study, questions about the social value, scientific validity, and quality of the consent. Some claim that not providing penicillin to these men was not an ethical violation. After all, it is argued, given the appalling health system in the South for blacks, most, if not all, of the participants would not have obtained penicillin treatment anyway; withholding penicillin thus did not deprive them of something they would otherwise have received, and therefore was not unethical. However, that these black sharecroppers would not have received penicillin in the regular health care system because of racism does not make withholding penicillin in a research trial ethical. In a just health care system in the United States, these black men, at least in the 1950s, would have received penicillin, just as more fortunate whites did. This entitlement—not the actual practices at the time—determines what care they should have received in clinical medicine and, therefore, as part of the research trial.

When dealing with clinical research in developing countries, the question is not just a matter of domestic justice, but also one of global justice. Deciding what health care services should be provided to research participants in a trial depends upon knowing what services they would be entitled to in a just global order. This is how the standard of care issue is inescapably linked to claims about global justice. Each of the positions in the standard of care debate reflects established theory about global justice.

There is one important exception to the requirement that participants receive the services they are entitled to that are relevant to the condition under study. If withholding the services would not increase participants' risk of suffering death, irreversible morbidity, disability, reversible but serious harm, or severe discomfort, then it need not be provided as part of a research trial. Another way to put this exception is that if it is a reasonable therapeutic option for patients to forego the medical service they are entitled to, then that medical service can be withheld from participants in the control arm as part of a research trial.[30,31] For example, there are effective interventions for migraine headache, yet many reasonable patients forgo these interventions without suffering a risk of death, irreversible disability, or even reversible harm; the discomfort is transient and many people tolerate it. If many reasonable patients forego these treatments, despite an entitlement to them, and if foregoing the treatments does not cause the irreversible harms listed, then a clinical trial can ethically withhold the medical services from the control group. This is why it is ethical to conduct placebo-controlled trials for new migraine medications even though there are many standard clinical drug treatments. Similarly, reasonable people routinely forego standard medical treatments for gastroesophageal reflux (GERD), joint pain, prostatic hypertrophy, hair loss, and many other conditions without exposing themselves to serious risks. In these conditions, despite standard therapies that patients are entitled to, it is ethical to conduct placebo-controlled trials. (The actual conduct of a trial must meet other ethical requirements as well; participants must be informed about the availability of these standard treatments and that these treatments will be withheld, but that there will be no chance of death, irreversible morbidity, disability, etc. The discussion above is about the *design* of the trial and the control arm.)

For most research trials in developing countries, however, this exception is irrelevant. The trials involve serious diseases—HIV, TB, malaria, pneumonia, diarrhea, etc.—in which use of placebo or treatment with less than the worldwide best standard of care exposes participants to an increased risk of death, irreversible morbidity, disability, or reversible but serious harm. Hence, in deciding what interventions the participants in the control arm of a trial in a developing country should receive, the fundamental ethical question is: What treatments are participants entitled to in a just global order?

EGALITARIAN COSMOPOLITANISM AND
THE STANDARD OF CARE

Underlying the view of the critics of the short-course AZT trials and supporters of the Declaration of Helsinki position is an egalitarian cosmopolitanism. This view begins with moral egalitarianism. All people are fundamentally equal regardless of what country they were born or reside in: "every human being has a global stature as the ultimate unit of moral concern."[32] This moral egalitarianism naturally leads to cosmopolitanism in matters of justice.[33] In this view, since a person's country of birth and residence is arbitrary from a moral point of view, it should not determine entitlement to basic rights, opportunities, or resources. There is no domestic justice, only global justice.

There are many different ways of justifying egalitarian cosmopolitanism. A utilitarian approach contends that each person is a locus of utility.

For utility maximization, national borders are irrelevant, and the utility of every person should be equally considered. Further, since individuals in developing countries are starting from a low level of utility, increases in their utility are likely to be substantially higher for each expenditure of a unit of resources. We should focus our resources on increasing their entitlements because this will maximally increase utility. This appears to be Peter Singer's position.[34,35]

A Rawlsian contractarian approach, like that articulated by Charles Beitz, can also justify egalitarian cosmopolitanism.[36] Society is a system of social cooperation; the principles of justice define how the benefits and burdens of that cooperation will be distributed. Since all people are free and equal, the starting point of justice is that these benefits and burdens of social cooperation should be distributed equally. Deviations from the equal distribution of opportunities and resources need to be justified, and are justified when they benefit the least advantaged. This reasoning applies to domestic society because citizens share a national community and a common conception of justice. But, as Beitz argues, this same logic should apply globally. The interdependent relationships across borders resemble the relationships within one country that justify the Rawlsian principles for domestic justice. As the world becomes more interdependent, global justice becomes more like domestic justice, and so the same principles should apply:

> States participate in complex international economic, political, and cultural relationships that suggest the existence of a global scheme of social cooperation. As Kant notes, international economic cooperation creates a new basis for international morality. If social cooperation is the foundation of distributive justice, then one might think that international economic interdependence lends support to a principle of global distributive justice similar to that which applies within domestic society.[37]

This is especially true since the non-voluntary economic, social, and political structures of the international order greatly affect the life prospects of each individual in the world, and it is these structures that are the focus of the principles of global justice. Thus, Rawls' three principles of justice—the equal distribution of the basic liberties, the fair equality of opportunity, and the difference principle for the distribution of resources—should apply not just within one country but across all countries regardless of boundaries.

Another justification comes from Pogge's view that developed countries have exploited developing countries in many ways, including unfair international agreements.[38] To remedy this exploitation would require a fairer distribution of the world's resources along the lines of egalitarian cosmopolitanism.

For entitlements to health care services, the implications of egalitarian cosmopolitanism are not straightforward. Daniels tries to extend Rawls' theory to health care. He argues that health is important because it secures an individual's opportunity:

> [D]eviations from normal health and functioning are important because they compromise people's opportunities: If an individual's fair share of the normal range is the array of life plans he may reasonably choose, given his talents and skills, then disease and disability shrink his share from what is fair.[39]

Thus, people are entitled to health care as part of the guarantee of fair equality of opportunity. Health care preserves functioning, prevents illness, and mitigates natural disabilities and accidents of fortune, thereby guaranteeing individuals the opportunity to pursue their own conception of the good life.

Unfortunately, the effort to go from this principle to delineating what health care services people are actually entitled to as a matter of justice has been a failure. This principle is insufficiently specific to answer the question, or even to help policymakers decide what resources should be devoted to health care. All that can be said is that people are entitled to some health care services. Expanding this process to a global scenario will not provide a determinate answer to health care entitlements, either.

This ambiguity or uncertainty about the content of actual health care entitlements might seem to undermine the relevance of egalitarian cosmopolitanism for the standard of care debate. After all, if we cannot determine what specific services people are entitled to in general, then how can we determine what services people are entitled to in the control arms of research trials in developing countries?

However, this ambiguity may be less problematic in practice than it appears in principle. Egalitarian cosmopolitanism suggests that everyone, in every country, is entitled to the same health care services, whatever those services turn out to be. This would support the idea that everyone in the world is entitled to the same medical tests and treatments. By examining what health care services people in developed countries are entitled to, we can determine what everyone else in the world should be entitled to. And this in turn determines what tests and treatments a research participant in a control arm of any study conducted anywhere in the world—including developing countries—should receive.

Obviously, this reasoning makes a big assumption: that people in developed countries actually receive medical services close to what they are entitled to. Most probably do receive health services close to those that they are entitled to, although it is possible that the health care services are more than they are entitled to.

Nevertheless, what people in developed countries actually receive can serve as an important benchmark, because in a practical sense it defines what services research participants in the country should receive in a trial.

The only reasonably successful effort to justify a situation in which people in developed countries receive more services than they are entitled to is based on cost-effectiveness analysis. On this view, people are entitled to those health care services that are cost-effective—that is, services that cost approximately $50,000 (or about £33,000) per quality-adjusted life year (QALY). This would eliminate from the pool of entitled services some health care services that people in developed countries routinely receive, say mammography for women less than 50 years of age. However, almost all the services that might be provided in clinical trials in developing countries—from highly active antiretroviral treatment to the Directly Observed Treatment Short Course (DOTS) protocol for controlling TB, and from all available immunizations to malaria treatments—certainly falls beneath this usual cost-effectiveness cut-off.

Thus egalitarian cosmopolitanism provides a relatively simple test that can be easily applied by researchers, institutional review board (IRB) members, and external observers to determine what services should be provided to participants in a developing country in the control arm of a trial: whatever participants in the control arm of a similar trial in a developed country would receive.

This egalitarian cosmopolitanism seems to provide the normative justification of the Declaration of Helsinki's paragraph 32. It means that everyone in the world who participates in research is entitled to the same set of health care services. Participants in developing countries should receive the care that people in developed countries are entitled to, namely "the best available services." This egalitarian cosmopolitanism also makes sense of the "double standard" charge. A person who participates in a research trial in a developing country is entitled to the same services a participant would receive in a developed country. Providing control-arm participants fewer services would be to let factors that are not morally relevant (e.g., country of residence) influence what health care services they receive.

If we strictly follow this egalitarian cosmopolitan inspired framework, then the only ethically acceptable trials are those in which the intervention is *ex ante* thought—or known—to be equivalent to or better than current treatment. There would be no justification for designing a research study in which the intervention might *ex ante* be thought—or known—to be less effective but cheaper or logistically easier to provide. Similarly, on this view, the idea that we could do a trial on cardiac interventions in developing countries without coronary care unit services if they are provided in developed countries is simply false. Whatever people in developed countries receive as a matter of entitlement must be provided to people in developing countries as a matter of justice. Thus, conducting a research trial in a developing country with a new vaccine that might be bivalent instead of tetravalent but substantially cheaper, or not require refrigeration but be somewhat less effective, would clearly be unethical on this view. It would deny people services or provide services that are less than those to which individuals in developed countries are entitled.

STATISM AND THE STANDARD OF CARE

If egalitarian cosmopolitanism suggests that people in developing countries are entitled to the same health care services as people in developed countries, Rawls' version of global justice, articulated in his book *The Law of Peoples*, takes the opposite position, namely that people in developing countries—even if not all are just or even decent—are (largely) entitled to the health care services they are actually receiving.[40] In other words, most participants in the control arms of trials in developing countries— except those in burdened societies—should receive the type of medical care they actually are receiving. This is the position that some defenders of the short-course AZT trials, particularly the FDA, invoked to justify using a placebo.

Rawls' view, which might be called statism, begins not with moral egalitarianism of individuals but with states or, as Rawls would have it, peoples. For Rawls, global justice is qualitatively different from domestic justice. People in one state share a common history, political tradition, and civic life, and a common set of cultural, educational, and social institutions. As a people, their primary goal is to preserve their political independence. The fundamental question regarding global justice is not what liberties, opportunities, and resources every person in the world is entitled to, or what resources people in developed countries owe to people in developing countries, but rather "what the foreign policy of a reasonably just liberal people should be."[41] The objective of global justice is to see how states should treat other states.

Next, Rawls argues that the goal of foreign policy is not to improve the well-being of people throughout the world. While the ultimate concern of a cosmopolitan view is the well-being of individuals, the ultimate concern of Rawls' statist view is the justice of societies and the international order.[42] That is, the goal of global justice is to bring about states that are "fully just and stable for the right reasons;" the goal is to permit countries to develop into liberal states, or at least decent non-liberal societies in which institutions are reasonably just.

Third, for most states that are not even decent, Rawls believes the major shortcoming is not the economic well-being of the population but the lack of political liberties. Rawls argues that a just, stable state need not be particularly well off. The possibility of being or becoming a liberal or decent state is not linked to the economic well-being of a country. Rich states can be illiberal and unjust—witness Saudi Arabia. Conversely, poor states can have just institutions and be liberal or decent states—witness the Indian state of Kerala, or Botswana. There is no necessary link between a state's economic well-being and whether the state is just. The key determinant of a state's being just, according to Rawls, is its political culture and institutions. Political culture and institutions, he argues, are not necessarily linked to economic well-being, cannot be imposed from the outside, and cannot be facilitated by external, development aid:

> [T]he crucial element in how a country fares is its political culture—its members' political and civic virtues—and not the level of its resources, [hence] the arbitrariness of the distribution of natural resources causes no difficulty.[43]

Thus, although for cosmopolitanism the focus of global justice is the guarantee of political rights and economic well-being of individuals throughout the world, for statism it is institutions and political culture that are just within each state. And for statism the main difficulty is not poverty—except in burdened societies—but the lack of a political culture that respects political and individual liberties.

Rawls acknowledges that some minimal level of economic well-being is necessary for the establishment and stability of just institutions in liberal and decent states. Some states—which he calls burdened states—have such grinding poverty and deprivation that they cannot establish just social and political institutions. These states need economic aid, and the world community has a duty of assistance to provide it. However, there are two limits to the duty of assistance. The world community has a duty only to get states to the minimal level necessary for them to establish just institutions; there is no ongoing, open-ended commitment to poor countries no matter who they are. As Rawls puts it, there is a target or threshold—economic well-being necessary for a just state—and a cut-off. The duty of assistance between states is a temporary necessity, not an ongoing obligation of justice. Above this minimal threshold, global justice does not demand redistribution of resources between states. Consequently, wealthy states do not have obligations of economic assistance other than to very poor states. The economic inequality among states and among individuals residing in different states is not unjust; it would be unjust only when it precludes establishment of just liberal or decent states. Because economic well-being is not a good in itself, and because there is no necessary link between economic well-being and having the political institutions to ensure liberal or decent states, large economic differences are permissible, are not unjust, and are not a particular concern of global justice.

What are the implications of statist views like Rawls' for the standard of care debate in international research? Within just societies, whether liberal or decent, there is some mechanism to establish entitlements to basic liberties, opportunities, and resources, and this is part of what makes them just. Since health care services are a component of guaranteeing opportunity, a just state would establish an entitlement to health care services. In liberal or decent states, this entitlement establishes what people enrolled in research should be guaranteed in a trial. Thus, under this view, we look at what people actually receive in the country where the research is being conducted to determine what people enrolled in the control arm of the research trial should receive.

This view of statism seems to justify the use of placebos in the control arms of the short-course AZT trials. On Rawls' view, the people in these developing countries, as a matter of fact and policy, were not receiving any interventions, including AZT, to prevent maternal–fetal HIV transmission prior to the trials. Thus they were not entitled to any intervention, much less the 076 regimen, for preventing maternal–fetal HIV transmission. Even if we have some questions about the justice of some of these countries, these questions mostly had to do with their political liberties; some of them at the time of the trial, such as Thailand and Uganda, were decent and not burdened. Furthermore, women and their infants actually received better care as part of the trial than they would have without the trial—better prenatal care, better delivery care, and better follow-up monitoring. Thus, to provide them placebo would not deprive them of health care services they were entitled to; and to enroll them in the trial would actually provide them better care than they would otherwise be entitled to, whichever arm of the trial they were in.

While this approach may work for most states, Rawls does not think it works for so-called burdened states. These states are so deprived that they need external development assistance, otherwise they cannot develop into liberal or decent societies. People in these states are entitled to more than they are actually receiving. Hence, looking at what health care services people in burdened states actually get is not a suitable way to determine what they are entitled to and what care should be provided to participants in the control arm of a research study. The entitlement to health care is likely to increase with the assistance provided by other states. By how much?

Here, there is a problem of what constitutes a burdened society. One approach is to identify the level of economic well-being necessary to have just institutions. This can be established by examining the amount of resources in poor states that are nevertheless liberal or decent states. These states are decent or liberal and thus, by definition, have sufficient resources not to be "burdened" in Rawls' sense. This would provide a relatively simple and practical way to establish entitlements to health care. The entitlement is what poor but liberal or decent states provide their citizens.

Examining the list of least-developed and then the list of low-income countries reveals a problem with this approach (Table 8.1). Some countries on the least-developed list are decent if not liberal societies. Consider Mali, with a per capita income in 2009 of just $680 (purchasing power parity). While it may not qualify as a liberal society, it certainly seems a decent society, with a peaceful transfer of presidential power, free press, no military oppression, and reasonable parliamentary system. And certainly some countries on the low-income list, such as Ghana, with a per capita annual income in 2009 of just over $1,000, are decent if not liberal societies. Many others on these lists—even some in the middle-income group, Syria, Cuba, Gabon—are clearly neither decent nor liberal, but are totalitarian or dysfunctional societies. Thus, strictly following Rawls' position, it would seem that even a least-developed country and certainly a low-income country could have enough resources to be a decent society.

More relevant for the HIV maternal–fetal transmission studies begun in the mid-1990s, some of the countries in which placebo-controlled trials were being conducted were lower-middle-income states and reasonably liberal or decent states. For instance, Thailand was one research site. In 1995, 17.5% of Thailand's population was in poverty and per capita income was over $2,690. In the mid-1990s a placebo controlled trial was conducted there because the 076 regimen was not an intervention people were usually provided or entitled to. Thus, if we are trying to determine what the control group in these short-course AZT studies were entitled to as a matter of justice, the Rawlsian view would seem to indicate that they were not entitled to the 076 long-course AZT regimen. Placebo was a reasonable control arm. Even if burdened states were entitled to assistance and additional resources, and the entitlement of their citizens to health

Table 8.1. GROSS NATIONAL INCOME OF LEAST-DEVELOPED AND
LOW-INCOME COUNTRIES

Least-Developed Countries[47]	Per Capita GNI[48] (2009)	Low-Income Countries	Per Capita GNI (2009)
Afghanistan (2008)	$310	Armenia	$3,100
Angola	$3,750	Azerbaijan	$4,480
Bangladesh	$580	Cameroon	$1,190
Benin	$750	Congo	$2,080
Bhutan	$2,030	Côte d'Ivoire	$1,070
Burkina Faso	$510	Georgia	$2,530
Burundi	$150	Ghana	$1,190
Cambodia	$610	India	$1,180
Central African Republic	$450	Indonesia	$2,050
Chad (2008)	$540	Kenya	$760
Comoros	$870	Kyrgyzstan	$870
Dem. Republic of the Congo	$160	Mongolia	$1,630
Djibouti	$1,280	Nicaragua	$1,000
Equatorial Guinea	$12,420	Nigeria	$1,190
Eritrea (2008)	$270	Pakistan	$1,000
Ethiopia	$330	Republic of Moldova	$1,560
Gambia	$440	Senegal	$1,040
Guinea	$370	Tajikistan	$700
Guinea-Bissau	$510	Turkmenistan	$3,420
Haiti	$	Ukraine	$2,800
Kiribati	$1,830	Uzbekistan	$1,100
Lao People's Dem. Republic	$880	Viet Nam	$930
Lesotho	$980	Zimbabwe	NA
Liberia	$160		
Madagascar (2008)	$420		
Malawi	$280		
Mali	$680		
Mauritania	$960		
Mozambique	$440		
Myanmar	NA		
Nepal	$440		
Niger	$340		
Rwanda	$460		
Sierra Leone	$340		
Somalia	NA		
Sudan	NA		
Timor-Leste (2008)	$2,460		
Togo	$440		
Uganda	$460		
United Rep. of Tanzania	$500		
Yemen	$1,060		
Zambia	$970		

Based on United Nations definition of Least Developed Countries and World Bank figures for gross domestic income.

care services increased, it is unlikely it would have included entitlement to the 076 regimen, since citizens in Thailand were not entitled to this treatment.

On the Rawlsian view, the entitlement to health care services will vary not just among countries of different economic well-being, but across research trials and over time. What health care services research participants in developing countries are entitled to will vary depending upon what citizens of poor but liberal or decent countries are entitled to for the condition under study. This could change over time. As drug prices come down, interventions become easier to perform, or the impact of an intervention is shown to be much greater than previously expected, people in developing countries may acquire an entitlement to new health care services that would change what participants in control arms should receive.

A MIDDLE POSITION: PRAGMATIC COSMOPOLITANISM

As its advocates acknowledge, the egalitarian cosmopolitan position is an aspirational ideal of global justice rather than a view that can be practically implemented today. However, espousing egalitarian cosmopolitanism creates a problem for clinical researchers who need to conduct trials in the least-developed and low-income countries today under non-ideal conditions. What health care services should be provided to participants in the control arm in the actual non-ideal world? Should the services specified by the aspirational ideal of justice be followed in research trials conducted today? If a researcher deviates from the aspirational ideal, is he or she acting unjustly or unethically?

Applying the egalitarian cosmopolitan position to the maternal–fetal HIV transmission studies in the mid-1990s would have meant that studies using placebo would have been unethical. Indeed, in a point not widely recognized, these studies would also have been unethical using the 076 regimen in the control arm. The short course that was tested in the intervention arm was predicted to be less clinically effective than the 076 long-course AZT regimen in preventing HIV transmission. Consequently, research participants in the intervention arm would have been given something known to be worse than what they were entitled to. That is, using the short-course AZT regimen would put participants at risk of serious harms from contracting HIV. According to egalitarian cosmopolitanism, this is unethical.

To emphasize an important point about the egalitarian cosmopolitan view: designing a trial with the 076 long-course regimen as the control arm and the short-course AZT regimen as the intervention arm actually violates the Declaration of Helsinki and the egalitarian cosmopolitan view. This was the position advocated by critics, such as Lurie and Wolfe and Angell, of the actual trials. But such a trial would be unethical because the people in the intervention arm would be denied the best available treatment and would be given a treatment *ex ante* known to be inferior to the standard 076 regimen. More generally, any trial testing a treatment thought—or known—to be inferior would be unethical.

Adhering to the Declaration of Helsinki would have prevented development of the subsequent interventions, such as single-dose nevirapine, proven in the HIVNET trial in 1997–1999 in Uganda as partially effective for preventing maternal–fetal HIV transmission. This intervention is much less expensive and simpler to administer than the 076 AZT regimen, but also clinically less effective at preventing HIV transmission. Without the short-course AZT or nevirapine trials, there would have been a period of many years, maybe as much as a decade, during which pregnant HIV-infected women in developing countries would have been getting no treatment to prevent maternal–fetal HIV transmission.

This conclusion creates a dilemma for egalitarian cosmopolitans. In the name of the aspirational ideal of egalitarian cosmopolitanism, there can be no research that uses treatments that are known—or *ex ante* predicted—to be clinically inferior, but potentially cheaper or logistically more feasible, than treatments used in the most-developed countries. This would seem to greatly limit research in developing countries and to preclude many incremental improvements in health care services within those countries. Requiring that the entitlement of citizens in the most-advanced developed countries be used to determine the control arms for research trials in developing countries would preclude trials that might lead to interventions that are cheaper or easier to implement and that would mitigate health problems in those countries even though they are less clinically effective than interventions people in developed countries receive. In the name of global justice to help bring the poor up to developed-country economic standards, the poor would be left right where they are. The ideal seems to be the enemy of the good, or at least of amelioration and incremental improvements.

For researchers who participated in the placebo-controlled short-course AZT trials and the HIVNET nevirapine study to prevent mother-to-child transmission, this is no idle concern. They were severely criticized as unjust and immoral. How should we handle the problem of the role of aspirational ideals in the non-ideal world? Is there an alternative approach that can fulfill some of the ideals of egalitarian cosmopolitanism without preventing incremental progress?

Consider what I will call pragmatic cosmopolitanism. It recognizes moral egalitarianism—the equality of individuals wherever they are born or reside; this is the cosmopolitan aspect. But it also recognizes that there has never been, nor is ever likely to be, a time in which every person around the globe has an equal entitlement to resources. The pragmatic aspect of pragmatic cosmopolitanism recognizes that the key outcome is not equivalent entitlement to resources, such as health care services; the key outcome is reasonably similar health outcomes. What we care about is that people should have good enough health so they have the opportunity to live a complete life and pursue a reasonable range of life plans.

How can we translate into policy "good enough health so they have the opportunity to live a complete life and pursue a reasonable range of life plans"? One crude measure of "good enough health" is a reasonable life span—a complete life. This does not capture every facet of a healthy life; for example, it excludes rates of disability. However, people who live a long time are likely to be relatively healthy. This is especially true in societies that devote fewer resources to health care. It may be possible to live a long time with chronic health problems or serious disabilities, such as kidney failure or diabetes, in developed countries, if the system devotes a substantial amount of resources to keeping such people alive and somewhat functional. But it is difficult to live a long time with serious disabilities or chronic health problems in developing countries that devote few resources to health care. People with chronic kidney failure or diabetes just do not live that long in countries that have fewer health care resources. The pragmatic portion of pragmatic cosmopolitanism suggests that people everywhere in the world should be entitled to the amount of health care services that provide a complete life—that is, a reasonable life expectancy of between 70 and 75 years of age.

Thus, the pragmatic cosmopolitan view is cosmopolitan by recognizing that people everywhere, regardless of the country in which they reside, are entitled to have resources to pursue a reasonable range of life plans. The pragmatic side is that this does not require the kind of equality of resources that would be associated with a worldwide utilitarianism or a global difference principle and fair equality of opportunity on a global scale. Instead, it is pragmatic by determining relevant outcomes, such as a life span of 70 to 75 years, and defining entitlements by the minimum resources sufficient to realize this life span—a complete life.

Some might object that using a reasonable life expectancy may be consistent with a cosmopolitanism justified by worldwide utilitarianism, but not with a cosmopolitanism justified by a global difference principle. The utilitarian may focus on averages rather than distributions within a population, but a global difference principle is concerned with the least well off, not with averages. Consequently, ignoring how the disabled or people with serious illness fare by focusing on the average life span would seem inconsistent with cosmopolitanism. However, even the global difference principle must utilize some aggregation; the "least well off" is not an individual but some aggregate of individuals sharing some characteristics, and within this aggregation some are likely to fare better than others with the resources given. Then the question is how we characterize the least well off. It does not seem unreasonable for pragmatic cosmopolitanism to characterize them as the people in a society with an expected life span below 70 years of age, rather than people with a specific medical condition.

ANOTHER MIDDLE POSITION: BENEFICENT STATISM

For many people, some aspects of Rawls' statism may be appealing, but its almost exclusive emphasis on just institutions and minimization of the importance of economic well-being seems troubling. First, it is unclear what Rawls views as burdened societies and how much assistance they should receive. Rawls seems to suggest that hundreds of millions of people require economic assistance in "our world as it is with its extreme injustices, crippling poverty, and inequalities."[44] In many countries, the annual per capita income and expenditures on health are nothing short of pitiful. The challenge for Rawlsian statism is that some of these countries, Mali and Bangladesh for instance, seem to have functioning, if not fully entrenched, democracies that are decent if not liberal. Furthermore, the threats to stability and entrenchment of democracy do not seem directly related to economic well-being but to political culture; tribal, ethnic and cultural conflicts; and instability from problems in neighboring states. The poverty experienced by many citizens of these countries seems to demand redress not just as a supererogatory act but as a moral obligation. Yet these countries would not seem to qualify under Rawls' criteria for restricting development aid to burdened societies.

Second, Rawls' position on economic well-being seems too dismissive. Morally, what matters is not merely that there are just political institutions and that individuals have the opportunity to effectively participate in these institutions. That people are free and equal also implies that they have the opportunity to pursue a reasonable range of life plans and to realize diverse human capacities. While economic well-being past a minimal threshold may have little to do with whether a country has decent political institutions that ensure reasonable political rights, it certainly has an important impact on the ability of people to pursue their life plans. While states may have a duty of assistance that extends to helping other states rise to the level where they can establish just political institutions, beneficence would seem to argue for going beyond this minimal level and ensuring sufficient economic well-being so that individuals in those countries can pursue a reasonable range of life plans within a decent political order. In Rawls' conception of global justice, this might not constitute an obligation of justice, but it might be an obligation of mutual aid. Furthermore, this obligation might not be limited to individual citizens but might extend to states. Not only might individuals have obligations to help others in need, when the assistance imposes no substantial burdens, but states might have such obligations, too. Developed countries have an obligation of mutual aid to contribute to poor states—say the low-income states—insofar as the contribution can make a substantial improvement in the well-being of individuals in those poor states. The amount owed should fit with our common sense notions of mutual aid; the contribution must substantially improve the well-being of the individuals in a poor state while imposing only minimal burdens on the citizens of the contributing states.

The relevant measure of well-being is whether people can pursue a reasonable range of life plans. Since our concern here is with health care, we can say that states have an obligation to provide resources to poor states to ensure those health care services that can raise the average life expectancy of 70 or 75 years. This would allow individuals in those poor states to live a complete life and to pursue a reasonable range of life plans. The goal is average life span in a state. Why? The distribution of resources within a state is a matter of domestic justice. On Rawls' statist view, donor states should not be in a position of evaluating, much less imposing, a particular distribution of resources on a poor state—for example, by judging the resource distribution between urban and rural or among ethnic groups or sexes.

HOW MUCH HEALTH CARE ARE PEOPLE IN DEVELOPING COUNTRIES ENTITLED TO?

Pragmatic cosmopolitanism and beneficent statism come to similar positions on what level of health care resources people in developing countries are morally entitled to as a matter of either justice or mutual aid. This is the level necessary to have an average population life expectancy of 70 to 75 years. Specifically, what does this level of health care require? There are diverse approaches to this question that imply slightly different levels of support.

One approach is the life expectancy approach. This approach begins by examining the important and enduring relationships between health and health care expenditures. This relationship reveals that most of the health gain from health services happens with very small expenditures, and that greater expenditures above a threshold provide only marginal improvements in average life expectancy (Fig. 8.1). There is a steep increase in average life expectancy with every additional dollar spent that plateaus at an average population life expectancy of 75 years or so at health care expenditures of about $200 per person per annum. Additional expenditures up to $4,000 or more per person per year add little or nothing to average life expectancy. Using this empirical relationship, we might say that people in all states should be entitled to about $200 per person per year of health care services that would allow them to live in a society with a life expectancy of 70 to 75. This is the level of health care services provided in Sri Lanka. It is substantial compared to the $10 or even less spent per person in many of the least-developed and low-income countries; but Sri Lanka's level of $200 is clearly less than what people in developed countries are entitled to or what people in developing countries would be entitled to as part of egalitarian cosmopolitanism.

How this $200 should be distributed—what precise set of services people would get or be entitled to—would vary by country. Doubtless states would vary their distribution of services by diseases that afflict people, what they consider the best approach to addressing the diseases, available infrastructure, etc. In some countries, emphasis might be on hospitals and surgical suites, while others might emphasize vaccines for cervical cancer. With growing chronic diseases some states might emphasize treatment for hypertension or diabetes over malaria.

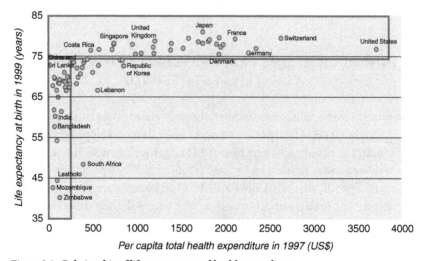

Figure 8.1. Relationship of life expectancy and health expenditures.

version of precedent curve. — Log scale,
*   is it accurate — query?*

The medical services approach considers those diseases that cause a disproportionate amount of premature mortality and morbidity in the low-income countries, thus reducing average life expectancy to below 70 to 75 years. In 2001, the WHO Commission on Macroeconomics and Health argued that:

> The health prospects of the poorest billion [people in the world] could be radically improved by targeting a relatively small set of diseases and conditions. The primary targets are:
>
> - HIV/AIDS
> - Malaria
> - Tuberculosis
> - Maternal and perinatal conditions
> - Widespread causes of child mortality including measles, tetanus, diphtheria, acute respiratory infection, and diarrheal disease
> - Malnutrition that exacerbates those diseases
> - Other vaccine-preventable illness [e.g., hepatitis B]
> - Tobacco-related disease
>
> These conditions are not the only health problems that afflict the poor. . . . However, these diseases account for a large proportion of avoidable mortality of the poor. . . .

Targeting the health care services for treating these diseases—and adding other modest interventions such as family planning and surgery for trauma and emergency problems—could increase the average life span in these low-income countries to close to 70 or 75 years. Clearly, more than money is involved. In many cases the problems are lack of managerial capacity, infrastructure, and trained health care personnel, along with logistical and other barriers. Nevertheless, development aid is critical. Even "development aid skeptics" have found that aid targeted at health care services generates excellent returns.[45]

This suggests that the basic health care services to which people in any country are entitled would address these diseases. They would include:

- DOTS treatment for TB
- Malaria prevention with bed nets and indoor spraying as well as treatment for acute clinical illness episodes
- Vaccinations against diphtheria, pertussis and tetanus (DPT), bacille Calmette-Guérin (BCG), polio (oral polio vaccine), measles, hepatitis B and *Haemophilus influenzae* B (HIB), pneumococcus, and rotavirus
- Antibiotics for acute respiratory infections, oral rehydration for diarrhea, and micronutrient supplements and other treatments for malnutrition

- Contraceptives
- Cesarean sections

The WHO Commission on Macroeconomics and Health estimated that providing these combined interventions for the least-developed and low-income countries would require a total of $38 per person per year. Other estimates of the cost for these essential services from the WHO and the International Monetary Fund (IMF) have arrived at similar figures, between $35 and $50 per person per year.

How burdensome would it be to provide these services? Would it surpass the minimal burden threshold? In 2009, there were about 837 million people living in the 49 least-developed countries.[46] These essential health interventions would cost between $29 billion and $42 billion per year. Assume the 30 high-income developed countries (as classified by the World Bank World Development Indicators) alone funded all this aid—ignoring for the moment that the low-income countries themselves already contribute a portion. The combined annual GDP of these 30 developed countries is approximately $39.5 trillion (2009). Hence the $42 billion amounts to 0.1% of GDP, or 1 cent for every $10. For a household at the median income in the United States this would amount to around $53 per year per family. By any standard this is a small burden for people in any developed country.

The life expectancy approach provides a health outcome that should be equalized for everyone in the world—life expectancy—and then derives an estimate of how much that would cost in many countries. The equalization of the substantive goal seems most consistent with pragmatic cosmopolitanism. The idea of a core set of medical services that everyone is guaranteed, with no further obligation above that level, seems more consistent with beneficent statism.

APPLYING PRAGMATIC COSMOPOLITANISM
AND BENEFICENT STATISM

Nine points summarize where we have come.

1. Ensuring that research has a favorable risk–benefit ratio is a core ethical principle of clinical research.
2. To fulfill this principle requires providing participants in the control arm of a research study all medical services related to the scientific objective that they are entitled to.
3. The medical services people are entitled to are not determined by what services they actually receive, but by principles of justice.

4. Egalitarian cosmopolitanism suggests that the medical services people are entitled to are the same regardless of country of birth or residence. To a good approximation, the services people are entitled to are determined by the best available services in the world, which are the services that people in the rich developed countries are entitled to and are receiving.

5. Statism suggests that the medical services people are entitled to are determined by domestic justice of each state. For liberal and decent states—excluding the very poorest, burdened societies—this is pretty close to the services they are actually receiving.

6. An alternative to egalitarian cosmopolitanism is pragmatic cosmopolitanism, which endorses the same core moral egalitarianism but asserts that the equality to be sought is not of resources but of some outcome, such as reasonable life expectancy. In the health sphere this suggests that people should have an average life expectancy of about 70–75 years and are entitled to health care services that achieve this.

7. An alternative to statism is beneficent statism, which suggests that rich states should provide aid not just to the level such that the recipient country can be a decent and just state, but to a level where there are substantial benefits to the population's well-being when the burden on the contributing developed country is minimal.

8. One way to achieve such middle positions, especially pragmatic cosmopolitanism, is the life expectancy approach. This is an empirical determination in which it is noted that average life expectancy is about 70–75 years when annual health expenditures are about $200 per person, and it does not substantially increase with higher health expenditures. Thus, every person in the world is entitled to medical services that are provided by countries that reach a life expectancy of 70–75 years.

9. Another way to effectuate these middle positions, especially beneficent statism, is the medical services approach. This defines specific medical interventions that target those diseases that cause the greatest morbidity and mortality among the world's poorest people. A high estimate of the price for the provision of these interventions is $50 per person, which would require a contribution of about $53 per year per median-income household in the United States (Table 8.2).

Given these points, what would people in poor developing countries have been entitled to regarding AZT for the prevention of maternal–fetal HIV transmission in 1995–1996, when the controversial studies were being initiated? Under the view of beneficent statism as represented by the $38 basket of essential services, it is fairly clear that they would not

Table 8.2 DIFFERENT APPROACH TO THE GLOBAL STANDARD OF CARE

Entitlement	Same services for everyone in the world	Services obtain to achieve an average life expectancy of 70 years	Basic services to address major causes of morbidity and morality	Services the population is already receiving
Theory of Global Justice	Egalitarian cosmopolitanism	Pragmatic cosmopolitanism	Beneficent statism	Rawlsian statism
Operationalization	Developed countries' standard of care (Declaration of Helsinki standard)	Average life expectancy	Basket of essential services	Standard of care in the host country

have been entitled to the 076 AZT regimen. Even in Thailand, which is a middle-income country, women were not entitled to the 076 AZT regimen at that time. This was probably true even for the life expectancy approach. Thus, only on the egalitarian cosmopolitan view would research participants in the original maternal–fetal HIV transmission trials have been entitled to the 076 regimen.

It is worth noting that both the life expectancy approach and the $38 basket of essential services imply that HIV-infected individuals in developing countries should be entitled to highly active triple-drug anti-retroviral therapy now that the price of such therapy is relatively low.

RESPONSE TO CRITICS

There are many potential objections to this way of analyzing the standard of care debate and the particular analyses of the theories of global justice.

First, why should biomedical researchers be responsible for providing people in developing countries with the medical services to which they are entitled in clinical trials? The obligation is derived from considerations of global justice and therefore seems to rest on countries rather than individual researchers trying to improve the conditions in the world.

Researchers are fulfilling the obligations of ethical research, not the obligations of global justice for their country. Researchers are obligated to conduct their research in an ethical manner. Having a favorable risk–benefit ratio is a critical requirement for ethical research. To fulfill this principle—that is, to ensure that research participants are not exposed to unnecessary risks from the research enterprise—researchers must ensure that participants are not denied services they should otherwise receive as a matter of justice. If participants are denied those services as part of the research enterprise, they are exposed to risks, and the risks are not minimized, and the favorable risk–benefit principle is violated.

It may be the case that, because of racism, sexism, corruption, or incompetence, people are not actually receiving medical services they are entitled to as a matter of justice. This does not permit researchers to conduct a trial using as a basis for the control arm the medical services these people are actually receiving. This would mean researchers could use cases of bad medicine to define the control intervention for a trial. On this logic, the men in Tuskegee could have been legitimately denied penicillin because they would not have received it in the normal course of events. Similarly, children receiving suboptimal asthma care could be randomized to receive bad care versus a new drug. The standard for evaluating the risk–benefit ratio is not what people are actually getting but what they are entitled to as a matter of justice.

Second, should researchers be required to provide participants in developing countries *all* the medical services they are entitled to, such as the $38 basket of essential medical services?

Researchers are not the agents for ensuring that global justice is fulfilled. Researchers are not responsible for providing all the medical services participants are entitled to. They must as a matter of ethics provide only those services relevant to the scientific objectives of their study, the services relevant to assessing the risks and benefits of the research interventions. It is wrong to believe that a researcher evaluating a new TB drug must provide participants a hepatitis B vaccine or malaria treatment that they are entitled to. Researchers must refer to entitlements specified by global justice to determine what interventions *relevant to the scientific objectives* of their research study must be provided, but this does not comprise all the medical interventions people might be entitled to.

Third, if health care entitlements based on requirements of justice are not being implemented, is this sufficient to render unethical a research study that aims to ameliorate actual conditions and diseases? In many developing countries, people are not close to receiving the services required by the beneficent statist or pragmatic cosmopolitan views of justice. In much of sub-Saharan Africa and Southeast Asia, people receive $5 to $10 of health care per year, far from even the $38 basket of essential medical services. What if a researcher is trying a new intervention to improve both the rate of initiating TB treatment and compliance? Why should this research be prohibited because, theoretically, people in the country are entitled to TB treatment but are not receiving it? Why is it ethical to have an unfulfilled standard of care prohibit important research that aims to actually improve the condition of people?

This is a clash between two principles related to research ethics—social value and favorable risk–benefit ratio. The research might have social value given the conditions of the people but would violate the principle that requires risks to be minimized. Such conflicts among principles are common. Favoring the social value principle while exposing participants to excessive risks is a clear case of exploitation of the participants for the good of society. To avoid this exploitation we should resolve such a conflict in favor of the risk–benefit ratio. If the social value is substantial, and the people could be assured that the intervention would be implemented expeditiously, it might be worth favoring the social value over the risk–benefit ratio. But this would have to be evaluated as an exception to the rules of research ethics.

CONCLUSION

For nearly 15 years, debate has raged over the ethics of conducting research in developing countries. On one side are people who, invoking the Declaration of Helsinki, one of the canonical documents in bioethics, require that research participants receive the best proven treatments available anywhere in the world. These commentators have been highly critical of existing research, condemning it as resting on a double standard. The other side has criticized this standard as idealistic, and ended up criticizing the Declaration of Helsinki as outmoded. Ironically, while the debate has implicated theories of global justice, the different parties have not really invoked these underlying theories to justify their positions. But this debate recapitulates a conflict between egalitarian cosmopolitanism and statism, the former justifying the Declaration of Helsinki and the latter a position that accepts the current system as the standard of care.

It is only by seeing this debate in research ethics as a manifestation of the deeper debate about global justice can we see that resolving it requires moving beyond the two standard positions in global justice. Egalitarian cosmopolitanism is an ideal theory that is difficult to apply to the real world. Statism seems to ignore other obligations we might have. Recognizing that both theories seem to have critical truths but also have limitations, as revealed in this debate about the standard of care, we tried to develop principled middle positions, namely pragmatic cosmopolitanism and beneficent statism. One recognizes that ideal theories can be the enemy of moral action in a non-ideal world; the other recognizes that states have moral duties, including mutual aid. These middle positions permit more reasonable and coherent resolutions of the standard of care debate.

NOTES

1. Centers for Disease Control and Prevention. (June 5, 1981). Pneumocystis Pneumonia—Los Angeles. *MMWR Weekly Reports* 30(21):1–3.
2. Kaiser Family Foundation. *Global HIV/AIDS Timeline.* Available at: http://www.kff.org/hivaids/timeline/hivtimeline.cfm
3. Serwadda D, Mugerwa RD, Sewankambo NK, et al. (1985) Slim disease: a new disease in Uganda and its association with HTLV-III infection. *Lancet* 2:849–852.
4. Fischl MA., Richman DD, Grieco MH, et al. (1987). The efficacy of azidothymidine (AZT) in the treatment of patients with AIDS and AIDS-related complex, a double-blind, placebo-controlled trial. *New England Journal of Medicine* 317:185–191.
5. Connor EM, Sperling RS, Gelber R, et al. (1994). Reduction of maternal–infant transmission of human immunodeficiency virus type 1 with zidovudine treatment, *New England Journal of Medicine* 331(18):1173–1180.
6. UNAIDS. (1998). *Report on the Global AIDS Epidemic, June 1998.* Geneva: UNAIDS. Available at: http://www.unaids.org/en/media/unaids/contentassets/dataimport/pub/report/1998/19981125_global_epidemic_report_en.pdf
7. World Health Organization. (June 23–25, 1994). *Recommendations From the Meeting on Mother-to-Infant Transmission of HIV by Use of Antiretrovirals.* Geneva: WHO.
8. Lurie P, Wolfe SM. (1997). Unethical trials of interventions to reduce perinatal transmission of the human immunodeficiency virus in developing countries. *New England Journal of Medicine* 337:853–855.
9. Angell M. (1997). The ethics of clinical research in the Third World. *New England Journal of Medicine* 337:847–849.
10. Annas GJ, Grodin MA. (1998). Human rights and maternal-fetal HIV transmission prevention trials in Africa. *American Journal of Public Health* 88:548–550.
11. Susser M. (1998). The prevention of perinatal HIV in less developed countries. *American Journal of Public Health* 88:547–548.
12. Bayer R. (1998). The debate over maternal-fetal HIV transmission prevention trials in Africa, Asia, and the Caribbean: racist exploitation or exploitation of racism? *American Journal of Public Health* 88:567–570.
13. Shapiro HT, Meslin EM. (2001). Ethical issues in the design and conduct of clinical trials in developing countries. *New England Journal of Medicine* 345:139–142.
14. Glantz LH Annas GJ, Grodin MA, Mariner WK. (1998). Research in developing countries: taking "benefit" seriously. *Hastings Center Report* 28:38–42.
15. London AJ. (2000). The ambiguity and exigency: clarifying "standard of care" arguments in international research. *Journal of Medicine and Philosophy* 25:379–397.
16. Glantz et al., *op. cit.*
17. Council for International Organizations of Medical Sciences. (2002). *International Ethical Guidelines for Biomedical Research Involving Human Subjects.* Geneva: CIOMS.
18. Participants in the 2001 Conference on Ethical Aspects of Research in Developing Countries. (2004). Moral standards for research in developing countries: From "reasonable availability" to "fair benefits" *Hastings Center Report* 34(3):17–27.
19. Participants in the 2001 Conference on Ethical Aspects of Research in Developing Countries. *Op cit.*

20. Participants in the 2001 Conference on Ethical Aspects of Research in Developing Countries. *Op cit.*

21. Lurie and Wolfe, *op. cit.*

22. Angell, *op. cit.*

23. *The Declaration of Helsinki*, September 1989, Paragraph II. 3.

24. Faden R, Kass N. (1998). HIV research, ethics, and the developing world. *American Journal of Public Health* 88:548–550.

25. Emanuel E, Wendler D, Killen J, Grady C. (2004). What makes clinical research in developing countries ethical? The benchmarks of ethical research. *Journal of Infectious Diseases* 189:930–937.

26. Connor, *op. cit.*

27. Shaffer N, Chuachoowong R, Mock PA, et al. (1999). Short-course zidovudine for perinatal HIV-1 transmission in Bangkok, Thailand: a randomised controlled trial. Bangkok Collaborative Perinatal HIV Transmission Study Group. *Lancet* 353:773–780.

28. Wiktor SZ, Ekpini E, Karon JM, et al. (1999). Short-course oral zidovudine for prevention of mother-to-child transmission of HIV-1 in Abidjan, Côte d'Ivoire: a randomised trial. *Lancet* 353:781–785.

29. Brocklehurst P, Volmink J. (2002). Antiretrovirals for reducing the risk of mother-to-child transmission of HIV infection. *Cochrane Database Syst Rev* (2):CD003510.

30. Emanuel EJ, Miller FH. (2001). The ethics of placebo controlled trials—a middle ground. *New England Journal of Medicine* 345:915–919.

31. Daugherty CK, Ratain MJ, Emanuel EJ, Farrell AT, Schilsky RJ. (2008). Ethical, scientific, and regulatory perspective regarding the use of placebo controls in cancer clinical trials. *Journal of Clinical Oncology* 26:1371–1378.

32. Pogge T. (2002). *World Poverty and Human Rights.* Cambridge, UK: Polity Press.

33. Beitz C. (1979). *Political Theory and International Relations.* Princeton, NJ: Princeton University Press.

34. Singer P. (1994). *Ethics.* Oxford: Oxford University Press.

35. Singer P. (1993). *Practical Ethics.* Cambridge: Cambridge University Press.

36. Beitz, *op. cit.*

37. Beitz, *op. cit.*

38. Pogge, *op. cit.*

39. Daniels N. (1985). *Just Health Care.* New York: Cambridge University Press.

40. Rawls J. (1999). *The Law of Peoples.* Cambridge, MA: Harvard University Press.

41. Rawls, *op. cit.,* p. 83.

42. Rawls, *op. cit.,* p. 119.

43. Rawls, *op. cit.,* p. 117.

44. Rawls, *op. cit.,* p. 117.

45. Easterly W. (2006). *The White Man's Burden.* New York: Penguin Press.

46. World Bank. World Development Indicators. Aggregate for Least Developed Countries: UN Classification. Population, 2009. Available at: http://data.worldbank.org.

47. United Nations Office of the High Representative for the Least Developed Countries, Landlocked Developing Countries and Small Island Developing Countries. Available at: http://www.unohrlls.org/en/ldc/related/62/

48. World Bank. Available at: http://data.worldbank.org/indicator/NY.GNP.PCAP.CD

CHAPTER 9

☙

International NGO Health Programs in a Non-ideal World

*Imperialism, Respect, and
Procedural Justice*

LISA FULLER

I n *Development as Freedom*, Amartya Sen remarks that "[t]he contemporary world is dominated by the West, and even though the imperial authority of the erstwhile rulers of the world has declined, the dominance of the West remains as strong as ever."[1] His comment captures the common intuition that the so-called "increasing globalization" of our world is actually a process of increasing Westernization. One troubling aspect of the West's global dominance is the propensity of Western beliefs, values, and practices to undermine traditional ways of life and distinctive religious and cultural belief systems. In this chapter, I propose to consider this phenomenon as it emerges in the context of development and humanitarian aid programs, particularly those delivering medical treatment, nutrition and access to clean water. (For brevity I will refer to these services simply as "health services.") I will argue that in order to avoid contributing to cultural imperialism, international nongovernmental organizations (INGOs) have a duty to ensure that they do not offer services in a way that requires their beneficiaries to choose between accessing essential health services and violating or otherwise undermining significant traditional norms and practices.[2] INGOs can accomplish this by means

how does this relate to ⌾

of an *iterated process of reciprocal negotiation*, in which both the INGOs' and the beneficiaries' deep values and concerns play a role. In essence, my claim is that employing such a process is a requirement of procedural justice, given the non-ideal conditions in which INGOs must operate.

Many people in the developing world access essential health services either partially or primarily through programs run by INGOs. Given that such programs are typically designed and run by Westerners, and funded by Western countries and their citizens, it is not surprising that such programs are regarded by many as vehicles for Western cultural imperialism. Despite the fact that many INGOs make efforts to employ local staff and to employ a "participatory approach" to program management and implementation, they are still commonly accused of harboring a predominantly Western, liberal agenda and being insufficiently engaged with, and responsive to, the communities they serve.[3]

To set the scene for our inquiry, suppose that an INGO has chosen a population to receive aid, and a group of expatriate staff has been sent to the selected area to set up a program. Typically, before a team is sent, the INGO will have either (i) sent an exploratory mission to determine the type and extent of unmet basic needs, or (ii) reviewed reliable reports from other groups present in the area on this subject. So, a team usually will not have been sent unless staffers have some idea of the situation on the ground and some idea of what types of services their organization intends to offer. In the early stages of a program, the INGO staff will usually confer with the relevant authorities (local elders, mayors, governors, or regional or national government representatives), as well as consult local staff hired to set up and run the project.

Suppose further that during these consultations, the objectives of the project or some element of its proposed implementation meets with considerable resistance from the people consulted. What should be done if: (a) community members do not feel that the chosen activities reflect their most urgent priorities, or (b) they want to change significant elements of the way the project is implemented, or (c) many refuse the proposed services altogether—and do so on the ground that the project as envisioned by the INGO is incompatible with some belief, value, or custom widely held or practiced by the community?

The problem here is that some community members may be very badly off and so may feel that they have little choice but to compromise their values in order to obtain the material benefits they need. This is also a problem at another level, because the provision of health services (under some conditions) may have the effect of contributing to the disappearance of valued practices. Given that INGOs' main purpose is to improve the welfare of vulnerable and impoverished people, they ought not to contribute to other types of loss and pain either for those people or for those who come after them. In addition, as a purely practical matter, many projects simply won't work efficiently—or at all—if they are designed in a way that conflicts with the dominant cultural beliefs and social conventions of the recipients. However, if INGO officials allow the religious, cultural, or value judgments of the recipient population to substantially govern project design, they might have to compromise their own deeply held convictions, and/or become involved in practices that they regard as morally wrong.   TENSION

In what follows, I argue that INGOs have a duty to structure their programs such that accepting aid does not require beneficiaries to violate their central moral and religious beliefs. I also claim that in cases where conflicts about factual or scientific beliefs arise, INGOs must take steps to convince beneficiaries of their truth before implementing programs. Finally, I suggest that providing services in certain ways may undermine cultural practices, and so may contribute to their eventual disappearance. Where these practices are still valued by the community, I take this to be an objectionable instance of cultural imperialism.[4]

TWO CASES OF CULTURAL CONFLICT

In order to concretely illustrate the type of difficulty I have in mind, I will briefly sketch out two examples. I will refer to these cases throughout my analysis, and will attempt to resolve them in the concluding stages of the chapter.

Feeding Center

Severe malnutrition in children under 5 years old is found to be prevalent in an area where drought has decreased the annual food supply significantly. While some members of other age groups are found to be malnourished or undernourished, these cases are relatively few, and young children are found to be much worse off as a group. The INGO decides to treat this group both because INGO officials think this is the right thing to do, and because they think this approach can do the most good overall (by preventing long-term deficits in brain and behavioral development). A therapeutic feeding center with a capacity of 150 is proposed, where severely malnourished and ill children can be admitted and treated. Local community members are happy to have a feeding center in their area but suggest that, at a minimum, it should be open to all age groups. Specifically, they suggest that heads of households and respected elders should have priority. They argue that heads of households must be able to work to provide for them and that it would be disrespectful for very young children to receive food and medical care while elderly parents and grandparents—who have done so much for their families and their community—are suffering. Some people imply that they would feel ashamed to bring their children to the proposed center for treatment if other groups were not also given the opportunity to be treated.

Traditional Healers

An INGO arrives prepared to set up an HIV/AIDS treatment center in a location easily accessible by patients from several villages. INGO staff plan to provide voluntary counseling and testing and antiretroviral therapy. They learn that local traditional healers typically prescribe herbal remedies that are known to interact negatively with antiretrovirals. Such healers are licensed by the government and are often very highly esteemed members of their communities. At the time the INGO arrives, these traditional healers are the primary source of health care for most people. Potential patients and healers alike reject the suggestion that people taking antiretrovirals should be required to cease using traditional remedies before being permitted to begin treatment. They also suggest that traditional healers should be made full treatment partners at the proposed center, since the patients already know, trust, and respect them.[5]

PRELIMINARIES: CULTURAL IMPERIALISM AND CULTURAL BELIEFS

Now that we have an idea of the type of conflicts at issue, we might do well to examine the idea of cultural imperialism in a bit more depth. While I don't want to get bogged down in a lengthy discussion of the merits of various definitions of culture, at least a working definition is necessary here. It seems reasonable to use a definition that takes in elements of several competing (although still quite standard) definitions commonly used in the social sciences. I therefore propose the following working definition: *culture* is that complex whole that includes knowledge, belief, art, law, custom, and any other institutions, habits, and signifying practices acquired by people as members of particular societies and passed on from generation to generation by means of education, the family, and tradition.[6] To avoid a similarly fraught discussion of the idea of imperialism, I will attempt to provide some explanation and analysis of the specific notion of imperialism at work in the context of international aid projects. These remarks are not intended to be exhaustive, but should provide at least a rough idea of what it is that INGOs are accused of perpetrating. *— by whom?*

The notion of cultural imperialism clearly has its roots in the colonial era, when Western nations set out to conquer and rule territories in North America, Asia, and Africa for the dual purpose of exploiting their natural resources and "civilizing" their inhabitants. In this context, cultural imperialism has three key features: (1) the perpetrators or "imperialists" *intend* to both dominate and reform the beliefs and practices of people from another culture, (2) they feel they are entitled to do this because they *believe* that Western culture is inherently superior to so-called "native" or "indigenous" cultures, and (3) they carry out their task *coercively*, with the aid of significant economic and military resources, as well as through political channels such as colonial judicial systems.

But these features do not appear—at first glance anyway—to apply in the context of delivering medical aid in the present day. Certainly INGO staff do not usually *aim* to dominate beneficiaries or obliterate their cultural practices. On the contrary, INGO workers typically accept what John Tomlinson calls the "sovereignty" of particular cultures—"the idea that 'how a life is lived' is a judgment to be made by the particular collectivity that possesses this culture, *and by no one else*."[7] Indeed, Tomlinson notes that criticism of cultural imperialism is often grounded on "respect for a plurality of ways of living," which is itself a central tenet of the contemporary liberalism to which most INGO workers broadly subscribe.[8] Further, INGO workers are not likely to believe that Western culture is inherently superior to others; instead, they are often vocal critics of the way in which the populations they serve have been oppressed by Western powers. Finally, INGOs have no obvious mechanisms of coercive force at their disposal, and must operate within the legal and political context of the countries where they work, rather than being able to occupy and run such institutions themselves.

While on the surface INGO medical aid does not *seem* to share in the pernicious elements of historical colonialism, some parallels are worth noting. First, the rise of INGO aid work over the last 50-odd years has occurred along with the growth of the phenomenon known as "globalization." One notable similarity between globalization and colonialism is that both appear to squeeze out particular sets of local values, beliefs, and practices and to replace them with another more homogenous set, which originated in the West. Typically, this "replacement" set of values is characterized by a strong emphasis on a few key values and practices, such as autonomy, individualism, secularism, and privacy.[9] Following Tomlinson,

Western values

we can understand this as a process of "cultural loss" for those non-Western people whose local, historical beliefs and practices are, in a sense, drowned by a flood of messages and interventions from the West.[10] Among other factors, this loss is said to be caused by the actions of the media, multinational corporations, foreign governments, INGOs, tourists, and intergovernmental organizations such as the United Nations. If INGOs are in fact contributing to cultural imperialism, then they will no doubt be doing so as part of this larger process of globalization.

A further parallel is that INGOs have the capacity to unduly influence beneficiaries into abandoning or contravening their own beliefs in their effort to secure material benefits. This influence, which I will argue can amount to coercion in certain circumstances, is not exercised by means of laws or military/police power, but rather is a function of the kinds of offers made by INGOs together with the background against which beneficiaries must consider them. In the following two sections, I will argue that while such coercion may be inadvertent and often invisible, the potential for it is nevertheless real.

Thus, there are some potentially troubling similarities between historical colonialism and the context in which INGO medical aid is currently delivered. But this is a concern only if we assume that contributing to cultural imperialism is somehow morally wrong. There are two main reasons why cultural imperialism may be morally objectionable. First, there can be no doubt that pressure to abandon, disregard, or violate valued practices and beliefs can cause distress. It can cause feelings of shame or pain or sadness in those who contravene traditional beliefs. It may also cause others in their communities to see those who abandon traditional beliefs as reprehensible or misguided, causing further suffering and difficulties due to social exclusion or criticism. Experiencing such distress makes people worse off. Second, the coercion associated with imperialism denotes a loss of choice or control over an individual's or a community's actions or way of life. Even when losing the ability to choose does not have a negative impact on people's welfare (and it often will), it is nevertheless insulting and disrespectful for choices about significant aspects of people's lives (and their society more generally) to be taken out of their hands. In these ways, contributors to cultural imperialism may wrong as well as harm others through their actions. As Sen notes, while not every way of life, belief, or practice may be judged worthy of preservation, members from all classes of the relevant society "should be able to be active in deciding what to preserve and what to let go."[11]

Finally, a preliminary word about cultural beliefs. Clearly, a variety of beliefs and views can be broadly described as "cultural" or somehow importantly influenced by culture. It is worth distinguishing some of them here. First, we should note that there is a difference between some-one's *factual* beliefs and his or her *moral and religious* beliefs. Factual beliefs are beliefs about causes and effects in the world, including beliefs about scientific and medical matters. While there may be some overlap between religious and factual beliefs in terms of specifying what is real, how the world came into existence, and so on, for the purposes of this chapter factual beliefs should be understood as those that relate to whether or not certain medicines and medical treatments are likely to make people better off by curing them, alleviating pain, or extending life.[12] The moral and religious beliefs I am concerned with are those that concern right and wrong—that is, which actions are morally prohibited (or shameful or blameworthy), and which are permissible. These will include beliefs about honor, modesty, and appropriate public and private behavior.[13] A second type of moral belief that is relevant here is someone's assessment of what would improve his or her welfare. This type of belief is basically a judgment about what would constitute a benefit for that person.

I have specified what counts as cultural imperialism and what types of beliefs might generate conflicts between recipients and INGOs. I now explain how INGOs can contribute to cultural imperialism through their work.

THE PROBLEM I: INFORMED CONSENT IS NOT ENOUGH

It might well be argued that the simple provision of health services by INGOs cannot be morally questionable, since this merely provides an opportunity for beneficiaries to improve their overall welfare. According to this type of view, all that is required for an INGO to work legitimately is for it to obtain informed consent from individual recipients of health services concerning the policies and procedures of the programs in which they wish to participate. After all, isn't this the ethical standard we apply in hospitals and clinics in the West? A supporter of this view might argue as follows: "It is nonsense to suggest that any beneficial services provided with the full and informed consent of the recipient can nonetheless be morally objectionable. As long as someone has the option to refuse, what could be wrong with offering someone health services?"

The plausibility of this view depends critically on what it means to have "an option to refuse." In one sense, a person who can either do X or refrain from doing X has such an option. Nevertheless, in what seems to me to be a more important sense, a person living in poverty or near poverty may not have a viable choice between accessing essential health services and, as in *Feeding Center,* respecting the society's normative requirement that elders' needs be placed first. Thus, a mother may have both malnourished children and parents, but if an INGO offers only a program for children, then she will naturally feel that she must make use of it. However, even though her choice set was constrained on the one hand by the decisions of the INGO and on the other by her economic circumstances, she likely will still feel ashamed at having done what she did.

The first sense of "having an option to refuse" implicitly relies on what might be called the "standard" view of coercion. The standard view holds that people are coerced only when they are threatened with sanctions, and not when they are offered benefits or inducements. David Zimmerman explains this view nicely as follows:

> P coerces Q only if he changes the range of actions open to Q *and* this change makes Q considerably worse off than he would have been [in the normal course of events]. Thus, in the standard highwayman case, P's proposal, "Your money or your life" counts as coercive (in part) because P changes Q's range of options and because Q is considerably worse off in the "threat situation" in which he has to make a choice between keeping his money and keeping his life than he is in the "pre-threat situation" in which he can have both.[14]

On this view, then, potential aid recipients are free to either accept or refuse the offer of health services. Since they are not threatened with a sanction that would make them worse off than they would normally be, they are not coerced.

By contrast, the second sense of "having an option to refuse" is grounded in an alternative theory of what constitutes coercion. This view holds that offers or inducements *can* be coercive, although in a different way from threats. Joan McGregor argues that one key difference between coercive threats and coercive offers is that with threats "the coercer puts his victim in a vulnerable position," whereas "with coercive offers the proposer finds his victim in a vulnerable position."[15]

We may draw a distinction here between *offers* and *gifts.* Offers come with a price, whereas gifts do not. This means that even accepting an offer that, on balance, makes me better off will usually require me to accept some conditions and/or carry out some specific acts. An offer is coercive when it exacts an unreasonable "price" in exchange for the benefits offered.[16] The "price" of accepting an offer can be explicitly stated, or it can be the result of complying with the system by which the benefits are distributed and organized. The "price" of accepting a coercive offer is also typically something that is regarded as an evil by the person making the choice. In McGregor's words, "[i]t is true of all acts of coercion that the victim *does not want to do* the demanded action, but he does have a motive for acting in the demanded way [italics added]."[17]

So, in *Feeding Center*, the price of obtaining benefits is both the shame a mother feels when she takes her children to the center and the fact that she has violated what she regards as a normative requirement. Once we recognize that even offers come with conditions attached (a price), then we can see that these conditions can often be adjusted by the person making the offer. In particular, the terms of the offer can sometimes be made easier for the beneficiary to accept.

Onora O'Neill is another proponent of the view that offers of assistance can be coercive. She argues that offers in which someone must abandon moral or customary obligations in order to obtain basic necessities (or perhaps even survival) are effectively "unrefusable." More specifically, she identifies several key features of such coercive offers and suggests that they are not meaningfully different from coercive threats. She notes that "[u]nrefusable 'offers' work because they link the choice of any but the compliant option to residual option(s) which the particular agent cannot survive or sustain.[18] An unrefusable 'offer' is not, indeed, one where non-compliance is made logically or physically impossible for all victims; it is one that a particular victim cannot refuse without deep damage to sense of self or identity."[19] Finally, O'Neill observes that it is easier to construct such "offers" when their recipients have meager resources and capabilities at their disposal, or when they have many commitments to others that they are obliged to fulfill. It is so easy, in fact, to construct and proffer unrefusable "offers" to vulnerable people that it is possible to do this without even being aware of it. O'Neill puts forward the following words of caution:

> [A]gents who seek not to coerce have to make sure that they do not inadvertently make unrefusable "offers." Any offers they make must not link options either overtly or covertly to consequences with which those to whom they make the offer cannot live. Like coercers, they will therefore need to take account of others' strengths and weaknesses, of their *specific* vulnerabilities and of the *actual* limits of their capabilities. In particular, they will have to be alert to the ease with which the weak can be coerced.[20]

It is worth noticing that it is the *structure of the offer* that is coercive here, and once the offer has been made, then the recipient cannot escape it. Of course, some of those to whom such "offers" are made might be willing to refuse the compliant option, and so suffer the severe consequences associated with doing so. However, even resisters of this type cannot restructure the offer itself and so cannot avoid the coercion inherent in it.

O'Neill remarks, "What exceptional people refuse when coerced is compliance with the option that coercers want. They do not and cannot refuse
the *'offer'*, of which that option is one disjunct. The *mark of coercion is the
unrefusable 'offer', not the unrefusable option* [emphasis in original]."[21] Thus,
even when people give their consent, they may still be constrained in a
way that looks morally objectionable. As Zimmerman argues, the offer is
objectionable because the recipient does not want to face this particular
"disjunctive choice"—and by structuring the choice in this way, the
offerer has undermined the recipient's freedom "to set [his or her] own
ends and to consider reasons for action which go beyond" this particular
choice situation.[22]

THE PROBLEM II: BACKGROUND CONDITIONS AND
NON-IDEAL THEORY

It now appears that INGOs could be inadvertently coercing their potential recipients, even when they think they are merely providing services or
offering benefits. By making coercive or "unrefusable" offers, INGOs can
cause recipients to reject local cultural norms and practices in favor of
Western practices or values. Thus, it follows that INGOs may contribute
to cultural imperialism *without intending to.*[23] Note that this analysis of
"inadvertent coercion" fits neatly into the earlier discussion of the unin-
tentional cultural imperialism characteristic of globalization (as contrasted with the intentional domination of historical colonialism). It does
not presuppose that the perpetrators *intend* to coerce the victims into
behaving according to an alternative set of cultural norms, and yet that is
what occurs. Given that INGOs control the (often life-saving) resources,
they have a duty to avoid making certain types of unrefusable "offers."
They cannot do this unless they first recognize the relevant background
conditions against which they offer their services.

The background conditions of which INGOs must take account are:
(1) the power differential between beneficiaries and INGOs; (2) the significance of a given cultural practice to (at least some of) their potential
beneficiaries; and (3) the non-ideal political context in which INGOs
must do their work.

Power Differential

O'Neill reminds us that it is the relative weakness of recipients vis-à-vis
those making the unrefusable "offers" that makes such "offers" possible in
the first place. The party in the more powerful position is the one who is

able to structure the "offer," and thereby to exert control over the choices available to the less powerful party. Indeed, McGregor sets out two advantages that those who have the power to coerce must have: (a) the weaker party must be in a relationship of "dependence" with regard to the stronger—that is, there must be few, if any, alternative sources from which the necessary benefit can be obtained, and (b) the stronger party must be able to "cause or prevent a positive evil to the weaker party."[24] INGOs frequently have both of these advantages.

Given that INGO staff are in a position of economic and administrative power relative to those people to whom they offer their services, it is clear that the character of the offers made by the former to the latter are affected by this inequality. As the stronger party, it is the responsibility of INGOs to ensure that their offers are not structured in a manner that gives recipients no option other than to violate a norm or undermine a cultural practice that has significance for them. They can do this only by relinquishing some of their hold over the structure of the choices that are presented to individual recipients. If INGOs simply open their doors and offer services without taking account of this difference in power, then they are irresponsible.

May not be homogenous – Some may be happy to violate cultural norms – eg. mothers of starving children.

OK

Cultural Values

The significance of a given norm to potential recipients is crucial in determining whether an offer that requires them to violate that norm is in fact coercive. Clearly if a norm or practice has very little or no value to potential recipients, then acting against it will not count as a significant evil. However, it seems clear that any act that recipients believe they are obligated not to perform, or that would cause them to feel shame, remorse, or regret, is an act they regard as an evil. Daniel Lyons helpfully points out that the usual mark of a coercive offer is that someone is "rationally reluctant" to accept it, even if it seems to offer a benefit.[25] On this view, people are "rationally reluctant" to accept an offer when they must perform act X in order to obtain benefit Y, but regard X as morally wrong or harmful for themselves (or someone they care about or are responsible for) in some non-trivial way. This type of reluctance seems to indicate that the cultural norm is significant in the relevant sense.[26]

Political Context

Finally, it is worth pointing out that INGOs, by definition, work in political contexts that are very far from approximating ideal conditions of either domestic or international justice.[27] Thus, the problems under consideration are problems in non-ideal theory, in both senses of "non-ideal" as outlined by Gopal Sreenivasan: (i) institutions where INGOs work "may be unjust or may not exist at all," and (ii) both domestically and internationally, individuals and states have failed to do their fair share for aid recipients within a scheme of distributive justice.[28] This is important for formulating appropriate procedures for determining how to benefit them.

Under non-ideal conditions, procedural justice will "look different" than it does under ideal conditions.[29] Under ideal conditions, where the background is a just society and a just world, recipients would have the option to refuse health services offered by INGOs on terms with which they disagree. This is because recipients would be able to access alternative sources of health care in a just world. A given INGO would not be "the only game in town." Recipients would also have various rights to religious and personal freedom that would be protected by domestic institutions, to which they could appeal in cases of conflict. In an ideal situation, then, making an offer of particular benefits under terms acceptable to the service provider would typically be ethically neutral.

By contrast, in non-ideal conditions, background institutions are often unjust or nonexistent and people are materially very badly off. This means that it is the responsibility of the party with more power and resources to avoid wronging the weaker party, even unintentionally. Therefore, procedures to protect the vulnerable from coercion and cultural loss must be put in place by INGOs themselves.

Further, it is the responsibility of an INGO, as the party in control of resources and wishing to help recipients, to make its best effort to help recipients *as they are* and not as INGOs (or other Westerners) would like them to be. George Sher notes that non-ideal theory "takes as its point of departure some problem that is raised either by the injustice of some past or present social arrangements or else *by some limitation of what people can be expected to know or do* [emphasis added]."[30]

I take it as given that people from all cultures are socialized such that certain norms and values are internalized by them. While these values can of course be altered, this is often a long and involved process, which even after much effort may result in only partial success. It does not seem appropriate to expect aid recipients to be able to shed all their ingrained socialization under desperate circumstances, just so that INGOs can organize projects in certain ways. If recipients cannot be expected to do this, then it follows that helping them requires INGOs to take seriously recipients' view of what is good for them and what they may permissibly do, rather than some view of these issues that they could or should have. It is no good offering those services that recipients *would value* if they were living under different institutions or subscribed to a different set of norms. It is also not appropriate to offer services *in ways* that recipients find objectionable but would think were fine if they held different beliefs. Real people are non-ideal. Their beliefs are not necessarily perfectly rational or consistent or in accordance with justice. Still, regardless of the cultural norms to which vulnerable people subscribe, they ought to be helped on terms that they can live with and protected from coercion and paternalism.

Based on aspects of the analyses provided by McGregor, Lyons, and O'Neill, we can say that if INGOs wish to avoid contributing to cultural imperialism, they ought to avoid making offers in which (a) the recipients are rationally reluctant to accept the offer because accepting would violate or undermine a cultural norm or practice that they currently value, but (b) recipients also have a strong motive to accept the offer because they need life-saving or essential health services for themselves or those close to them, and (c) they cannot access equivalent services from an alternative provider, but (d) the INGO could offer these or other essential health services on easier terms—that is, terms that would not require recipients to violate or undermine a cultural norm or practice that they currently value.[31]

One important qualification needs to be noted. As in the traditional doctor–patient relationship, INGOs need not offer services that they think do not provide any medical benefit to recipients. This is only reasonable, since INGO staff must act in ways that do not violate their own deep ethical and professional commitments. However, they also should recognize that the people they are trying to help hold certain values and beliefs, and are in need of help even if these beliefs are sometimes objectionable.[32] If the mandate of an INGO is to improve the health of a given population, then it needs to do this in ways that *both* recipients and INGO staff can agree are beneficial, and without causing recipients to undergo distress, shame, or remorse in the process.[33]

OBJECTIONS

A critic might object that the foregoing analysis confuses or conflates the distinct notions of coercion and exploitation. Given that potential aid beneficiaries are already in a vulnerable position before the INGO arrives, one might well think that their situation is akin to the classic case of exploitation, in which the capitalist exploits the worker who needs a job or else will starve. In this case, the capitalist has not created the bad situation of the worker; he merely takes advantage of it in order to pay his worker lower wages and so reap more profits for himself.

What if both sets of values coincided?

It is true that INGOs *find* their beneficiaries in a vulnerable position rather than *making* them vulnerable by threatening to make them worse off. It is also true that in some cases of exploitation, as in the case of aid recipients, the more vulnerable party is made better off by the transaction. However, these resemblances are insufficient to support the claim that INGOs exploit, rather than coerce, their beneficiaries when they design projects with only their own values in mind.

While there are many and varied accounts of exploitation in the philosophical literature, it is characteristic of the great majority of them that in order to exploit someone, the exploiter must take advantage of him or her *for the exploiter's benefit*. As Robert Mayer notes, "those who exploit always extract a benefit from that which they exploit. They gain. Exploitation is thus a process of acquisition."[34] The moral wrong of exploitation is that the more powerful party takes unfair advantage of the less powerful, and so gains at the weaker party's expense. But this feature is missing from the INGO case. INGOs do not gain from offering beneficiaries access to health services, even when they cause beneficiaries to act in ways that go against the norms and practices that the beneficiaries value.

Rather, INGOs are trying to benefit their recipients. They intend to act for the good of those they wish to help, not to gain benefits for themselves. Coercion comes into the picture when they do not attend carefully enough to the background conditions, to their recipients' *own ideas* of what is good for them, and/or to those actions that are morally permissible and impermissible in their beneficiaries' eyes. In essence, either negligently or paternalistically, INGOs can end up coercing recipients for (what the INGO considers) their own good. It is true of both the standard and the alternative accounts of coercion that people can be coerced into doing something for their own good.

By contrast, it seems contradictory to say that "I exploited him for his own good." Acts of exploitation may benefit the exploited, but the object is always for the exploiter to gain. The exploited person has been harmed or cheated or treated unfairly *in order that* benefits will flow to the exploiter. Coercion does not have this feature. It is perfectly ordinary to coerce someone with only his or her interests in mind; in this case, the coercer gains nothing from the effort.[35] The wrong of exploitation is taking advantage of someone's vulnerable position to deprive that person of benefits that are rightly his or hers, in order to take the benefits for oneself. Exploitation is a species of wrongful gain, whereas coercion improperly limits the freedom of some person or group.

A second objection to my analysis might be that the main problem I raise is not specific to instances of cultural imperialism, but rather is widely applicable to circumstances in which the more powerful try to help the less powerful. It is surely true that unrefusable "offers" can be made in a variety of contexts and by a variety of agents. As such, we might simply note that some kinds of offers made by INGOs constitute a subset of this larger group.[36] However, it is also true that this type of practice has implications for cultural imperialism that it might not have in situations where both parties are from the same culture, broadly speaking.[37]

One implication is best explained if we envision the consequences of offering the kinds of coercive choices just sketched out to very many vulnerable people in a given community at once. The result, I would argue, is a particular kind of collective action problem. It may be beneficial for an individual to accept badly needed medical services even at the cost of failing to fulfill societal obligations, or feeling shame, or going against tradition. In an individual case, we might even say that the good done for the aid recipient outweighs the harm of being coerced. However, if many people make the same choice, over a substantial period of time, some customary ways of doing things could begin to disappear because they have no current practitioners. And the *reason* these practices will have disappeared is that INGOs encouraged their practitioners to defect from them on an individual basis by making unrefusable "offers."

Consider *Traditional Healers*. Given that the healers need a minimum number of patients to sustain a viable practice, if they were to lose sufficient support, then they would have to seek other ways of making a living, and might not endeavor to teach anyone else how to carry out their practices in future. (I assume that the healers had something to offer the community other than the treatment for HIV/AIDS. This does not seem far-fetched, since they were the only health care practitioners in the village before the INGO arrived.) Each individual who chose not to go back to his or her traditional healer for further HIV treatments might have thought that he or she was merely choosing between two services. All individuals, in a sense, expect to "free-ride"—they expect that the practice they have stopped supporting will still be there for them if they want to access it again. But in cases of widespread defection, the traditional healer may be gone.

THE SOLUTION I: NEGOTIATION, RECIPROCITY, AND PROCEDURAL JUSTICE

Having established that INGOs should guard against making coercive offers of the type identified above, I now address the question of how that might be accomplished. Cultural imperialism both can and should be avoided by INGOs, even in situations where difficult conflicts arise. It is my contention that INGOs can avoid contributing to cultural imperialism in instances of conflict by adopting an *iterated process of reciprocal negotiation*. Even against a background of injustice, INGOs can design and implement projects justly by making decisions within a particular procedural framework. The framework is intended to allow agreement to be reached concerning *who* should receive health services, and *how* they should be provided.

However, before any process of negotiated decision-making can succeed, it would seem that INGOs and recipients need—or need to work out—at least some shared beliefs and values, in order both to communicate effectively and to arrive at mutually acceptable projects. After all, if the worldviews of the two groups are completely at odds, it seems implausible that there could be any jointly satisfactory project designs.[38] Henry Shue rightly notes that this shared background need not exist *prior* to the actual discussions that lead to agreement. He argues, "[e]ven the shared 'premise' on which new-found agreement rests need not have existed *ex ante*, before the conversation occurred."[39] Jeremy Waldron suggests that, in the course of a conversation, the evaluations of various proposals by all parties involved can "themselves . . . constitute the background practice that gives them sense and substance."[40] This insight is at the heart of the decision-making procedure I am recommending here.

This process would work as follows: INGO staff and various representatives of the community (including potential beneficiaries) would sit down and discuss the plans for a new project. They would each present their own reasons for supporting or condemning the project.[41] Then, as negotiations proceed—as comments and suggestions are put forward and considered—it will become clear which commitments each side has that it is unwilling to compromise, and which ones are not as deeply held and so admit of some "wiggle room." Indeed, it might happen that some concerns are shared by both parties and so can constitute a shared end for the project, which then needs to be spelled out in terms of concrete implementation mechanisms, which are then also subject to deliberation and negotiation.

Waldron's point is that after several iterations, the line between the reflective evaluations of one group and the other will blur, such that it will not be not entirely accurate to assign each belief or principle to one group rather than the other. Even if no deep or ultimate shared concern is uncovered, they will have a mutual interest in reaching agreement. In addition, they will have some idea of the parameters within which they must work once the non-negotiable commitments are on the table. This framework of deep commitments can act as their background in the way Waldron suggests.

[handwritten margin note: Who's sitting at the table ?]

To be clear, I am *not* suggesting that INGOs and the communities in which they work should somehow "just compromise and get on with it." The danger of a forced settlement is that one or more parties may feel morally compromised by the resulting agreement. Michele Carter and Craig Klugman note that in cases of cultural conflict, the requirement that each side compromise "can have a negative connotation, implying situations where an individual [or group] may be asked to forfeit a cherished belief or practice."[42]

It should not be necessary for either INGOs or their recipients to violate cherished or central beliefs or values in order to agree upon the ends and implementation mechanisms for aid projects. INGO projects are necessarily limited in scope, and so do not affect every aspect of recipients' lives. Disagreement, and the accompanying need for mutual concessions, need only affect a few elements of the cultural and moral framework within which each party operates. Only partial overlap is actually needed here. In addition, we should recognize that such overlap does not always have to occur at the level of fundamental moral principle (although of course it may). As Shue (following Rawls) points out, "ultimate premises [may] differ but certain intermediate premises [can] work equally well."[43]

In contrast to mere compromise, the model I am suggesting here calls for *reciprocal negotiation* or engagement. According to Lawrence Becker, one central element of reciprocity is that the parties are disposed "to return good for good."[44] This does not mean that recipients must make a return in kind, but rather that responses (to proposals, to benefits offered, or to concessions made) should be both "fitting and proportional."[45] For a response to be fitting, it must be in line with the special purpose of the institution, practice, or relationship of which it is a part.[46] For it to be proportional, responses must not demand much more, or give much less, than is reasonable relative to what has been offered or conceded by the other. The reciprocal negotiation must also be characterized by equal respect and standing, in the sense that each participant must have the right to respond to proposals and make suggestions, and each party's point of view must be considered seriously and responded to without hostility or intimidation. The requirement that each participant negotiate on equal footing reflects the validity and relevance of each autonomous person's view of what is good for him or her and for the community. As such, reciprocal negotiation under conditions of equality is an expression of respect, and is demanded by the equal dignity of all persons.[47] It is precisely by affording recipients this respect that INGOs can avoid coercive offers.

I argue that a process of this type must be used by INGOs if they wish to avoid becoming agents of cultural imperialism. In effect, I am suggesting that an iterated process of reciprocal negotiation under conditions of equality is a requirement of procedural justice for INGOs.[48]

However, given the importance of negotiation and reciprocity, what should be done when a community and an INGO come to a genuine impasse? In my view, it is better to agree not to proceed than to carry out projects that violate the deep commitments of either those who participate in them or those who are responsible for running them on a day-to-day basis.[49] I recognize that certain benefits will necessarily be foregone as a result, and some money will inevitably be wasted. But I regard this as acceptable given the alternatives: (1) the INGO designs and runs the project with its own values in mind and does not alter it in the ways requested by the community, or (2) the INGO contravenes its own principles and gives in to requests that it views as ethically unacceptable.

I have already argued at length for why option (1) is objectionable. But it is also objectionable to use the good will of INGOs against them, as in option (2), by suggesting that the delivery of medical treatment is inherently oppressive so long as *any* consideration of INGOs' (Western) values has informed decisions about which services to offer and how to offer them. Aid agencies and individual practitioners have to justify themselves to the wider world and must be able to "live with themselves." They should not have to participate in projects or activities that they regard as unethical. It is not reasonable to accuse INGO officials of participating in cultural imperialism simply because they will not give over total control to the community or its representatives—again, because this approach is entirely one-sided and does not demonstrate a concern for reciprocity. Reciprocity is required because it is a concrete demonstration of respect for those with whom one is engaged. It is the recognition that one is negotiating with self-determining individuals who possess the capacity for both ethical reflection and reasoned judgment; and so it cuts both ways.

THE SOLUTION II: TWO CASES REVISITED

To sum up the foregoing analysis, INGOs have a negative duty to avoid making unrefusable offers, as well as a positive duty to create the conditions under which is possible to engage in reciprocal negotiation with recipients, because this process constitutes a concrete expression of respect for recipients as self-determining beings. I now return to the examples I introduced at the beginning of the chapter. While the kind of model I have been advocating here does not really admit of application in the absence of the actual affected parties, I will nevertheless allow myself to speculate about some possible resolutions to the various conflicts.[50]

Feeding Center

In this case, the INGO has two main value concerns at stake: efficiency and distributive justice. It chose to target children under 5 years old because that strategy would produce the most medical benefits per patient treated and would help remedy the effects of a cultural bias within the community that works to distribute a disproportionate amount of essential resources to (usually male) heads of households and elders. But INGO officials are also committed to the equal value of each individual, and presumably they would not think that serving malnourished members of any group was a waste of money. The community also has two main concerns: community members think elders should have priority (on the basis of desert or respect), and that, practically speaking, everyone will be better off if those who are working are well fed and able to perform their duties. Thus, both groups are concerned with efficiency, but they are calculating benefits at different levels. To accommodate their other concerns, each group could agree to sacrifice some amount of efficiency. The outcome of this would be that the INGO would make the feeding center open to all on an equal basis.

This arrangement would not be a violation of the INGO's main commitments, and it would avoid making a coercive offer. The INGO could agree to see family groups rather than individual patients and to treat any family member who meets the criteria for malnutrition. This would be mainly children in any case, and so this policy would lower the level of good outcomes that the program is able to generate—or raise the cost—by only a relatively small amount. However, even if the efficiency lost in terms of health outcomes was more significant, it is the responsibility of health practitioners to help their patients in ways that do not wrong or harm them, even under resource constraints. If fewer people are helped as a result, then at least the help has been provided in a morally defensible way.

The INGO might also decide to engage in supplementary feeding for undernourished family members, thus preventing them from becoming worse and acknowledging the need for heads of households to remain well for the sake of the whole family. Certainly an INGO would agree that it is better for the community generally for its working people to be healthy.

By treating everyone equally, the INGO risks giving more resources to privileged community members who already have an unequally large share within the family and community. Thus, it could be reinforcing existing inequalities. But if the INGO merely agrees to evaluate everyone and to provide lesser food supplements to those who are not technically malnourished, then it would still be concentrating its efforts on those who need help the most. Granted, this approach would widen the recipient group to include all of the sickest (rather than just those who are under 5 years old), but this arrangement would preserve the INGO's concern for treating the most vulnerable while at the same time accommodating the community's priorities to some extent.

Traditional Healers

In the *Traditional Healers* case, the INGO's concern is to provide the best-quality medical care for its patients; the community's concerns are to treat these trusted members of the community with due respect and to maintain access to traditional healers in the face of new "competition." Community members are also concerned, of course, to get the benefits that the INGO is offering, such as antiretroviral therapy for people with HIV/AIDS. Both sides agree on who should be helped and the benefits to be provided, but they disagree on the conditions attached to accessing that help. This conflict admits of three potential resolutions:

1. The INGO could hire the traditional healers as center staff and have them work directly with the INGO doctor(s). In this way they could collaborate on a case-by-case basis in determining the best treatment for each patient.
2. As above, but in addition INGO physicians and staff members could ask the traditional healers what it would take to convince them that the remedy they have been prescribing is not beneficial in combination with antiretroviral therapies. The INGO staff members could then try to demonstrate this to the satisfaction of the traditional healers, which might induce the traditional healers to abandon treatments that interact negatively with Western medications. Since this is a disagreement about facts, the INGO only has to provide evidence to the relevant parties that its view is true.
3. To avoid contributing to cultural imperialism, the INGO could move on to another community of people who need HIV/AIDS treatment just as badly but do not have sets of beliefs and practices that conflict with the project as originally envisioned.

Option (1) is objectionable from the INGO point of view because it does not ensure that the traditional healers will forgo prescribing the remedies to which the INGO objects, thereby lowering the standard of patient care. It does not address the INGO's concerns, but simply ensures that the disagreement will be out in the open. It does, however, respect the value commitments of the community and avoids making coercive offers.

To my mind, option (2) is the best resolution, but it is not entirely without drawbacks. It will cost time and money to attempt to persuade the traditional healers that the remedies at issue actually interact negatively with antiretrovirals. While there is no way to say hypothetically what would ultimately prove persuasive, nevertheless we might imagine that both sides could appeal to people they trust in the Ministry of Health, or in respected nearby hospitals, to conduct experiments or review literature and then report back their opinions. It might be difficult to find people trusted by both "camps" who also have the scientific expertise required to assess the evidence, but this doesn't seem clearly impossible.

Option (2) also promises to deliver the best scenario for patient care. Provided that the traditional healers could be persuaded of the truth of the INGO's claim, then patients would cease taking the relevant traditional remedies. They would also benefit from being treated both by someone they trust and who knows them intimately, as well as by someone who has specialized knowledge of HIV/AIDS. Close cooperation between the INGO and local healers also might significantly improve patient compliance. Further, by collaborating with the traditional healers, the INGO would avoid making patients choose between the two services, and would ensure that it is not contributing to the demise of a local practice that is valued by some members of the community.

Alternatively, option (3) looks like an acceptable resolution under the reciprocal negotiation model, because at least it would not impose unacceptable choices on community members, possibly undercutting their traditional practices. It also would not "waste" money demonstrating scientific facts that are not disputed in the Western medical community.

From another perspective, however, option (3) seems *very* imperialistic, and so looks like the least attractive resolution. This type of decision-making would no doubt send the message to communities in the developing world that those populations that are most similar to INGOs in terms of values are more likely to receive aid. Once this becomes known, it would create an incentive for communities to alter practices in advance in order to appear more compatible with the work of certain organizations. This would then become simply a large-scale version of the unacceptable choice situation discussed earlier, and so would also be an objectionable course of action for an INGO to take, especially as a matter of ongoing policy.

OBJECTION

No doubt my (admittedly tentative) resolutions of these cases will be quite unsatisfying to many people who might be described as holding broadly liberal values or convictions. Some proponents of universal human rights and some liberal feminists would certainly object that these resolutions either reinforce existing injustices or at least do nothing to rectify them. Such critics also might suggest that when those with more power—such as elders, heads of households, and traditional healers— support a policy choice, it is highly likely that the policy will work to the detriment of the less powerful, such as women, children, young people, and the very poorest. According to this view, INGOs should set an example by insisting on human rights and equal treatment, rather than accommodating cultural practices and beliefs; if their insistence contributes to the elimination of certain unjust practices, then so much the better.

There can be no question that if INGOs consult only the people in power, any agreement reached will merely reinforce existing social inequalities. This difficulty can be rectified by seeking out those who have less power in the community and including their views in the discussion, as well as by paying special attention to whether certain individuals are intimidated into supporting certain views, and whether certain individuals are not genuinely trying to reach agreement based on shared principles but are, say, pressing for a resolution that personally benefits them (for instance, financially). By assessing these aspects of the claims made within negotiation, INGOs can choose to build an agreement that does not merely reflect the interests of the powerful, but that reflects the legitimate concerns of all participants—including INGOs' own staff and likely patients from different parts of the community.

To some, this initial response will seem to take an overly rosy view of the possibilities that the model allows for. In particular, Susan Moller Okin suggests that "interactive" approaches based around dialogue and agreement (such as the one I have proposed) have limited value in contexts where "oppressed people have internalized their oppression."[51] She claims that "we are not always enlightened about what is just by asking persons who seem to be suffering injustice what they want."[52] This is because oppressed people's view of what they want or what is good for them is often distorted by "adaptive preferences." This phenomenon occurs when oppressed people learn to desire things that are in fact harmful to them, or at least to accept their inferior social position. They may also adjust their hopes and expectations such that they desire only small improvements in their overall condition, rather than demanding more robust fulfillment of their human rights.[53] Therefore, according to Okin, engaging in reciprocal negotiation—even with a varied group of community members, and even taking into account possible intimidation and conflict of interest—will inevitably favor the (unjust) status quo. This means that the only just way for INGOs to determine project designs and goals is to make sure they are in line with the demands of human rights and equality.

It is important to remember at this point that there are dissenters in virtually every cultural context. As Ann Cudd notes, "where there is a request for information or debate from within the culture, there is an opportunity for respectful and rational dialogue."[54] Indeed, it does not seem reasonable to assume that all people with power will espouse the same views, or that (some of) those with less power will not see perfectly well that they are at a disadvantage. This takes some of the sting out of Okin's objection, since her view seems to be premised on the idea that most members of a cultural group are not themselves critical of it. *[may not be vocally so]*

Okin does admit that sometimes oppressed people criticize the institutions and practices that oppress them. However, I agree with Alison Jaggar when she says (referring to Okin and Martha Nussbaum) that "their practice of raising questions about adaptive preferences and false consciousness only when confronted by views that oppose their own encourages dismissing the views of the poor and oppressed without considering them seriously."[55] Considering such views seriously entails understanding the significance of the beliefs and practices at issue from the point of view of the person who holds the view, and, in an empathetic manner, trying to see what violating certain beliefs or norms would mean for that person in terms of social consequences, personal distress, guilt, and shame.[56] It cannot be reasonable for INGOs to take a dismissive attitude toward recipients' important moral and religious beliefs; after all, the aim of the exercise is to benefit them. And one worthwhile way to benefit recipients is to improve their health in ways *they deem to be important*, without also imposing extra costs on them.

Granted, this may not be the most efficient way to help them; it also may not abolish those cultural practices that, from a Western perspective, are pernicious. I do not mean to make light of either the concern for wasted money and time, or the necessity to work against injustice. Nevertheless, it seems unnecessary to trade off treating people with respect (no matter what their beliefs are) in favor of maximizing efficiency, when there are many ways to improve the health and welfare of beneficiaries that are consistent with mutual respect.[57] If an INGO is satisfied that a project designed through reciprocal negotiation can provide real health benefits at a reasonable cost, then it should choose this instead of a more efficient but coercive (and so imperialist) alternative. In effect, the duty to avoid coercive offers should act as a constraint on what INGOs view *as* viable alternatives in the first place.

Finally, when considering how to provide health care in developing countries, we would do well to keep two things in mind: (1) In the West, we do not generally think it is the province of health care providers to reform the desires and beliefs of their patients, even where this would create better health outcomes for less cost; and (2) we are not in agreement in our own societies about how to prioritize health care for different groups under resource constraints. If we are willing to admit that there is continued controversy in our own societies around what health services should be offered, to whom, and on what basis, then we should not be surprised to find that this is also an issue where INGOs may not see eye to eye with beneficiaries from other cultures. The right solution to these difficult issues—both at home and abroad—is to engage in debate and try to reach mutually acceptable arrangements, rather than to make paternalistic assumptions about the interests of others, or to make coercive offers based on only the providers' assessments of health needs.

ok, but only up to a point

CONCLUSION

It is the argument of this paper that it is inappropriate for INGOs to impose themselves on beneficiaries in ways that result in distress, coercion, and (perhaps) cultural loss. That said, the reciprocal negotiation model does not require INGOs to *endorse* beliefs, values, or practices that they find morally suspect, and it does not prohibit them from trying to persuade people of their views by reasoned argument. Reciprocal negotiation merely requires INGOs to take steps to ensure that their recipients are not forced to choose between upholding their values and obtaining the health services that they desperately need.

ACKNOWLEDGMENTS

I would like to thank audiences from the Department of Philosophy at the University of Sheffield and the Department of Medical History & Ethics at the University of Washington for their feedback on earlier drafts of this paper. I would also like to thank the Social Sciences and Humanities

Research Council of Canada for funding the postdoctoral fellowship I held from 2006–2008, when this paper was originally drafted. Finally, I extend my thanks to the editors of this volume for their valuable comments and support.

NOTES

1. Sen A. (1999). *Development as Freedom*. Oxford: Oxford University Press, p. 240.
2. "INGOs" should be understood to refer to those organizations that originate in the West and provide health services directly to beneficiaries. I am not here referring to either those nongovernmental organizations that originate in the developing world (and so typically serve the communities in which they originate) or those concerned only with monitoring and reporting, such as Human Rights Watch or Amnesty International.
3. For a nice discussion of the various forms this type of charge can take, see Ibhawoh B. (2007). Human rights INGOs and the North–South gap. In D.A. Bell and J. Coicaud, eds. *Ethics in Action*. Cambridge: Cambridge University Press, pp. 79–97. He points out that it is no longer plausible to insist that criticisms of INGOs generally originate with opportunistic members of the ruling elite, since even those people who accept that INGOs play an important role in delivering vital services have concerns about their priorities and methods.
4. Where such practices are grossly unjust, it might seem implausible to argue that it is objectionable to contribute to their disappearance. As I will point out later, however, if the practice is valued, then it is up to the community itself and not to outside agencies to decide whether the value the local community places on it justifies sustaining the practice, or whether it ought to be abolished.
5. Both examples have their basis in real situations that have been faced by INGOs, although of course they have been considerably fictionalized. INGOs that deliver health services often run into some of the most serious issues of intercultural conflict. The reason for this (as has been discussed widely in the bioethics literature) is that matters concerning the body (especially women's bodies), birth, death, and the endurance of suffering are likely to be surrounded by a host of important customs and traditions. In addition, these are matters about which people usually have strong beliefs, and to which they may have profound emotional reactions.
6. This is my own definition. It combines the commonplace "complex whole," "semiotic," and "particular way of life" understandings of the concept. These elements are all cited by John Tomlinson in his 1991 book *Cultural Imperialism: A Critical Introduction*. London: Pinter, pp. 4–7. It also borrows heavily from a definition formulated by E. B. Tylor, a prominent 19th-century British anthropologist. This definition also includes the important notion that practices are transmitted to each new generation by the last and so can usually be expected to continue over many generations. For instance, see Fairchild HP. (1966). *Dictionary of Sociology*. Totowa, NJ: Littlefield, Adams and Co., pp. 80–81.
7. Tomlinson J. (1991). *Cultural Imperialism: A Critical Introduction*. London: Pinter, p. 6.
8. Ibid., p. 6.
9. This characterization of the dominant values of the West is largely owed to Segun Gbadegesin. (1998). Bioethics and Cultural Diversity. In Kuhse H and Singer P, eds. *A Companion to Bioethics*. Oxford: Blackwell, p. 30.
10. Tomlinson, 1991, p. 173.
11. Sen, 1999, p. 242.

12. While this is admittedly a very rough account of what constitutes factual beliefs, I think it is adequate for the purposes of my analysis.

13. I will not distinguish between those normative requirements that are said to originate from God or religious scripture, since in both the moral and religious cases, those who hold these beliefs understand them as required, and so violating them would be wrong from their point of view. Whether these beliefs rest on religious authority or something more secular is not strictly relevant to my discussion here.

14. Zimmerman D. (1981). Coercive wage offers. *Philosophy and Public Affairs* 10, p. 124. Note that in this passage, Zimmerman is merely articulating the standard view but does not himself subscribe to it; he allows that some offers can be coercive.

15. McGregor J. (1988–89). Bargaining advantages and coercion. *Philosophy Research Archives* 14, p. 45.

16. Both Virginia Held and Daniel Lyons note that offers have "prices." According to Held, gifts do not have conditions attached, but offers characteristically do. She notes that "[i]t seems fair to say that gifts are among the rarest of entities in the political and public realms." See Held V. (1972). Coercion and coercive offers. In Pennock JR and Chapman JW, eds. *NOMOS XIV: Coercion.* Chicago: Aldine, p. 57, and Lyons D. (1975). Welcome threats and coercive offers. *Philosophy* 50, p. 427.

17. McGregor, 1988–89, p. 35.

18. O'Neill O. (2000). Which are the offers **you** can't refuse? *Bounds of Justice.* Cambridge: Cambridge University Press, p. 93. The "compliant option" here refers to the one the person making the offer wants the recipient to choose. For example, in *Traditional Healers,* the INGO wants patients who could benefit from antiretrovirals to cease taking certain traditional remedies and accept the antiretrovirals, as opposed to (b) refusing the antiretrovirals and continuing with the traditional remedies. Even though the patients might be better off all things considered by opting for the compliant option, it might nevertheless be "unrefusable" on O'Neill's terms and so objectionable. O'Neill, 2000, p. 91.

19. Ibid., p. 91.

20. Ibid., p. 93.

21. Ibid., p. 91.

22. Zimmerman, 1981, pp. 128, 130.

23. Not everyone who agrees that offers can be coercive also agrees that coercive offers can be made *unintentionally.* Nevertheless, O'Neill's intuition that this is possible seems plausible to me.

24. McGregor, 1988–89, p. 35.

25. Lyons, 1975, pp. 428–429.

26. Note that it is not important whether recipients have reason to feel remorse, regret, etc. or whether they merely think (mistakenly) that they do. Either way they are coerced since they are put in the position of choosing (by their lights) the lesser of two evils.

27. At least for the purpose of this paper, I take an ideally just society to be one that approximates the requirements of Rawls' theory of justice. See Rawls J. (1971). *A Theory of Justice.* Cambridge, MA: Harvard University Press.

28. Following Rawls, Sreenivasan calls these two branches of non-ideal theory "partial-compliance theory" and "transitional theory" respectively. Sreenivasan G. (2007). Health and justice in our non-ideal world. *Politics, Philosophy & Economics* 6, p. 220.

29. For instance, it is widely recognized that college admissions procedures that favor historically disadvantaged groups are just, given the unjust circumstances in which they are applied.

30. Sher G. (1997). *Approximate Justice: Studies in Non-Ideal Theory.* New York: Rowman and Littlefield Publishers, p. 1.

31. Clearly more needs to be said both about what happens when no easier terms are available or acceptable to the INGO, and so it cannot offer easier terms, and what happens when the other essential services that can be offered do not provide equivalent benefits. The requirement to avoid coercion has the most force when either the same or other roughly equally beneficial services can be provided on terms acceptable to beneficiaries. I take it that this will usually be possible, given that aid beneficiaries tend to be very badly off.

32. To be fair, some INGOs do recognize this, at least in some cases. However, they are also sometimes criticized for treating people without regard to their beliefs. For instance, an INGO may be criticized by donors and others for helping people from the "wrong" or "undeserving" side of a conflict. This type of criticism puts pressure on INGOs to refrain from helping those who have "objectionable" beliefs or histories.

33. Where this can be avoided. Sometimes people feel shame or distress simply because they are sick, or because the treatments that can help them are invasive or debilitating in the short term. Clearly these feelings can't always be avoided by restructuring the way services are delivered or by offering different health services.

34. Mayer R. (2007). What's wrong with exploitation? *Journal of Applied Philosophy* 24(2), p. 139.

35. We do this all the time when we pass laws that require everyone to, say, wear a bicycle helmet or seatbelt.

36. I am not suggesting that intra-cultural coercion would be a trivial matter, just that cultural imperialism is distinct from this in certain ways.

37. I say "broadly speaking," because it has been suggested to me that Oxford-educated charity workers working with the urban poor in Britain might encounter similar difficulties in terms of setting project goals for recipients who have ideas about what is in their interests that differ from the charity workers' ideas. My response to this scenario is twofold. Either what is going on in this case is that the parties have differing opinions from within the same culture of what constitutes the recipients' interests, or there is an element of "class culture" in play here that is analogous to the INGO case.

38. I don't think it is very likely that any INGO staff and a group of potential beneficiaries would be so different as to have no common ground at all from which to begin discussion. However, there may well be extreme cases where the area of overlap is quite small. I am partly responding to what is essentially a worst-case scenario here.

39. Shue H. (2004). Thickening convergence: human rights and cultural diversity. In Chatterjee DK, ed. *The Ethics of Assistance: Morality and the Distant Needy*. Cambridge: Cambridge University Press, pp. 217–241.

40. Waldron J. (1987). Nonsense upon Stilts? —a reply. In Waldron J, ed. *Nonsense upon Stilts: Bentham, Burke, and Marx on the Rights of Man*. London: Methuen, pp. 151–209.

41. These conversations typically take place with the aid of translators, which of course makes the whole enterprise more difficult and time-consuming. However, this added dimension need not prevent the process of dialogue from working in the long run. Also, it might not be either appropriate or possible for all representatives to meet at the same time, which would require the INGO to go through several iterations of the proposed design *with each group* until one could be accepted by all. This again would add to the cost of the process but does not present an insurmountable challenge to obtaining agreement.

42. Carter MA and Klugman CM. (2001). Cultural engagement in clinical ethics. *Cambridge Quarterly of Healthcare Ethics* 10:20.

43. Shue, 2004, p. 232.

44. Becker L. (1986). *Reciprocity*. New York: Routledge & Kegan Paul, p. 89.

45. Ibid., p. 105.

46. Ibid., p. 106.

47. My account draws on Seyla Benhabib's characterization of both "universal moral respect" and "egalitarian reciprocity" in her 2002 book *The Claims of Culture*. Princeton, NJ: Princeton University Press, pp. 19, 107, 115–120. However, ultimately my view differs from hers substantially, since the contexts in which we are discussing cultural conflict are importantly different. She is concerned to articulate a deliberative view appropriate for members of multicultural democratic states, which presupposes a certain set of background conditions not present in the INGO case.

48. I try to spell out what this process might look like with respect to my two examples in the next section.

49. Another reason why such an impasse is unlikely is that within communities there are differing views, and usually it will be the most extreme ones that generate such intractable conflicts. However, since usually not everyone in the community holds these views, it is likely that some agreement can be reached that satisfies most people.

50. The resolutions I discuss in this section should be taken as merely speculative because the model I have proposed is predicated on the actual act of negotiating in good faith between real people with real interests, beliefs, and commitments. It is also worth noting at this point that the model does not suppose that it is enough for an INGO to consider what it would be "reasonable" for a population to accept, without engaging that population in the reciprocal negotiation process. This is because it is not sufficient justification for interfering with someone's community and his or her fundamental choices that the person doing the interfering thinks it is reasonable to do so.

51. Okin SM. (1994). Gender inequality and cultural differences. *Political Theory* 22, p. 19.

52. Ibid., p. 19.

53. As Alison Jaggar notes, this is sometimes characterized as "false consciousness" by Western feminists influenced by Marxist critique of ideology. Jaggar A. (2005). 'Saving Amina': Global justice for women and intercultural dialogue. *Ethics and International Affairs* 19, p. 58.

54. Cudd AE. (2005). Missionary positions. *Hypatia* 20, p. 171.

55. Jaggar, 2005, p. 69. Jaggar's point here seems particularly true of Okin when the view in question has its origin in religion. See Okin. (2003). Poverty, well-being and gender: what counts, who's heard? *Philosophy and Public Affairs* 31, p. 313. See also Nussbaum M. (1992). Human functioning and social justice: in defense of Aristotelian essentialism. *Political Theory* 20(2), p. 204.

56. Whether or not beneficiaries' moral and religious beliefs are true is beside the point here, as it is when treating patients in the West. What someone regards as a benefit should also be taken seriously, although some questioning is appropriate in cases where the procedure in question is life-threatening or seriously debilitating. Compare, for instance, the common practice of male circumcision in North America, which is not usually questioned and yet is also thought to have limited (if any) health benefits. I have already noted that what an INGO regards as mistaken factual beliefs should be dealt with in a different manner.

57. This is a reiteration of Lyons' point, noted earlier, that INGOs can usually offer the same or other, roughly equally beneficial health services under easier terms than those put forward in a coercive offer. Why not do this when it can no doubt be done in the vast majority of cases, and at the same time it circumvents the danger of cultural imperialism?

CHAPTER 10

༄

Global-Health Impact Labels

NIR EYAL

INTRODUCING GHILS

This chapter proposes a labeling system that rewards corporations for promoting global health. What I call a *global-health impact label* (GHIL) would be an accreditation, certifying that some company, product, service, or brand had been rigorously reviewed by an independent body and was found to have a sufficiently favorable impact on *global health*, in a specific way or overall, and was decently ethical otherwise. "Global health" herein designates the basic health interests of the global poor.

A GHIL could recognize, for example, a South African construction company that provides generous HIV education, testing, and medication to its employees;[1] a Namibian–German beer producer that annually funds training for more than a certain number of clinical officers and anesthetists to conduct cesarean sections in Namibia or otherwise demonstrably helps cut Namibian maternal mortality by more than a certain percent; a Chinese toy producer that meets International Labor Organization workplace recommendations on safe noise levels, as shown by its fulfillment of International Organization for Standardization benchmarks; a British–U.S. pharmaceutical company that does *not* lobby for devastating patent protections in developing countries. GHILs could be awarded by international organizations, by nonprofits (as I shall assume), or by still other types of organizations.

This section explains what GHILs would be, how they might promote global health, how they would differ from somewhat similar mechanisms, and what forms they might take. The second section notes the advantages of GHILs over other avenues through which ordinary citizens may attempt to promote global health. The third section describes a specific GHIL designed to reduce the contribution of medical tourism to critical shortages of health workers in the public sectors of some countries. The fourth section defends GHILs against potential objections.

Let me start then by elaborating on what GHILs would be. A GHIL would assess the impact of a company, or of a specific brand or product, on global health, according to the certifiers' assessment. Different GHIL certifiers might assess global health differently and reach divergent judgments. Certifications could also take different forms—for example, pass/fail, a point rank, platinum/gold/silver/uncertified, or a complex index. Later I explain how we can tell good GHILs from bad ones, and coordinate between good ones. A certified company could display its GHIL on products. Certifiers could state a company's rating on their websites and, through downloadable apps, on online shoppers' computer screens and smart phones. They could add detailed explanations of the grounds for the company's rating.

GHILs could improve global health primarily because some people care about global health enough to want to consume and invest in ways that promote global health. They want, specifically, to encourage companies to promote global health. But it is hard to tell which companies promote global health and which do not, and therefore, to decide which companies to reward and which to penalize, for instance, by buying or not buying their products. Good GHILs would provide that information. Professional and independent certifiers that monitor companies' real impact on global health, and make it transparent to the public, would increase consumers' and investors' knowledge and companies' accountability. That could help motivate even purely profit-seeking companies to promote global health. Receiving a GHIL could separate them from their competition in the minds of ethically inclined consumers and investors, as well as regulators, workers, and business partners. It could also help them avoid negative consequences by saving them from consumer boycotts, investor activism, union protest, and hostile regulations and taxes (although GHILs would not provide legal immunity).

Consumer and investor activism can be powerful tools for social change, as evidenced by the Montgomery bus boycott, the Shell *Brent Spar* boycott, and divestment from apartheid South Africa. Such activism is on the rise. Sales of fair trade-certified products grew an average of almost 40% per year in the 5 years from 2005 to 2010, with exceptional growth even during the global recession.[2] Socially responsible investments are now valued at an estimated $2.71 trillion, more than a tenth of the U.S. investment marketplace.[3]

The main way in which GHILs could advance global health, then, would be by mobilizing and focusing consumer and investor activism, to help consumers and investors advance global health more effectively. In addition, clever certification criteria could educate people and companies about the value of global health and on effective ways to promote it. For example, a GHIL for companies that helped train mental health workers in countries where they are in severe scarcity could draw attention to the need to train and hire more such workers. What is more, the publication of major GHILs' certification criteria could serve to improve coordination among different initiatives for global health. For example, if a large, successful GHIL accreditor said that it would start certifying companies for hiring anesthetists in the small kingdom of Lesotho, that might give other organizations the assurance they need to focus on other countries.

While GHILs could work alongside other interventions for global health, there are some important differences between GHILs and other mechanisms through which corporations and consumers may seek to improve global health. First and foremost, GHILs would assess impact, and do so independently. By separating the entity assessing the impact of a practice from the entity engaged in the practice, GHILs would offer consumers more reliable information on which to base decisions regarding companies in both developing and developed countries. For example, in the developing world, studies show that "philanthropy generally gets an even higher priority as a manifestation of [corporate social responsibility than in Europe and the United States]," and many companies offer medical aid in their areas of operation, ranging from money and medication to free-immunization camps and telemedicine services.[4] But the aid is rarely monitored—externally *or* internally.[5] Interventions that help a few sick children are frequently pursued even when other, less visible but arguably more fruitful interventions, such as lobbying a reluctant government to fight AIDS harder, would have had higher public health impact.[6]

In rich countries, corporate social responsibility and charitable efforts are increasingly systematized and reviewed. For example, giant retailer Walmart's 2009 ethical standards for its suppliers around the world cover employees' health issues. They ban child labor and forced labor, discrimination against the disabled, long work hours, and unhealthy and hazardous sub-contractor facilities. However, compliance audits are run by Walmart itself, which does not assess the broader impact on global health.[7] Walmart's 2010 guidelines on cutting carbon footprints have wide-ranging global health implications, but audits remain internal.[8] Corporate partners in the (Red)™ network like American Express, Apple, Starbucks, Nike, and Gap donate a portion of profits from some regularly priced products to the Global Fund to Fight AIDS, Tuberculosis and

Malaria—reportedly $170 million as of September 2011. However, it is hard for ordinary consumers to verify reports and to tell how companies like Nike, Gap, and Apple, often accused of subcontracting sweatshops, affect global health overall.[9] Some fair trade label requirements promote the health of some of the global poor by requiring occupational safety and a premium for development in workers' communities. But insofar as the monitoring organization sells certified products, certification is not entirely independent.[10]

Apart from providing a clearer separation between producers and certifiers, GHILs would be distinguished by what they certify companies for: global health impact. When certification standards do not bear on health, or on global health, they are not GHILs. As an example, environmental certification standards that neglect the health impact of pollution, say, or assess it mainly inside developed countries, and do not touch on non-environmental determinants of global health, are not GHILs.[11] Certifications for international charities that focus on procedural or formalistic matters like organizational democracy and transparency, not primarily on global health, are not GHILs. Likewise, as the terms for these practices suggest, some *ethical* investment and *fair* trade certifications focus primarily on companies' ethics, not on their global health impact. Many focus exclusively on companies' treatment of their own employees' health, presumably on the assumption that companies have special ethical responsibilities to employees. GHILs would focus primarily on global impact, paying only limited and instrumental attention to procedural aspects and to company ethics.[12]

Some new rankings for international charities, for example, GiveWell's, focus on reviewed charities' impacts on global health (usually assessed in disability-adjusted life years).[13] GHILs would play a similar role for the non-charitable sector, where independent review is arguably more needed, and less likely to "crowd out" genuine altruism.[14]

GHILs could take a number of different forms. For example, a GHIL could reflect the certifiers' assessment either of a company's overall global health "footprint" or of a select component, such as workplace safety, or the company's impact on a single Millennium Development Goal in one district. The selected component might encapsulate what is most important and easily verifiable, what certifiers' expertise enables them to assess, or what is most likely to motivate consumers and investors. It would be possible to focus on specific success indicators—proxy factors that are easy to inspect, quantify, and assign responsibility for, such as the degree of workplace noise-level reduction, the number of mosquito nets supplied (a measure available before health impact becomes available for assessment), or the reduction in maternal mortality in a randomly selected district where a company volunteered trained personnel (necessarily a retrospective assessment).[15]

In addition, different GHIL-certifying organizations might use different methods for health impact assessment, divergent both technically and substantially. For example, the assessment could take in, or ignore, the equality of health increase distribution, whether health increases for the young or the old, and the varying economic effects of improved health.

In assessing a company for certification, GHIL examiners could evaluate the company straightforwardly, or relative to its means, to what it charges for certified products, or to industry norms. They could assess only companies' recent health impact, or continue to penalize companies' earlier violations. They could check whether a company has a certified supply chain, or evaluate the company on its own, so to speak. GHIL certifiers could be very rigorous, requiring frequent and detailed reports and using random audits, complaint hotlines, and the frequent denial of certifications; or they could be somewhat more lenient, which might preserve some companies' genuine altruism and help small companies obtain GHILs. They could use the same criteria over many years, or gradually elevate the requirements as the industry meets earlier requirements. GHILs could be issued by independent nonprofits (as I shall assume), by governments, or by for-profit companies or trade associations. They could serve exclusively private consumers and investors, or also public agencies.

Finally, second-order GHILs could reflect assessments of the quality of other GHILs and their certifying organizations, to monitor the monitors and, specifically, to prevent companies from creating in-house GHIL certifiers that produced whitewashed reports. The existence of such second-order GHILs could also be designed to thwart industry attempts to disparage other GHILs unjustifiably, or to otherwise upset the GHIL system. Assessments by prestigious second-order GHIL certifiers that a company abused the system would warrant penalties like blacklisting and the concerted denial of any GHIL for that company—for indirectly undermining global health. The financial independence or the overall quality of second-order GHIL organizations could be monitored either by an independently prestigious body, or by a broad base of Internet users (see "Defending GHILs" below).

AN INITIAL CASE FOR GHILS

In 2003, a United Nations sub-commission unanimously adopted draft norms that demand, as a matter of "obligation," that:

> Transnational corporations and other business enterprises shall . . . contribute to . . .
> [civil, cultural, economic, political, and social rights'] realization, in particular the
> rights to development; adequate food and drinking water; the highest attainable

standard of physical and mental health; adequate housing; . . . and refrain from
actions which obstruct or impede the realization of those rights.[16]

The draft norms make clear that the obligations are subject to external
verification and monitoring mechanisms to ensure that they are being
met. Recent United Nations reports express similar ideas.[17] GHILs can be
seen as such external verification and monitoring mechanisms. As I now
argue, they also enjoy certain advantages over some other ways for ordinary citizens to promote global health.

First, GHILs would be relatively easy to create, but could have high
impact. Being especially feasible, they would give ordinary citizens a
greater opportunity for global health promotion than do direct investment
and work in global health, political activity, and legal action. Unlike starting a medical aid organization, for example, a GHIL system could be easy
to fund. Offering medical services costs a lot of money; issuing a label that
motivates others to do so could cost almost nothing. While it does cost
money to monitor companies' operations and to build a reputation, it costs
much less than providing large-scale health services. In the certification
industry, expenses are usually built into certification fees. Fees from large
companies that seek accreditation, owed regardless of whether their peti-
tion is accepted,[18] could potentially cover expenses for evaluating small
companies, and non-cooperating companies for blacklisting. GHILs could
also be funded through donations or government subsidies. In any event,
their relatively low cost would give GHILs an edge over Thomas Pogge's
"health impact fund," which would create incentives for pharmaceutical
companies to promote global health by offering them money to do so. To
offer a lot of money costs a lot of money; to offer a label can be cheap.[19]

Ordinary citizens could initiate and run GHIL-awarding operations.
To be sure, states can and should defend global health through many
channels: donations, diplomacy, domestic and international legislation,
perhaps even sanctions and military intervention. However, from an ordinary citizen's perspective, GHILs would compare well with trying to
influence government to do something. They would increase the efficacy
of citizens' grassroots action, bypassing the wait for busy politicians,
powerful interest groups, or the general electorate to take the basic health
rights of the global poor seriously.[20] GHIL denial could rapidly and
painlessly prod corporations to promote global health. Here, GHILs
stand in sharp contrast to court cases which, in both developing and
developed countries, can be long, tedious, and expensive (recall that a
GHIL would not provide legal immunity). GHILs stand a chance of
affecting company profits and incentive structures very soon after their
inception. By comparison, altering international laws and codes requires
especially long pleading and haggling in stagnant international forums,

where even existing legislation on social and economic health rights is rarely enforced.[21]

GHILs also have the potential to eliminate collective action problems that can plague other efforts to improve global health. In general, a single consumer, investor, or company has little impact on global health, which poses a potential collective action problem for ethically inclined consumers, investors, and companies.[22] By contrast, starting a GHIL program could move many consumers and investors who (irrationally?) overlook their small individual impacts, and that makes starting a GHIL perfectly rational. Of course, trusts, unions, and pension funds also represent many individuals, and they can exercise considerable power. But many such bodies have potentially conflicting legal duties, say, to maximize investor profits—from which GHILs are free.

State intervention can resolve many collective action problems. But it does not resolve all. When a developing country government regulates or sues companies to protect local health, companies will sometimes move to other countries. For example, legislation to make sweatshops safer in Bangladesh pushed producers to China.[23] A GHIL operation could apply globally, monitoring and responding to a company's practices in many countries—so there would be no escaping it. It therefore could also potentially overcome intergovernmental collective action problems.

Finally, given the potential impact of GHILs, they would be surprisingly difficult to suppress. Across legal systems, there are few legal constraints against issuing, selectively assigning, and promulgating a certification, provided that outright defamation is avoided. Nor are there usually laws forbidding consumers, investors, and activists from taking a certification into account in their decisions.[24] It is normally draconian to regulate what we express and how we reach our decisions. Certainly international law, which primarily binds states, not individuals, would not restrict GHILs.[25]

Notwithstanding these advantages, we do not have to choose between GHILs and more conventional ways to promote global health. GHILs could work in concert with other measures. They could enhance a general pool of incentives to promote global health. Should they ever conflict with other measures that work better in some areas, GHILs could be introduced in those areas where they would work best.

A GHIL FOR THE MEDICAL TOURISM INDUSTRY

Let me illustrate with a sketch of a specific GHIL idea. Medical tourism—travel across international borders to obtain health care—is expanding rapidly. Some experts now predict that the global revenue will reach

US$100 billion as early as 2012.[26] The general expectation is that both public and private Western insurers will increasingly encourage their patients to travel abroad for cheaper care, so the medical tourism industry is projected to continue to grow. India's medical tourism sector is expected to grow 30% annually from 2009 to 2015.[27] Currently, India and Thailand are the two global leaders in supplying services to medical tourists.

There are many good things about medical tourism, including the provision of good-quality affordable health care to tourists, and much-needed income for middle-income countries. But for countries like India and Thailand there is also a dark side. Providing services to tourists probably exacerbates the already critical shortages of health professionals in the underserved sectors of these countries. In India, for example, "There are fears . . . that medical tourism could worsen the internal brain drain and lure professionals from the public sector and rural areas to take jobs in urban centers."[28] Private urban hospitals offer very high salaries by local standards, so physicians, nurses, and other health workers leave or avoid the public sector—including rural clinics and hospitals—and work in private establishments in cities. Low health worker density is a major bottleneck to delivering essential medical interventions ranging from vaccinations to antiretrovirals; it is known to increase maternal, infant, and child mortality.[29] In India, for example, "One of the major factors responsible for poor performance in hospitals is the absence of personnel."[30]

Can this problem be remedied? On the face of it, the prospects seem grim. For what could be done about a perfectly legal trend, driven by Western tourists' demand for health services and tourist hospitals' supply? There is nothing illegal about paying health workers handsomely, nor about accepting a high salary. Destination country governments might have offset this problem if they invested the tax revenue from tourist hospitals in public and rural, health. But expecting that they do so is not always realistic (see "Political equality" below).

Imagine, however, the following alternative system. Suppose all tourist and other private hospitals in countries like India and Thailand paid their health workers 12 months a year, but expected them to spend a quarter of that period offering expertise and services in resource-poor settings in the region. An internist or a nurse might staff a remote rural clinic for 3 months every year or provide similar services via telemedicine. A cosmetic surgeon might re-train to perform cesarean sections, to function as a general practitioner, or to coordinate services in a public urban hospital for a week every month. Many variations are conceivable. The fundamental idea is that these health workers would help alleviate their region's critical health worker shortages, but would keep their existing salary levels so as to keep them motivated to stay.

So far so good. But why would tourist hospitals accept such an arrangement? Suppose that a national government tried to force hospitals to accept it. Competitors in other countries might then be able to offer tourists better deals. The country might lose out in the global competition against providers in India, Thailand, Malaysia, the Philippines, Cuba, Dubai, Germany, Israel, Belgium, and elsewhere. Perhaps for this reason, trade ministries tend to oppose regulating tourist hospitals,[31] and Indian regulations requiring tourist hospitals to reserve beds for the local poor in return for generous subsidies are seldom honored or enforced.[32] A collective action problem exists, then: client demand tends to advantage the destination country that regulates its providers the least. Therefore, even if all destination countries support regulation, there is a danger that none will do so.

Coordinated international regulation might have resolved this collective action problem between developing countries, but international regulation tends to be toothless. In recent history, even a wholly voluntary draft code of practice on international health worker recruitment has encountered great difficulty in international forums.[33]

Surprisingly, a GHIL could drive many hospitals to implement this arrangement. Let me explain how.

Tourist hospitals the world over seek quality certifications from Joint Commission International (JCI), an affiliate of the U.S. nonprofit Joint Commission, which certifies America's hospitals for the quality of their medical services.[34] JCI certification is key for securing Western referrals. What would happen if, in regions containing pockets of critical health worker shortages, JCI added to private hospitals' hundreds of certification requirements (including a requirement to admit patients in an emergency, even if they cannot pay) the requirement that medical staff spend a quarter of their salaried time working in the region's underserved sectors? That would turn JCI certification into a GHIL, and might thereby effect the arrangement just described. Since losing JCI certification could mean losing substantial Western referrals, such altered standards could in principle transform hiring policy. After all, other JCI requirements have certainly instilled expensive quality-control measures.

Even if the relevant GHILs were not woven into regular JCI accreditation, a separate GHIL, offered by JCI or by an independent certifier, could assess hospitals' impact on health worker availability in underserved communities. It would thereby also somewhat affect referrals.

One way to encourage formation of GHILs for the medical tourism industry would be to link them with government health payment systems in tourists' source countries. The personal cost to a voter or administrator of requiring systems like Britain's National Health Service, once they start sending patients abroad *en masse*, to work solely with certified providers is small—potentially just casting a vote. Here, a GHIL could serve as a focal point for patients' and citizens' campaigns to turn public systems ethical.

But GHILs for the medical tourism industry could work even in a private insurance framework. They could work even though consumers behave far less ethically than they claim to do (see "Defending GHILs" below), and even if most medical tourists are indifferent to the impact of their choices on poor communities. GHILs could tackle other points on the supply chain, moving many private insurers to work only with certified hospitals, even if most of their clientele is willing to go to uncertified hospitals. GHIL certification—or decertification—could influence these insurers' *general* reputation, which affects much more than their medical tourism departments. Many insurers would find it advantageous to sacrifice the small direct profit from working with cheap uncertified providers and middleman companies, if working only with certified ones shielded them from damage to their general reputation. Such damage could come, for example, from competitors' advertisements accusing them of increasing child and maternal mortality by hiring uncertified suppliers. Damage to general reputation would affect many operations of an insurer, not just its medical tourism operations.[35]

Consider an analogy. Even if many Marks & Spencer shoppers would buy the best-tasting tea over fair trade tea, Marks & Spencer, perhaps considering its *general* reputation, now sells only fair trade tea and coffee, and that presumably transforms the incentives for middlemen and for producers. Likewise, even if few Walmart consumers mind whether the dairy product they purchase was made from cows treated with artificial growth hormone, and with negative environmental externalities, Walmart itself—perhaps wary of upsetting other potential consumers and encountering hostile activism and hostile regulation—is leveraging its market power to transform its own, and America's, dairy supplier practices.[36]

In addition to GHILs for tourist hospitals, there could also be GHILs for middleman organizations that work solely with certified hospitals, and for Western insurers and Western physicians who refer patients solely to certified middlemen and certified hospitals. If even a single link on this potentially long supply chain is heavily affected, the entire chain could be transformed. While a market niche would remain for cheap, uncertified hospitals, heavy impact on even a single link could prod most suppliers to try to secure a GHIL. This is essentially how Kimberley Process certification is said to have quickly stemmed trade in "conflict diamonds" worldwide, sharply decreasing Africa's natural resource-related conflicts. The diamond industry and consumers were evidently willing to trade in conflict diamonds; but enough governments, concerned about diamond-driven conflicts and about the bad reputation of having refused to cooperate, came to certify conflict-free diamonds and to ban trade in uncertified diamonds. Because governments were an essential link in this particular supply chain, this reportedly cut down the share of conflict diamonds from up to 15% of world diamond supply in the 1990s to nearly zero in some later years. GHILs that tackled essential links on corporate supply chains could in principle have similar impact on industries.[37]

It is thus possible that carefully designed and tested GHILs for the medical tourism industry would decrease the contribution of that industry to critical health worker shortages in tourist destination countries.

DEFENDING GHILS

I now present some possible objections to my proposal to promote global health through GHILs, and defend the proposal against them.

Efficacy

Scholars' skepticism about the efficacy of corporate social responsibility is profound, and it might be thought to cast doubt on GHILs as well. GHILs might be seen as attempts to reassure corporations that global health-related social responsibility on their part is likely to yield profits. In a related vein, several years ago, Kofi Annan, then the Secretary General of the UN, assured American business representatives regarding a central element of global health: "by joining the global fight against HIV/AIDS, your business will see benefits on its bottom line."[38] However, as Joshua Margolis and James Walsh point out, to be responsible, Annan's assurance required at least some positive association between corporate financial performance and corporate social responsibility. Their extensive literature review calls such association into question.[39] David Vogel is also skeptical about the profitability of corporate social responsibility: "A company's degree of responsibility or irresponsibility has rarely affected its sales, its attractiveness to potential employees, or its access to capital."[40] With regard to ethical investment, Eric Becker and Patrick McVeigh give a stark summary: "The historical performance of socially screened mutual funds supports the thesis that social screening does not affect the performance of stock portfolios."[41]

If corporate social responsibility isn't truly profitable, it might be argued, how can GHILs offer businesses enough assurance to alter their performance on global health? For example, how could a medical tourism GHIL give hospitals sufficient incentives to provide expensive services in underserved areas?

These questions rest on a mistake. In themselves, GHILs would not be examples of corporate social responsibility, and they cannot be dismissed on the ground that social responsibility has so far had little impact on consumers and investors or on company profits. If anything, GHILs would be intermediary tools that might make corporate social responsibility profitable. Put differently, the future correlation between corporate action for global health and corporate profit is partly up to us. Initiating a GHIL could strengthen that correlation, by introducing new sticks and new carrots for corporate performance on global health. A GHIL would alter the course of action that a corporation must take in order to maximize profit. The currently limited profit generated by corporate social responsibility may therefore provide all the more reason to introduce GHILs.

In fact, GHILs could partly address some of the reported barriers to higher profit for social responsibility. David Vogel writes, "One of the problems for [socially responsible investment] is that the information is often given by the firms themselves, and subject to much manipulation or even deceit with impunity."[42] In other words, some companies that currently ask for our custom on the ground that they are socially responsible are in fact quite irresponsible. Consumers and investors, suspecting this to be the case, remain indifferent to companies' claims to social responsibility. Even genuine disclosure and transparency, complete or "targeted,"[43] would not suffice for effective monitoring by investors and shoppers who lack the time and the expertise to gather and analyze all relevant data.[44] But a good GHIL would involve rigorous third-party monitoring; independent experts would gather and analyze data, restoring reliability. They would translate their conclusions into clear operational signals for busy consumers and investors—for example into simple green, amber, and red labels, displayed on the certifiers' website, on collaborating shopping and consumer report websites, on the iPhones of shoppers who use special "apps" to scan products' barcodes, or, perhaps with government enforcement, on products themselves.

Were GHILs to prove effective, it is possible that some companies would create in-house GHILs to send false signals about company efforts.[45] Second-order GHILs would thwart this development only to some degree, because companies could also create in-house second-order GHILs that would commend their in-house first-order GHILs and vilify independent critics. This is a serious complication, but it could be resolved in two ways.

In what we might call the *supervisor model*, an independent, reputable body—such as WHO, the UK Department for International Development

(DFID), or the Carter Center—would assess GHILs' independence from commercial interests. Such a trusted body would accredit GHIL certifiers either directly or by accrediting the higher-order GHIL overseers that monitor them. To gain accreditation in this system, GHIL certifiers would have to establish both that their standards and practices are sensible and their funding is truly independent from relevant commercial interests, and that they expect similar independence from any GHIL certifiers that they monitor. The supervisory body's accreditation would thus signal a GHIL certifier's quality and independence from relevant commercial interests. Naturally, companies certified by accredited certifiers would advertise their status and its importance. After some years, being accredited as a reputable GHIL certifier could become necessary for a GHIL certifier and the labels it issues to have traction with consumers and investors. This might dissuade corporations from creating in-house GHILs in the first place.

The supervisor model may seem to burden the supervisor with the task of monitoring thousands of GHILs, but the burden would remain limited. First, corporations would be wary of getting caught supporting fake in-house GHIL panels, so they would set up fewer of them; therefore the reliability of the average GHIL to be monitored would be higher. As the system proved more and more reliable, less elaborate monitoring would be needed. Second, the supervisory body could depend on higher-order GHIL panels to do the bulk of the monitoring work. It would need only to certify the quality and the commercial independence of the highest-order GHIL in the pyramid of downstream certifiers. Lower-order GHIL certifiers would need to establish their financial independence to a GHIL organization that the supervisory body approved.[46]

An alternative to the supervisory model is the *collaborative model,* according to which grassroots Internet users would thwart commercial interests and enhance the quality of influential GHILs without centralized supervision. On this model, all users of the web would be invited to rank GHILs, not only on financial independence but on many proxies for positive impact on global health. This e-collaboration would offer its second-order certification to decently ranked GHIL certifiers and, automatically, to any products, brands, and companies that those certifiers rank highly. Users could publicly rank and discuss each other's rankings. Crucially, individual users would be encouraged to delegate their "votes" to a user whom they trust more than themselves. For example, a user could order the website to add his or her vote automatically to any vote cast by trusted fellow user Doctors Without Borders. If many users did so, then Doctors Without Borders and any GHILs that it ranked highly (or delegated *its* votes to) would enjoy greater influence than users that had not established good reputations. Industry-funded users would find it hard to establish consistently high reputations.[47]

The collaborative model would face challenges similar to the ones faced by collaborative recommender systems and collaborative filtering systems in general. On eBay, Amazon, Wikipedia, Zagat, YouTube, and Consumer Reports, users' recommendations inform the system's later recommendations. The GHIL collaboration system could be designed to do the same. It would mobilize many Internet users so that together they would filter out relevant commercial interests. Collaborative recommender systems also face the challenge of potential manipulation by industry, through industry-sponsored reports that disguise themselves as independent. While in principle commercial interests stand to gain a lot from drowning these reputation systems with fake reports, "reputation systems appear to perform reasonably well."[48] The GHIL collaboration could potentially curtail commercial manipulation using the same methods that are being used or considered for these reputation systems. For example, the collaboration could demand that recommenders use their real names and addresses, or single lifetime pseudonyms. It could also adjust pricing in ways that decrease fake reporting, use moderators, and scan the recommendation base for suspicious patterns: a user who makes hundreds of recommendations in the course of one day, or a pattern in which A recommends B who recommends C who recommends A. GHIL collaboration systems could therefore make use of existing expertise on how to contain commercial manipulation of collaborative recommender systems.[49] Moreover, abuse attempts would identify holes in any GHIL system and would become more sophisticated. Therefore, a good GHIL system should remain flexible to become suitably sophisticated. It cannot be fully detailed in advance.[50]

Both corporate social responsibility and GHILs may seem to face a deeper problem. What if one reason that corporate social responsibility is not very profitable is that consumers and investors do not really care about it? When surveyed, consumers claim to buy ethically. For example, 84% of Americans say that they would be likely to switch to a brand associated with a good cause if price and quality were similar. However, experts believe that only 10% or less of the population actually buys ethically.[51] In a 2004 European study, 75% of respondents indicated that they would modify their purchases on the basis of social and environmental criteria, but only 3% did so.[52] Nevertheless, GHILs could also affect what consumers and investors care about. This is because beyond facilitating action on existing preferences, GHILs would prompt new basic and derivative preferences. For example, initially indifferent medical tourists might come to demand GHIL certification if certified hospitals and insurers bombard them with advertisements on the public health consequences of reckless hiring practices. To make a difference for good, GHILs would not require full compliance on consumers' part, only enough compliance on the part of consumers, investors, or regulators to make some such difference.

Research on the effectiveness of corporate social responsibility could indicate *which* forms of GHIL would work best. For example, we now increasingly suspect that negative measures (such as boycotts and investor activism) tend to mobilize companies more than positive measures like "buycotts" (deliberate purchase of virtuous companies' products).[53] This could support devising GHILs in some ways rather than others. For example, it might support publishing blacklists of products and companies that certifiers would *not* accredit. Likewise, skeptics concede that social responsibility does tend to enhance profit in particular niches—for specific products, producers, and consumer types. Hybrid cars, dolphin-free tuna, free-range eggs, the Body Shop, Britain's Co-Operative Bank, and American universities' sweatshop-free apparel and environmentally friendly buildings are cases in point.[54] It is perfectly open to certifiers to identify and focus mainly on such niches.[55] Certifiers could also focus on tapping groups of consumers and investors who are currently underutilized for ethical consumption and investment—for example, the earnings that Western workers defer for a secure retirement.[56]

In brief, where skepticism about contemporary corporate social responsibility is warranted, such skepticism could actually bolster the case for GHILs and help us identify ways to maximize their impact.

Fairness to Consumers

Thomas Pogge raised the concern that the proposed GHIL for the medical tourism industry might impose unfair financial burdens on uninsured Americans. It might prevent these often impoverished patients from receiving cheaper medical services abroad. Pogge argued that GHILs might thus be impermissible because of their negative effects on these third parties.[57]

While there may always remain a market niche for cheap, uncertified hospitals, and they could serve uninsured Americans, my basic response is that uninsured Americans are on average far more privileged than India's and Thailand's underserved rural populations. GHILs for the medical tourism industry that successfully promoted the basic health interests of these populations would therefore enhance net fairness. Generally put, a successful GHIL would enhance a basic interest of some of the world's worst-off populations. Even if it negatively affected the interests of some better-off consumers, that would be a price morally worth paying. In the rare cases in which a consumer is worse off than a GHIL's beneficiaries, certifiers could take the potential damage to that patient's health into account in deciding how much a product or a service promoted global health overall.

If priority for worse-off populations were conceded, it might be replied that other conceivable schemes would equally help the global poor without harming uninsured Americans. For example, if rich nations put enough money down to remunerate tourist hospitals that assist under-served communities, without raising prices for tourists, that would be even fairer.[58] My response is that, unfortunately, rich country govern-ments are unlikely to put down that money. I propose GHILs not as an ideal, but as a good feasible option. The appropriate comparison set for the choice to set up a GHIL comprises only those options that are open to the people who could set up a GHIL. Among these options, a system that includes GHILs does not seem to exclude any morally superior alternative.

In sum, assuming that global health is overwhelmingly important, and that there are no feasible alternatives that promote global health nearly as much without negative impact on consumers, any negative impact on them is very unlikely to trump the case for introducing a GHIL.

Market Freedom

Economic conservatives sometimes oppose corporate social responsibil-ity as incongruent with the nature and purpose of free enterprise. Could similar arguments be mounted against GHILs?

One imaginable conservative objection is that GHILs would violate market freedom. This is precisely what GHILs would not do. Third-party accreditation is a familiar market tool for signaling product quality. A GHIL is similar, only it signals global health impact. The proposed medical tourism GHIL would intentionally put pressure on producers in order to manipulate their conduct, potentially violating negative freedom,[59] but certainly not market freedom. Needless to say, the fact that consumer and investor altruism, and not greed, drives GHILs does not breach market freedom either; consistent libertarians support voluntary charity.[60]

If a government instructed its own agencies—or, through legislation, required all citizens—to buy only ethically certified products, that would violate market freedom. But GHILs need not be implicated in coercive government market intervention. For example, a medical tourism GHIL might be run by civil society actors and might accredit only private health insurers and private hospitals. Even GHILs that are implicated in coercive government intervention would not necessarily increase the extent of that intervention, merely its efficacy. Libertarian protest not-withstanding, governments already subsidize or, alternatively, tax or ban the purchase of many products; GHILs would only help governments identify the right products. Consistent libertarians mind the existence and the quantity of restrictions on citizens, not whether these restrictions promote global health.

Economically conservative opponents of GHILs might also cite Milton Friedman, for whom corporate acts of social responsibility betrayed obligations to prioritize stockholder returns.[61] But this objection would not apply to GHILs either. A GHIL could actually help conscientious executives promote stockholder returns and, at the same time, global health. An influential GHIL would transform the economic playing field. It would change the calculation of which policy would best promote stockholder returns, aligning profit with global health more closely. GHILs would enable executives to sidestep Friedman's worry.

A final economically conservative worry would be that, while a GHIL would not interfere with any of the liberties that make up a free market, it would nevertheless risk undermining a major free market benefit. The market generates collective wealth, and often consequent improvements in health, longevity, and infant survival, not only *despite* private selfish-ness, but often *thanks* to it. Take Adam Smith's butcher and baker, who produce tasty meat and tasty bread without any inherent regard for the interests of each other, only to attract each other's business.[62] If they became altruistic, they might wind up eating unsavory food: an altruistic butcher would buy bread not in the light of something he knows very well—namely, which bread *he* likes best—but in the light of speculation on what bread the baker would like him most to buy. But the butcher has little knowledge of that. Certainly in the global economy, with its numer-ous remote actors and multiply interconnected interests, knowledge of everyone's interests is limited. Even governments and certifiers lack that knowledge. As Ludwig Von Mises and Frederic Von Hayek famously held, the market's price system can facilitate information flow, resolving this economic calculation problem.[63] One may worry that the market is much less capable of resolving it when consumer preferences regard remote others. On this objection, the type of altruism expressed in GHILs under-mines efficiency.

One response to this worry is that, when it comes to basic global health, we know a lot about what would help the other people. No intimate famil-iarity with a community is necessary in order to know that decreasing toxic emissions in its drinking water and increasing the number of available health workers would tend to improve the community's health. Regional experts and local consultations can further increase certifiers' understanding of particular communities' needs. While there will always remain gaps in that understanding, the classical free market would usually do worse. In a purely egoistic market, the needs of the global poor are severely underrepresented, because of their tiny buying power. GHILs may represent their needs inaccurately, but they would still represent those needs far more accurately than the purely egoistic market does.

Personal Autonomy

When someone has no decent alternative to accepting an otherwise bad offer—for example, an offer to work in a sweatshop—we often find the situation troubling, *inter alia* on grounds of personal autonomy. While the situation need not involve literal coercion, threats, or interference with this person's liberty, he or she can be said in English to be "forced" or "compelled" to accept the bad offer, and some would call the offer "coercive."[64] A related worry arises about GHILs. They may be thought to invite investors and consumers to place formidable pressure on producers and leave them no choice except to do as certifiers demand. For example, imagine that requirements surrounding medical tourist industry hiring were woven into JCI's general standards; many Indian and Thai medical service providers might then lack economically viable alternatives to conforming to these requirements. Hence, the worry goes, they would effectively be forced to conform.

Conservatives who worry that "human rights conventions purport to regulate details of family life and basic operations of labor markets"[65] may also worry about the oppressive potential of GHILs. It is commitment to global health, not to privacy and the strictly personal, that defines GHILs. Nothing would seem to stop GHILs from laying down requirements that intrude into private lives and demand of individuals more than local and international law can or should.

These concerns notwithstanding, the effect on personal autonomy would likely remain ethically acceptable, for a number of reasons. First, while putting pressure on people might be inherently ethically problematic, putting pressure on corporations is not. The liberal worry surrounds the real freedom of people to say no, usually not the freedom of corporations, since corporations lack inherent autonomy rights.

It is true that when a corporation lacks acceptable alternatives to acting in some way, this sometimes means that its owners and employees lack acceptable alternatives to acting in certain ways. But that is not always the case. If tourist hospitals shut down, their owners, some of the wealthiest families in India and Thailand, as well as American investors, would not go hungry. They have perfectly acceptable alternatives, like starting a different business with their own capital. The sweatshop worker analogy does not extend to these people.

GHILs might affect the options available to some poorer people, too. For example, the medical tourism GHIL would change health workers' daily schedules, and some businesses that would have no option except to surrender to certifiers' pressures are small family-owned enterprises. I propose that certifiers take that danger into account and normally refrain from devising extremely demanding and intrusive standards, especially for small businesses. Consultation with relevant stakeholders could help identify and prevent excessive intrusion. However, regular market operations put much more formidable pressures on rich and poor owners and workers, including health workers. We rarely complain, for instance, when a hospital expects staff to act in particular ways in order to keep their jobs (e.g., "Go see the patient in room X," or "To continue working with us, start working in city Y, where we have opened a new branch"). If we do not object when market pressures affect small, family-owned companies for purely economic reasons, why object when they do so to promote global health?

Not only would GHILs not violate producers' autonomy, there are also reasons to believe they could enhance it. GHILs would help producers resolve a collective action problem that currently limits their autonomy by making it unaffordable for them to do what they truly want. Take the proposed medical tourism GHIL. Most health workers and hospital managers would probably be happy to assist medically needy compatriots more than they actually do. Many went into health care, rather than other interesting and lucrative professions, partly because they wanted to treat the serious medical problems of their compatriots.[66] By and large, far higher salaries in tourist hospitals are what lure health workers to serve the rich, not the poor; and the need to offer tourists competitive prices and services makes it impractical to pay staff to assist the poor on hospital time. The willingness of all to deprive disadvantaged poor compatriots of basic medical services is thus usually conditional on the link between helping the poor on the one hand, and working many more hours for much less pay on the other. GHILs for the global medical tourism industry could sever that link, and give individual tourist hospitals and their staff what they truly want: a kinder way to make nearly as much money. By placing similar pressure on competing tourist hospitals, the GHILs could make it affordable for all hospitals not to cut salaries for staff who assist the poor part-time. In this way, GHILs would enhance most service providers' autonomy, even if they leave each hospital no affordable choice except to adhere to certification requirements. They would bind providers to do what each would rather do but is currently under financial pressure not to do. Any negative impact on providers' autonomy from such a process would be superficial.

Political Equality

The power of certifiers, Internet users, consumers, and investors to manipulate social realities, sometimes in remote countries where a democratically elected government does not welcome that manipulation, raises a democratic concern. GHILs, like fair trade and socially responsible investment practices, would allow affluent people and rich nations to translate their superior buying and investment power into increased political power. Insofar as consumers and investors make a difference, their impact rests on their financial standing. In Michael Walzer's terms, GHILs might thereby extend the tyranny of money.[67] That may be considered problematic, either for associating more closely two spheres that should remain as separate as possible—money and politics—or for exporting the highly unfair distribution of money to another sphere. Compare GHILs to kickbacks that would buy support for global health from members of the U.S. Congress. Even if they served a good cause, such kickbacks would unfairly prioritize the agendas of rich donors over those of other voters, in breach of political equality.

Interestingly, on this point, libertarians, who approve of private donations in politics, may see to eye to eye with Walzerians, who usually do not. Libertarians may allege that GHILs would impose the will of individual certifiers on a majority, just as Milton Friedman thought corporate social responsibility imposes the will of businessmen who run companies on fellow citizens. For Friedman, advocates of corporate social responsibility must argue,

> that the problems are too urgent to wait on the slow course of political processes, that the exercise of social responsibility by businessmen is a quicker and surer way to solve pressing current problems . . . this argument must be rejected on grounds of principle. What it amounts to is an assertion that those who favor the taxes and expenditures in question have failed to persuade a majority of their fellow citizens to be of like mind and that they are seeking to attain by undemocratic procedures what they cannot attain by democratic procedures.[68]

Another way to express the Walzerian-Friedmanian worry is to say that GHILs give certifiers, or the people with the money for politicized consumption and investment, disproportionate power. For example, while U.S. citizens have the power to alter India's health policies, most Indian citizens cannot alter U.S. health policy. They could only fantasize about improving the health of U.S. minorities through consumer boycotts on the unhealthy junk food that impairs health in the United States.

Setting aside broader questions about corporate social responsibility in a democracy, what should we make of the specific worry that GHILs would extend national and international power imbalances? The worry is far from straightforward: by definition, GHILs would aim to have a positive impact on the basic health interests of the global poor. But better health—a longer life; better ambulatory, discursive, and cognitive capabilities; fewer worries about survival and more energy for political engagement—tends to increase political voice, standing, and impact, so much so that, for some thinkers, the reason to ensure fairness in health resource allocation is its positive impact on democratic equality.[69] Whether or not this is the primary reason to promote basic health, promoting it certainly tends to enhance democratic equality. Hence genuine GHILs would usually tend to level the political playing field.[70]

Moreover, the fact that GHILs would empower the affluent to promote global health may mean simply that disenfranchised populations would acquire an indirect political voice: thanks to GHILs, richer consumers would act as advocates for a section of the global poor. That these consumers count twice because they have money may sound awful, but it is quite benign. By analogy, there is nothing outrageous about making a woman hospitalized after an auto accident, whose husband was struck unconscious in the same accident, decide twice about consenting to possibly risky treatment—once for herself and once for her spouse, who otherwise would lack even approximate representation. While GHILs would not literally appoint consumers and investors as proxy decision makers for the global poor,[71] they would allow the rich to operate as advocates for the poor. The global poor often need that external advocacy. For example, India is an impressive democracy on many counts, but India's poor rural populations marshal little political impact, as is evidenced by public health expenditure itself. While India spends a similar percentage of GDP on health care as other developing countries, at least until recently it was a consistent outlier in spending a particularly low portion of this amount in the public sector.[72] This assessment is shared by both friends and foes of its medical tourism industry.[73]

Putting these points together, consider again the proposed medical tourism GHIL. A GHIL that would enhance the health, and consequently the political power, of India's and Thailand's disenfranchised poor would enable locals and foreigners to mobilize their economic power to advocate for populations that currently do not enjoy enough power.

If GHILs would in one way empower the poor, and in another way empower richer consumers and investors, why do I assume that, overall, GHILs would improve the balance of power? Partly because affluent people and rich nations' *failure* to buy and invest in accordance with a GHIL would not prevent the rich from having excessive political power. They could use their money to promote less progressive agendas instead. The initiators of GHILs would observe the existing imbalance of power, and appeal to the economically powerful, but would not thereby augment that power. GHILs would primarily divert the existing power of the rich toward a good cause. Overall, then, far from exacerbating power imbalances, GHILs would reduce them.[74]

Of course, GHILs are far from being perfect solutions. They would enable consumers and investors to advocate for the disenfranchised without having first asked the disenfranchised for their permission. But unappointed advocacy is likely to be more beneficial than no advocacy at all.

Good Advocacy

This brings me to a final objection: that the advocacy that GHILs would offer the poor is of such questionable quality that it might have negative effects. Bad advocacy can be worse for someone than no advocacy. Compare the structural adjustment programs of the 1980s. They were ostensibly put into place for the benefit of the global poor, but have been widely criticized for exacerbating global poverty. In the present context, the fact that the relevant accreditors, consumers, investors, and Internet users would never have been elected might initially raise the suspicion that they often would not be up to the task. No public, professional expert or stakeholder need have reviewed accreditors' certification criteria or their practical application. Nonetheless, the GHIL certifiers' decisions might control such matters as the daily schedules of Indian and Thai health workers, and upset the policy priorities of legitimate regimes.

This worry is not merely academic. To begin with, nothing prevents certifiers, consumers, investors, and Internet users from screening products and companies on the basis of moral and factual errors, a sense of cultural superiority, or vested interests. Consider the 2003 consumer boycott of French wines in America following France's refusal to support the U.S.-led attack on Iraq. At the peak, a "conservative estimate is that French wine sales would have been 27% higher if there had been no boycott."[75] More generally, "Although boycotts often are associated with liberal causes, they can come from the right as well as the left and are not unique to Western democracies."[76] GHILs would *not* by definition enhance global health, they only *purport* to enhance it. One can easily imagine certificates that promote anti-liberal global health goals—say, certificates for hospitals that refuse to perform abortions[77] or to admit patients of certain denominations.[78]

Additionally, the interests of rich consumers and investors are sometimes opposed to those of the global poor. Because certifying organizations depend on cooperative consumption and investment, these organizations might be considered especially prone to turn into what Richard Falk calls "hegemonic human rights organizations"—ones that side with the global rich against the global poor.[79] In some cases, consumers' and investors' vested interests will shape their values unconsciously, and choices that seem ethical to them might actually harm the global poor. For example, ethically minded consumers often feel obliged to support domestic products, and leading scholars have encouraged U.S. union pension funds to redirect investment to workers' geographical jurisdictions.[80] Such ostensibly ethical consumption and investment would reduce U.S. consumption of developing country products, arguably harming the global poor and their health.

Even people who are genuinely committed to promoting global health through consumer and investor action can fall prey to bias and ignorance. Causes deemed newsworthy, like tsunami and torture, move us to donate far more money than does ongoing famine, which takes more lives.[81] After its successful 1995 Shell *Brent Spar* boycott, Greenpeace admitted that one of the allegations that had drawn widespread consumer support during the boycott was false, and had to apologize to Shell.[82] Most social responsibility investment funds continue to screen out all nuclear energy and alcohol production, perhaps reflecting ignorance about their potential roles as checks on global warming and heart disease.[83] Although physicians continue to doubt the health risk, and despite the potential gains against malnutrition, fair trade certifications forbid producers from using genetically modified organisms either in the production or the processing of products,[84] presumably reflecting European consumer anxieties.[85]

In short, the agents who might champion GHILs would not be screened for good will, impartiality, or relevant skills. Are GHILs therefore really likely to get things right, even most of the time? Or would they empower bad advocates for vulnerable populations? When well-meaning proxy decision makers are appointed to make medical decisions on, say, end-of-life care, they have a hard time second-guessing what patients would have wanted. Wouldn't unelected strangers do much worse? Wouldn't GHIL certifiers, consumers, and investors often bring more harm than good?

I would argue that GHILs' advocacy would usually be far better than the alternative—having less advocacy overall. For one thing, the prospect of abuse of power by people and organizations who claim to promote global health but promote agendas contrary to global health—a problem in "humanitarian" interventions that can make interveners rich—is less relevant for GHILs. The potential for commercial abuse of the GHIL system could be curtailed through second-order GHILs and the supervision and Internet collaboration models mentioned above.

A different worry surrounds well-meaning consumers' and investors' potential ignorance of the correct way to promote global health. However, this worry is often exaggerated. Returning to the analogy of proxy decision making for medical care, what people need or want at the end of life varies considerably across patients. But people's needs and wants vary much less when it comes to the basic care that the global poor require. For example, no intimate acquaintance with a community is needed in order to know that polluted drinking water is usually bad for the community, or that a rural Thai community without a physician would benefit from having one. It is true that, very rarely, even matters like these are hard for non-locals to figure out. Perhaps the locals have a compelling religious need to keep their sacrosanct land free from strangers, even physicians, which may trump basic health. Certifiers ought to try to identify such rare cases by consulting with communities and relevant experts. But the potential for occasional error should not stand in the way of interventions that commonly produce important benefits.

It is true that in some areas of global health promotion, high factual and normative uncertainty and disagreement persist. A GHIL system's plurality and consequent flexibility would help address this challenge. Some GHIL certifiers probably would assess companies' overall global health impact retrospectively, and such assessments would often be difficult or contentious. But other GHIL certifiers might assess only components or indicators that were easy to measure and agreed upon. Other things being equal, it is likely that the GHILs that survive over a long period would tend to be those that focus on areas of relative professional agreement, where, for example, WHO and professional societies have issued clear guidelines, and experts agree on the facts. Such areas would command the broadest support both from authoritative Internet users like Doctors Without Borders and from cross-country coalitions of second-order GHILs, consumers, and investors.[86]

The vast majority of consumer and investor action that purports to serve global health succeeds to some degree. While it might not get things exactly right, and activists can be misinformed and even biased, consumer boycotts gone awry are the rare exceptions. Even when consumer and investor activism misses its target, it rarely misses it by as long a way as consistent neglect would.[87] In general, giving too much for tsunami victims and too little for famine is better than putting down nothing at all. The collaborative model mentioned above could further enhance the quality of influential GHILs' monitoring and dramatically cut ignorance among influential accreditors. Admittedly, most Internet users lack sufficient expertise on global health promotion to decide between GHILs. But many would have enough understanding to tell which users are decently knowledgeable and reliable—to realize, for example, that Doctors Without Borders is more credible than Joe Schmoe.

Even if I am right that GHILs are more likely to do good than harm, certifiers should nevertheless take precautions to minimize their own bias and error. It is important, for example, to involve both health policy experts and affected stakeholders in devising, honing, and operating certification standards. In general, GHILs could follow some international nonprofits that are in touch with communities and businesses in their areas of influence and forge alliances with social movements there.[88]

GHILs that do so might wind up having requirements that differ for different countries (like the JCI informed consent requirements),[89] or very minimal requirements that enjoy broad expert and stakeholder consensus across countries (like some International Organization for Standardization requirements).[90] This is a good thing. Normally, certifiers should not lay down universal standards based on views on global health that the certifiers hold with limited confidence. Instead, certification standards should aim for standards that any informed and well-meaning advocate for the global poor would support, and that tend to involve broad consensus.

Presumably, something along the lines of the proposed GHIL for the medical tourism industry would meet that standard. Despite open empirical questions and possible disagreements about details, the basic idea should draw broad support. It may remain unclear whether medical tourism at present is bad or good on balance (given the foreign income and skills it brings);[91] it may even remain unclear whether the impact on health worker shortages is bad or good on balance (given health workers' remittances to families).[92] But there is no serious controversy that medical tourism coupled with rural service and roughly the existing levels of foreign income would be an improvement upon the current system.

CONCLUSION

GHILs could fruitfully supplement the global health promotion tool kit. They have advantages over other tools, and they withstand a host of initial objections. A proposed GHIL for the medical tourism industry could help alleviate critical health worker shortages in countries with internal brain drains to the medical tourism sector.

ACKNOWLEDGMENTS

For their helpful comments, the author is grateful to Yukiko Asada, Nick Beckstead, I. Glenn Cohen, Will Crouch, Alnoor Ebrahim, Marc Fleurbaey, Archon Fung, Kaveri Gill, Axel Gosseries, Noa Havilio, Nien-hê Hsieh, Patricia Illingworth, Emily Jones, Yaron Klein, Larry Lessig, Azi Lev-On, Toby Ord, Jonathan Marks, Jeff McMahan, Joseph Millum, Toby Ord, Thomas Pogge, Eric Posner, Matthias Risse, Wolf Rogowski, Bettina Scholz, Prakash Sethi, Mildred Solomon, Daniel Star, Gilad Tanay, Sridhar Venkatapuram, Daniel Viehoff, Alex Voorhoeve, Katharine Young, Leif Wenar, and Dan Wikler, as well as to participants of the July 2009 workshop on Markets, Governance and Human Development in Cambridge University, a 2010 Safra Ethics Center seminar, a 2010 Harvard University Program in Ethics and Health seminar, and the 2010 session of Academics Stand Against Poverty (ASAP) at Yale University.

NOTES

1. Until recently, this practice was uncommon in the construction industry (Dickinson D and Innes D. [2004]. Fronts or front-lines? HIV/AIDS and big business in South Africa. *Transformation: Critical Perspectives on Southern Africa* 55:28–54.
2. Fairtrade Labelling Organisations International. www.fairtrade.net.
3. Social Investment Forum staff, "Socially Responsible Investing Facts," www.socialinvest. org/resources/sriguide/srifacts.cfm. Cowton, Christopher J., and Joakim Sandberg. (forthcoming). Socially Responsible Investment, in Ruth Chadwick (ed.), *Encyclopedia of Applied Ethics*, London: Academic Press.
4. Visser W. (2008). Corporate social responsibility in developing countries. In Andrew Crane, Abagail McWilliams, Dirk Matten, Jeremy Moon, and Donald S. Siegel (eds.), *The Oxford Handbook of Corporate Social Responsibility*. Oxford: Oxford University Press, p. 490.
5. As an example, in South Africa, "While nearly all large companies report having HIV/ AIDS policies, many have not conducted research on what to base these policies, programmes are often incomplete when measured against codes of good practice, there are gaps in the provision of antiretroviral drugs to full-time employees, and there is little independent monitoring of policy implementation" (Dickinson and Innes, 2004, p. 47).

6. Visser, 2008. For example, it has been suggested that if South African businesses lobbied Thabo Mbeki's government to fight AIDS, policy could have changed, as it did when South African businesses had lobbied for an end to apartheid (Dickinson and Innes, 2004).

7. Wal-Mart Stores, Inc., *Standards for Suppliers* [January 2009 Update], http://walmartstores.com/download/2727.pdf. See reported abuses at Foxvog, Bjorn Claeson and Liana. 2011. *Bangladesh Labor Leaders Win One Case; Ten More Cases Still to Go*, June 20, 2011 [cited September 9 2011]. Available from http://laborrightsblog.typepad.com/international_labor_right/2011/06/-bangladesh-labor-leaders-win-one-case-ten-more-cases-still-to-go-.html#more.

8. Wal-Mart Stores, Inc., *Sustainability* [May 2010 Update], http://instoresnow.walmart.com/Sustainability.aspx?povid=cat14503-env172199-module042610-lLink_wnsus; Rosenbloom S. (Feb. 25, 2010). Wal-Mart unveils plan to make supply chain greener. *New York Times*, p. B3; Vandenbergh MP and Cohen MA (2009). Climate change governance: boundaries and leakage. *NYU Law Review*.

9. (Red)™ Staff. *(Red)™ Results*. www.joinred.com/red/#impact_134. See also Cary, John, and Courtney E. Martin. 2011; *Apple's philanthropy needs a reboot*, September 5 2011 [cited September 6 2011]. Available from http://www.cnn.com/2011/OPINION/09/03/martin.cary.apple.charity/index.html?iref=allsearch; Barboza, David. 2010. After Suicides, Scrutiny of China's Grim Factories. *New York Times*, June 7, 2010, A1.

10. Fairtrade Labelling Organisations International. (December 2005). *Generic Fairtrade Standards for Small Farmers' Organisations*. www.fairtrade.net, p. 2; Balineau, Gaëlle, and Ivan Dufeu. 2010. Are Fair Trade Goods Credence Goods? A New Proposal. *Journal of Business Ethics* 92 (April):331–345; Blowfield, Michael, and Catherine Dolan. 2010; Fairtrade Facts and Fancies: What Kenyan Fairtrade Tea Tells us About Business' Role as Development Agent. *Journal of Business Ethics* 93 (supplement 2):143–162.

11. For example, *Green-e* is a certification system, but its exclusive focus is production in the United States and Canada (Green-e Staff. *Welcome to Green-E*. www.green-e.org). Internationally, Global Reporting Initiative, International Organization for Standardization, and other monitoring bodies cover environmentally affected health, among other things (GRI Staff. *GRI—the Global Reporting Initiative*. www.globalreporting.org/Home; ISO Staff. *ISO—International Organization for Standardization*. www.iso.org/iso/home.htm). Most of these entities do not provide labeling. For a review of disclosure systems for companies' carbon footprint, an issue that will affect global health, see Vandenbergh and Cohen, Climate change governance.

12. Even on the non-consequentialist position that we have agent-relative duties beyond improving impact, it is rarely our duty to ensure that other agents abide by their own agent-relative duties. So even non-consequentialists can endorse a near-complete focus on impact in the incentives that GHILs give other agents.

13. GiveWell Staff. *GiveWell: Real Change for Your Dollar*. www.givewell.net; Singer P. (2009). *The Life You Can Save: Acting Now to End World Poverty*. New York: Random House; Ord T. *Giving What We Can*. www.givingwhatwecan.org.

14. Onora O'Neill warns that the recent drive towards maximal accountability in many areas of public life might crowd out altruistic motivation (O'Neill O. [2002]. *Autonomy and Trust in Bioethics*. Cambridge: Cambridge University Press, p. 39).

15. Gruskin S and Ferguson L. (2009). Using indicators to determine the contribution of human rights to public health efforts. *WHO Bulletin* 87(9). It could also combine the use of substantive goals and technical indicators, in the following way: corporations would usually be expected to meet success indicators but there would be an appeal or interpretation-request process allowing the corporation to establish that it serves the broader goal or intent of the certification without meeting the technical indicator.

As Noa Havilio informs me, the environmental housing certificate LEED combines indicators and goals in this manner.

16. UN Sub-Commission for the Promotion and Protection of Human Rights. (2003). *Norms on the Responsibilities of Transnational Corporations and Other Business Enterprises with Regard to Human Rights.* Draft E/Cn.4/Sub.2/2003/12/Rev.2, Preamble, Articles §12, 16. Article 1 expresses a similar idea. Articles 7 and 13 focus on honoring health rights (as opposed to promoting them). See also Slomanson WR. (2007). *Fundamental Perspectives on International Law* (5th ed.). Belmont, CA: Thomson Publishers, p. 566.

17. Ruggie J (Special Representative of the Secretary-General on the issue of human rights and transnational corporations and other business enterprises). (2010). Human Rights Council, Fourteenth session. Agenda item 3: Promotion and protection of all human rights, civil, political, economic, social and cultural rights, including the right to development. *Business and Human Rights: Further steps toward the operationalization of the "protect, respect and remedy" framework.* United Nations A/HRC/14/27; Hunt P (Special Rapporteur on the right of everyone to the enjoyment of the highest attainable standard of health). (2009). Human Rights Council, Eleventh session, Agenda item 3: Promotion and Protection of all Human Rights, Civil, Political, Economic, Social and Cultural Rights, Including the Right to Development—*Annex: Mission to Glaxosmithkline.* United Nations A/HRC/11/12/Add.2. I am grateful to Patricia Illingworth for these references.

18. Admittedly, payment by certified companies risks creating incentives for GHILs to provide friendly reports that could gain them future business—a problem that corrupted several of America's credit rating agencies in recent years (Krugman P. [April 25, 2010]. Berating the raters. *New York Times*). As Larry Lessig pointed out to me, the fact that there was corruption when huge fees were at stake wouldn't automatically entail corruption when small issues are at stake; payment by certified companies only sometimes gives rise to such corruption; and some of the most reliable certifiers are funded primarily through that channel (see, e.g., UL Certification Staff. *About UL Mark Product Certification.* www. ul.com/global/eng/pages/corporate/certifications), still, the potential for corruption could be curtailed further. For example, a central regulator could receive certification requests and decide which GHIL would certify which company, thereby limiting GHILs' incentives to be soft on companies. Parallel solutions have been proposed for the credit rating industry. See *ibid.*; Richardson M and White L. (2008). The rating agencies: is regulation the answer? *NYU Stern White Papers*, available at http://whitepapers.stern. nyu.edu/summaries/ch03.html. Funding GHILs by selling their reports to consumers, as *Consumer Reports* does, would probably not work. However, the information may be of interest to other agents.

19. Hollis A and Pogge T. (2008). *The Health Impact Fund: Making New Medicines Accessible for All.* New Haven, CT Incentives for Global Health. yale.edu/macmillan/igh/hif_book. pdf. One other difference between GHILs and the proposed Health Impact Fund is that, unlike the latter, the GHIL system would not focus only on the pharmaceutical industry. Of course some GHILs would focus exclusively on that industry, as in a proposal by Nicole Hassoun, who unbeknownst to me has worked on a health impact label for drug manufacturers for some years: see her *Globalization and Global Justice: Shrinking Distance, Expanding Obligations* (under contract with Cambridge University Press), Chapter 7. But limiting GHILs to such a focus would have been arbitrary, since many non-medicinal determinants affect health, and since any company may affect health, through donations. Another difference is that, while the Health Impact Fund emphatically assesses impact retrospectively, some GHILs, recognizing the infeasibility of reliable retrospective impact assessment in some areas of global health, would use advance proxies. For example, where many companies, the environment, and political factors affect deaths from malaria, and individual contributions are hard to disentangle, the relevant GHIL might use proxies,

such as how many mosquito nets a company supplied. That said, GHILs could exist alongside the Health Impact Fund, or even certify companies for their donations to it.

20. Compare Vandenbergh and Cohen, 2009. "Climate Change Governance."
21. Guariglia O. (2007). Enforcing economic and social human rights. In Pogge T, ed. *Freedom from Poverty as a Human Right*. Oxford: Oxford University Press, pp. 356–357.
22. Cowton and Sandberg, "forthcoming."
23. Brooks E. (2005). Transnational campaigns against child labor: the garment industry in Bangladesh. In Bandy J and Smith J, eds. *Coalitions Across Borders: Transnational Protest and the Neoliberal Order*. Lanham, Md.: Rowman & Littlefield. Likewise, as states begin to regulate domestic carbon emissions, companies may move operations abroad (Vandenbergh and Cohen, "Climate Change Governance").
24. Vandenbergh and Cohen, 2009. "Climate Change Governance."
25. I am grateful to Katharine Young for clarifications on these matters.
26. PTI. 2010. Medical tourism to become USD 100 billion industry by 2012: Report. *Times of India*, December 11.
27. Deloitte Staff. (2009). *Medical Tourism: Update and Implications—2009 Report*, pp. 1–15.
28. Chinai R and Goswami R. (2007). Medical visas mark growth of Indian medical tourism. *Bulletin of the World Health Organization* 85(3). See also de Arellano ABR. (2007). Patients without borders: the emergence of medical tourism. *International Journal of Health Services* 37(1).
29. WHO. (2006). *World Health Report 2006: Working Together for Health* (Geneva: WHO); Institute of Medicine. (2007). Committee for the Evaluation of the President's Emergency Plan for AIDS Relief (PEPFAR) Implementation, *PEPFAR Implementation: Progress and Promise*. Washington, DC: National Academies Press.
30. Quoted in Singh N. (2008). Decentralization and public delivery of health care services in India. *Health Affairs* 27(4), pp. 991–1001.
31. Blouin C. (2007). Trade policy and health: from conflicting interests to policy coherence. *Bulletin of the World Health Organization* 85(3); Chinai and Goswami, 2007.
32. Overdorf J. (2009). The dark side of India's medical tourism. *Global Post*. http://www.globalpost.com/dispatch/india/091013/the-dark-side-medical-tourism.
33. WHO Secretariat Executive Board. (Dec. 3, 2009). *International Recruitment of Health Personnel: Draft Global Code of Practice. Provisional Agenda Item 4.5 (Eb126/8)*, 126th Session.
34. Timmons K. (2007). The value of accreditation. *Medical Tourism Magazine*; JCI staff. *JCI—Joint Commission International*. http://jointcommissioninternational.org.
35. In the United States, medical tourism could be driven by self-insured companies (Cohen IG. [2010]. Protecting patients with passports: medical tourism and the patient protective-argument. *Iowa Law Review* 95(5)). These companies would also have a reputational interest beyond their employees' health care department. The proposed solution would have the least impact on uninsured Americans, who fund their own travel and care abroad, but the April 2010 health care reform law should decrease the number of uninsured Americans.
36. McFarland J. (March 22, 2008). Wal-Mart move "tipping point" for non-hormone milk. *Toronto and Mail*; Rosenbloom, 2010. I thank Mildred Solomon for informing me of the related changes in America's dairy industry. On other externalities induced by Wal-Mart, and on several U.S. cheap food chains that now sell only fair trade coffee, see Vandenbergh and Cohen, "Climate Change Governance."
37. Kimberley process staff. *The Kimberley Process*. www.kimberleyprocess.com. But concerning recent failures, see Sethi, S. Prakash, and Olga Emelianova (2011). The Kimberley Process Certification Scheme (KPCS): a voluntary multigroup initiative to control trade in conflict diamonds. In S. Prakash Sethi, ed. *Globalization and self-regulation: the crucial role that corporate codes of conduct play in global business*. New York City: Palgrave Macmillan, pp. 213–248.

38. Quoted in Margolis JD and Walsh JP. (2003). Misery loves companies: rethinking social initiatives by business. *Administrative Science Quarterly* 48.

39. *Ibid.*, pp. 273, 277–278. See also Vogel D. (2005). *The Market for Virtue.* Washington DC: Brookings Institution Press, p. 29.

40. Vogel, 2005, p. 46. See also his Chapters 1 and 2. Other times Vogel concedes that, currently, corporate social responsibility "does not make firms *less* profitable [and that] means that it is possible for a firm to commit resources to [corporate social responsibility] without becoming less competitive." If that is an accurate assessment, then ramping up corporations' health promotion incentives even a little, as surely some well-designed GHILs could do, might break the rough tie. It would make it positively profitable for corporations to promote global health—which is enough to motivate even the ethically indifferent among them.

41. Becker E and McVeigh P. (2001). Social funds in the United States: their history, financial performance, and social impacts. In Fung A, Hebb T, and Rogers J, eds. *Working Capital: The Power of Labor's Pensions.* Ithaca, NY: ILR Press Books, p. 52.

42. Vogel, 2005, pp. 39, 67ff.

43. Fung A, Graham M, and Weil D. (2007). *Full Disclosure: The Perils and Promise of Transparency.* Cambridge: Cambridge University Press.

44. For similar approaches to disclosure, in a variety of areas, see *Ibid.*; Moore DA, Cain DM, and Loewenstein G. (2005). Coming clean but playing dirtier: the shortcomings of disclosure as a solution to conflicts of interest. In Moore DA, et al., eds. *Conflicts of Interest: Challenges and Solutions in Business, Law, Medicine, and Public Policy.* Cambridge: Cambridge University Press; Manson NC and O'Neill O. (2007). *Rethinking Informed Consent in Bioethics.* Cambridge: Cambridge University Press; and Lessig L. (Oct. 9, 2009). Against transparency: the perils of openness in government. *The New Republic.*

45. On such in-house certifications for friendliness to the environment, see O'Connell V. (April 24, 2010). "Green" goods, red flags. *Wall Street Journal.*

46. The involvement of a supervisory body such as WHO or DFID may seem incompatible with the aspiration of the GHIL system to enable ordinary citizens to act for global health. However, it would not make decisions about content, just technical decisions on independence from relevant commercial interests.

47. A further advantage of the collaborative model is that it could enable busy ethical consumers to check simply whether users like Doctors Without Borders approve of the GHILs that approve of a product, instead of which of a potentially confusing bevy of GHILs that product won. Such confusion currently poses a problem for labeling systems (Vandenbergh and Cohen, "Climate Change Governance"). Additional ways to diminish this confusion are to use the best social science on effective labeling, or even for different GHILs to unify their labeling systems, following the example of European fair trade labels.

48. Resnick P, Zeckhauser R, Friedman E, and Kuwabara K. (2000). Reputation systems. *Communications of the ACM* 43(12), pp. 45–48.

49. See, e.g., *Ibid.*; Resnick P and Sami R. (2007). The influence-limiter: provably manipulation-resistant recommender systems. in *Proceedings of the ACM Recommender Systems Conference.* Minneapolis, MN: ACM Conference On Recommender Systems; Riedl J and Konstan J. (2002). *Word of Mouse: The Marketing Power of Collaborative Filtering.* New York: Warner Business Plus; Lev-On A and Hardin R. (2008). Internet-based collaborations and their political significance. *Journal of Information Technology & Politics* 4(2), pp. 5–27, 20ff; Streitfeld, David. 2011. In a Race to Out-Rave Rivals, 5-Star Web Reviews Go for $5. *New York Times*, August 19, A1. I am grateful to Azi Lev-On for these citations and for a helpful conversation. Another source of relevant expertise could be found in quality control reporting, which faces similar pressures by industry.

50. I thank Jonathan Marks for related ideas.

51. Craig Smith N. (2008). Consumers as drivers of corporate social responsibility. In Crane A, McWilliams A, Matten D and Moon J, eds. *Oxford Handbook of Corporate Social Responsibility*. Oxford: Oxford University Press, p. 286.

52. *Ibid.*, p. 286.

53. E.g., Becker and McVeigh, 2001, p. 63ff; Craig Smith N, 2008, pp. 295–296. In the same vein, progressive nonprofits report that "naming and shaming" is often the most effective way to achieve results (Roth K. [2004]. Defending economic, social and cultural rights: practical issues faced by an international human rights organization. *Human Rights Quarterly* 26(1), pp. 67–68). In the context of global health, negative campaigns seem to have been key to companies' improved performance—for example, on child labor in Bangladesh (Brooks, 2005, pp. 133–134) and on antiretroviral patent protection lobbying (Gellman B. [Dec. 28, 2000]. A turning point that left millions behind; drug discounts benefit few while protecting pharmaceutical companies' profits. *Washington Post*).

54. Craig Smith, 2008, pp. 298–289; Burr AC. *LEED's Big Market Bias*. www.costar.com/News/Article/LEEDs-Big-Market-Bias/100646. Note also a skeptic's concession that, in parts of the business system, social responsibility is profitable: "there is a place in the business system for responsible firms, but the market for virtue is not sufficiently important to make it in the interest of all firms to behave more responsibly" (Vogel, 2005, p. 17).

55. Skeptics may retort that ethical "considerations do influence consumer behavior, but our understanding is limited of when, how, and why" (Craig Smith N, 2008, p. 299). However, it is open to certifiers to focus on the limited areas that we do understand, while continuing to search for other areas.

56. "[P]ension funds are the primary drivers of today's financial markets in the United States and around the world," according to Tessa Hebb (Hebb T. [2001]. Introduction: the challenge of labor's capital strategy. In Fung A, Hebb T, Rogers J, and Gerard LG, eds. *Working Capital: The Power of Labor's Pensions*. Ithaca, NY: ILR Press Books, p. 1), and yet, curiously, "pension fund assets are generally invested in almost the same manner as the funds of other large institutional investors" (Baker D and Fung A. [2001]. Collateral damage: do pension fund investments hurt workers? In Fung A, Hebb T, Rogers J, and Gerard LG, eds. *Working Capital: The Power of Labor's Pensions*. Ithaca, NY: ILR Press Books, p. 13). Pension funds and labor unions could be lobbied to use a greater share of investment to promote global health, in accordance with a GHIL standard. A few years ago, authors dismissed any form of social investment as unrealistic, noting simply, "below-market-rate investments, sometimes called social investments, undermine individuals' retirement security" (Hebb, 2001, p. 3) The suggested alternative approach was to make assets promote investments that at least benefit American workers. See also Baker and Fung, 2001; Calabrese M. (2001). Building on success: labor-friendly investment vehicles and the power of private equity. In Fung Λ, Hebb T, Rogers J, and Gerard LG, eds. *Working Capital: The Power of Labor's Pensions*. Ithaca, NY: ILR Press Books. But no ground was advanced in support of this pessimism about social investment, and currently, nearly one out of every nine dollars of assets under professional management in the United States is involved in socially responsible investment (Social Investment Forum staff. *Socially Responsible Investing Facts*). It now seems plausible that a significant portion of workers would tolerate a small setback to their retirement returns for the sake of global health.

57. Pogge, personal communication (prior to the passage of the April 2010 U.S. health law).

58. Pogge, personal communication.

59. It may meet Frederic Von Hayek's criterion for coercion, which for Hayek is a violation of freedom: "By 'coercion' we mean such control of the environment or circumstances of a person by an other that, in order to avoid greater evil, he is forced to act not according to a

coherent plan of his own but to serve the ends of an other" (Von Hayek FA. [1982]. *Law, Legislation, and Liberty*. London: Routledge, p. 20 ff). For Hayek, certain intentions and threats (with no moralized baseline) are the mark of coercion (*The Constitution of Liberty*. Chicago: University of Chicago Press, 1960, p. 133). Consumers and investors could certainly intentionally threaten to boycott companies unless they start promoting health.

60. Nozick R. (1986). *Anarchy, State and Utopia*. Oxford: Oxford University Press, pp. 265–268.

61. Friedman M. (Sept. 13, 1970). The social responsibility of business is to increase its profits. *New York Times Magazine*.

62. Adam Smith. (1976). *An Inquiry into the Nature and Causes of the Wealth of Nations*. In Campbell RH and Skinner AS, eds. *Glasgow Edition of the Works and Correspondence of Adam Smith*. Oxford: Clarendon Press, Book I, Chapter 2, p. 19.

63. Von Mises L. (1922/2010). *Socialism: An Economic and Sociological Analysis* (trans. Kahane J, 2nd ed.). Indianapolis, IN: Liberty Fund Inc, II.5.3, pp. 97–105; Von Hayek FA. (1944/2007). *The Road to Serfdom*. In Caldwell B, ed. *The Collected Works of F. A. Hayek*. Chicago: University of Chicago Press.

64. Cohen GA. (1979). Capitalism, freedom and the proletariat. In Ryan A, ed. *The Idea of Freedom: Essays in Honour of Isaiah Berlin*. Oxford: Oxford University Press; Wertheimer A. (1987). *Coercion*. Princeton: Princeton University Press; Fuller L, Chapter 9 in this volume.

65. Rabkin J. (1998). *Why Sovereignty Matters*. Washington DC: American Enterprise Institute, p. vii.

66. Eyal N, Hurst S. (2010). Coercion in the fight against brain drain. Shah R, ed. *The International Migration of Health Workers: Ethics, Rights, and Justice*. London: Palgrave Macmillan: 137-158.

67. Walzer M. (1983). *Spheres of Justice: A Defense of Pluralism and Equality*. New York: Basic Books, Introduction.

68. Friedman, 1970. Jeremy Rabkin uses similar arguments to warn against the growing influence of "unelected" international NGOs. For Rabkin, unelected international NGOs have led to human rights treaties that "bestow the status of 'human rights' . . . on many of feminist or liberal preferences" (Rabkin, 1998, 43). Invoking Friedman's further warning, that corporate social responsibility lacks democratic checks and balances, Rabkin notes that, while the U.S. Constitution "imposes a whole series of checks and balances on domestic law making," NGOs lack equivalent mechanisms (Rabkin, p. ix). Jennifer Rubenstein quotes many other conservative critics who express similar worries (Rubenstein J. [2009]. The ethics of INGO advocacy, or why it is OK that no one elected Oxfam. In *Democracy Seminar Series, Kennedy School of Government, Harvard University*. Cambridge, MA.

69. Daniels N. (2007). *Just Health: Meeting Health Needs Fairly*. Cambridge: Cambridge University Press.

70. I am grateful to Daniel Viehoff for a discussion of this point.

71. More generally, international nonprofits are best thought of as advocates for justice, not as (unelected) representatives for the poor. Rubenstein, 2009.

72. Singh, 2008; Kaveri Gill. (2009). *A Primary Evaluation of Service Delivery under the National Rural Health Mission (Nrhm): Findings from a Study in Andhra Pradesh, Uttar Pradesh, Bihar and Rajasthan*. Working Paper 1/2009—PEO. New Delhi: Planning Commission of India, e.g. pp. 17, 63; correspondence with Kaveri Gill.

73. Sengupta A and Nundy S. (2005). The private health sector in India. *British Medical Journal* 331; Bose A. (2005). Private health sector in India—is private health care at the cost of public health care? *British Medical Journal* 331; Economist Intelligence Unit. (March 16, 2009). Healthcare in India: rural development. *Economist*.

74. Nor are GHILs procedurally problematic in the way that extreme measures like using one's economic advantage to bribe officials for the sake of global health goals would be.

75. Chavis L and Leslie P. (2008 draft). *Consumer Boycotts: The Impact of the Iraq War on French Wine Sales in U.S.*

76. Craig Smith N, 2008, p. 285. The author gives an example: "Middle East sales of Danish dairy giant Arla ($430 m. annually) vanished almost overnight as a result of a boycott in early 2006, following the publication of cartoons that caricatured the Prophet Muhammad in the Danish newspaper *Jyllands-Posten*." Likewise, some NGO monitoring organizations, including ones lobbying for Israel, are ferociously anti-liberal and apparently prejudiced (Edwards M. [2006]. Foreword. In Jordan L and Van Tuijl P, eds. *NGO Accountability.* London: Earthscan, p. viii.

77. On his first day in office, President George W. Bush famously canceled all U.S. medical aid funding abortion. Consider also that social responsibility investment funds (SRIs) continue to screen out many "sin" products, historically defined along religious Puritan criteria; and that the Islamic Market Index, a leading SRI, screens out companies worldwide for violating Shari'ah law, even when secular norms permit company policies— for example, when companies collect interest (Becker and McVeigh, 2001, p. 52).

78. U.S. missionaries in Kenya recently offered small business loans that excluded Muslims (Stockman F. [Oct. 10, 2006]. For those excluded, loan program is no success. *Boston Globe*). Humanitarian aid was discriminatory on many additional occasions (Nardin T. [2003]. The moral basis for humanitarian intervention. In Lang Jr, AF, ed. *Just Intervention.* Washington DC: Georgetown University Press, p. 21).

79. Falk R. (2009). *Achieving Human Rights.* New York: Routledge, Chapter 2, pp. 25–38.

80. Calabrese, 2001, p. 95.

81. McMahan J. (2009). Humanitarian intervention, consent, and proportionality. In Davis NA, Keshen R, and McMahan J, eds. *Ethics and Humanity: Themes from the Philosophy of Jonathan Glover.* New York: Oxford University Press.

82. *New York Times* staff. (Sept. 6, 1995). Greenpeace apologizes to Shell Oil Company. *New York Times.* Greenpeace had alleged that the oil platform at sea still contained 5,500 tons of oil, which it did not.

83. Becker and McVeigh, 2001, pp. 59, 60; Vogel, *The Market for Virtue*, pp. 39–40.

84. Fairtrade Labelling Organisations International, 2005, p. 18. In a different example, the American consumers' 1990s protest that led to dramatic reduction in child labor in Bangladesh was later said to have been narrowly focused on whether children work, without providing them with alternative sources of livelihood or decent schooling, leading to many counterproductive results. Brooks, 2005.

85. "[C]onsumer pressures have virtually eliminated sales of genetically modified foods in Europe." Vogel, 2005, p. 51.

86. Besides, where true uncertainty exists, the existence of GHILs with incongruent criteria may be a *good* thing. Collaborative filtering, which we earlier compared to GHILs, works well in areas of persistent disagreement among experts. See Gladwell M. (2002). Foreword. In Riedl J and Konstan J, eds. *Word of Mouse: The Marketing Power of Collaborative Filtering.* New York: Warner Business Plus.

87. Indeed, even the cases I mentioned are not as extreme as their opponents make them out to be. For example, despite Shell's complaints, Greenpeace's opposition to Shell's oil dumping was never based solely on the matter on which it made false allegations—the presence of oil in the platform at sea.

88. Yamin AE. (2005). The future in the mirror: incorporating strategies for the defense and promotion of economic, social, and cultural rights into the mainstream human rights agenda. *Human Rights Quarterly* 27(4), p. 1239.

89. Timmons, 2007.

90. ISO staff. "ISO—International Organization for Standardization."
91. Brenzel L and Le Franc E. (2007). *Opportunities and Challenges for Expanding Trade in Health Services in the English-Speaking Caribbean.* Washington DC: World Bank; Rajan TD. (2005). Private health sector in India—let's not confuse the issues. *British Medical Journal* 331 p. 1339; Herrick DM. (2007). Medical tourism: global competitions in health care. In *NCPA Policy report;* Bookman MZ and Bookman KR. (2007). *Medical Tourism in Developing Countries.* London: Palgrave Macmillan; Cohen IG, 2008, e.g. p. 170.
92. Connell J and Brown RPC. (2004). The remittances of migrant Tongan and Samoan nurses from Australia. *Human Resources for Health* 2(1); Adams Jr., R. (2003). International migration, remittances and the brain drain: a study of 24 labor-exporting countries. *World Bank Policy Research Paper* 3069(5).

REFERENCES

Adams Jr., Richard. (2003). International migration, remittances and the brain drain: a study of 24 labor-exporting countries. *World Bank Policy Research Paper* 3069(5)

Baker, Dean, and Archon Fung. (2001). Collateral damage: do pension fund investments hurt workers? In Archon Fung, Tessa Hebb, Joel Rogers and Leo G. Gerard, eds. *Working Capital: The Power of Labor's Pensions.* Ithaca, NY: ILR Press Books, pp. 13–43

Balineau, Gaëlle, ahd Ivan Dufeu. (2010). Are Fair Trade Goods Credence Goods? A New Proposal. *Journal of Business Ethics* 92, pp. 331–345

Barboza, David. 2010. After Suicides, Scrutiny of China's Grim Factories. *New York Times,* June 7, 2010, A1

Becker, Eric, and Patrick McVeigh. (2001). Social funds in the United States: their history, financial performance, and social impacts. In Archon Fung, Tessa Hebb, Joel Rogers and Leo G. Gerard, eds. *Working Capital: The Power of Labor's Pensions.* Ithaca, NY: ILR Press Books, pp. 44–66

Blouin, C. (2007). Trade policy and health: from conflicting interests to policy coherence. *Bulletin of the World Health Organization* 85(3), pp. 169–173

Blowfield, Michael, and Catherine Dolan. (2010). Fairtrade Facts and Fancies: What Kenyan Fairtrade Tea Tells us About Business' Role as Development Agent. *Journal of Business Ethics* 93 (supplement 2), pp. 143–162

Bookman, Milica Z., and Karla R. Bookman. (2007). *Medical Tourism in Developing Countries.* London: Palgrave Macmillan

Bose, Amitava. (2005). Private health sector in India—is private health care at the cost of public health care? *British Medical Journal* 331, pp. 1338–1339

Brenzel, L., and E. Le Franc. (2007). *Opportunities and Challenges for Expanding Trade in Health Services in the English-Speaking Caribbean.* Washington DC: World Bank

Brooks, Ethel. (2005). Transnational campaigns against child labor: the garment industry in Bangladesh. In Joe Bandy and Jackie Smith, eds. *Coalitions Across Borders: Transnational Protest and the Neoliberal Order.* Lanham, MD: Rowman & Littlefield, pp. 121–139

Burr, Andrew C. *LEED's Big Market Bias.* www.costar.com/News/Article/LEEDs-Big-Market-Bias/100646

Calabrese, Michael. (2001). Building on success: labor-friendly investment vehicles and the power of private equity. In Archon Fung, Tessa Hebb, Joel Rogers and Leo G. Gerard, eds. *Working Capital: The Power of Labor's Pensions.* Ithaca, NY: ILR Press Books, pp. 93–127

Cary, John, and Courtney E. Martin. 2011; *Apple's philanthropy needs a reboot,* September 5 2011 [cited September 6 2011]. Available from www.cnn.com/2011/OPINION/09/03/martin.cary.apple.charity/index.html?iref=allsearch

Chavis, Larry, and Phillip Leslie. (2008 draft). *Consumer Boycotts: The Impact of the Iraq War on French Wine Sales in U.S.*

Chinai, R, and R. Goswami. (2007). Medical visas mark growth of Indian medical tourism. *Bulletin of the World Health Organization* 85(3), pp. 164–165

Cohen, Gerald A. (1979). Capitalism, freedom and the proletariat. In Alan Ryan, ed. *The Idea of Freedom: Essays in Honour of Isaiah Berlin.* Oxford: Oxford University Press, pp. 7–25

Cohen, Ivan Glenn. (2010). Protecting patients with passports: medical tourism and the patient protective-argument. *Iowa Law Review* 95(5), pp. 1467–1567

Connell, John, and Richard P. C. Brown. (2004). The remittances of migrant Tongan and Samoan nurses from Australia. *Human Resources for Health* 2(1), p. 2

Cowton, Christopher J., and Joakim Sandberg. (forthcoming). Socially Responsible Investment, in Ruth Chadwick, (ed.), *Encyclopedia of Applied Ethics,* London: Academic Press

Craig Smith, N, (2008). Consumers as drivers of corporate social responsibility. In Andrew Crane, Abagail McWilliams, Dirk Matten and Jeremy Moon, eds. *Oxford Handbook of Corporate Social Responsibility.* Oxford: Oxford University Press

Daniels, Norman. (2007). *Just Health: Meeting Health Needs Fairly.* Cambridge: Cambridge University Press

de Arellano, Annette B. Ramírez. (2007). Patients without borders: the emergence of medical tourism. *International Journal of Health Services* 37(1), pp. 193–198

Deloitte staff. (2009). *Medical Tourism: Update and Implications—2009 Report,* pp. 1–15

Dickinson, David, and Duncan Innes. (2004). Fronts or front-lines? HIV/AIDS and big business in South Africa. *Transformation: Critical Perspectives on Southern Africa* 55, pp. 28–54

Economist Intelligence Unit. (2009). Healthcare in India: rural development. *Economist,* March 16

Edwards, Michael. (2006). Foreword. In Lisa Jordan and Peter Van Tuijl, eds. *NGO Accountability.* London: Earthscan, pp. vii–ix

Eyal, N., and Hurst S. (2010). Coercion in the fight against brain drain. In Rebecca Shah, ed. *The International Migration of Health Workers: Ethics, Rights, and Justice.* London: Palgrave Macmillan: 137–158

Fairtrade Labelling Organisations International. www.fairtrade.net.

———. (Dec. 2005). Fairtrade Labelling Organisations International. (December 2005). *Generic Fairtrade Standards for Small Farmers' Organisations,* pp. 1–22. www.fairtrade.net.

Falk, Richard. (2009). *Achieving Human Rights.* New York: Routledge

Foxvog, Bjorn Claeson and Liana. 2011. *Bangladesh Labor Leaders Win One Case; Ten More Cases Still to Go,* June 20, 2011 [cited September 9 2011]. Available from http://laborrightsblog. typepad.com/international_labor_right/2011/06/-bangladesh-labor-leaders-win-one-case-ten-more-cases-still-to-go-.html#more

Friedman, Milton. (Sept. 13, 1970). The social responsibility of business is to increase its profits. *New York Times Magazine,* p. SM17

Fuller, Lisa. (2012). International nongovernmental organization health programs in a non-ideal world. In Joseph Millum and Ezekiel Emanuel, eds. *Global Justice and Bioethics.* New York: Oxford University Press

Fung, Archon, Mary Graham, and David Weil. (2007). *Full Disclosure: The Perils and Promise of Transparency.* Cambridge: Cambridge University Press

Gellman, Barton. (Dec. 28, 2000). A turning point that left millions behind; drug discounts benefit few while protecting pharmaceutical companies' profits. *Washington Post,* p. A1

GiveWell staff. *GiveWell: Real Change for Your Dollar.* www.givewell.net

Gill, Kaveri. (2009). *A Primary Evaluation of Service Delivery under the National Rural Health Mission (NRHM): Findings from a Study in Andhra Pradesh, Uttar Pradesh, Bihar and Rajasthan.* Working Paper 1/2009—PEO. New Delhi: Planning Commission of India

Gladwell, Malcolm. (2002). Foreword. In John Riedl and Joseph Konstan, eds. *Word of Mouse: The Marketing Power of Collaborative Filtering*. New York: Warner Business Plus, pp. vii–xv

Green-e staff. *Welcome to Green-E*. www.green-e.org

GRI staff. *GRI—the Global Reporting Initiative*. www.globalreporting.org/Home

Gruskin, Sofia, and Laura Ferguson. (2009). Using indicators to determine the contribution of human rights to public health efforts. *WHO Bulletin* 87(9), pp. 645–732

Guariglia, Osvaldo. (2007). Enforcing economic and social human rights. In Thomas Pogge, ed. *Freedom from Poverty as a Human Right*. Oxford: Oxford University Press, pp. 345–357

Hassoun, Nicole. (under contract). *Globalization and Global Justice: Shrinking Distance, Expanding Obligations*. Cambridge University Press

Hayek, Von-, Frederic August. (1960). *The Constitution of Liberty*. Chicago: University of Chicago Press

———. (1982). *Law, Legislation, and Liberty*. London: Routledge

———. (1044/2007). *The Road to Serfdom*. In Bruce Caldwell, ed. *The Collected Works of F. A. Hayek*. Chicago: University Of Chicago Press

Hebb, Tessa. (2001). Introduction: the challenge of labor's capital strategy. In Archon Fung, Tessa Hebb, Joel Rogers and Leo G. Gerard, eds. *Working Capital: The Power of Labor's Pensions*. Ithaca, NY: ILR Press Books, pp. 1–12

Herrick, Devon M. (2007). Medical tourism: global competitions in health care. In *NCPA Policy Report*

Hollis, Aidan, and Thomas Pogge. (2008). *The Health Impact Fund: Making New Medicines Accessible for All*. New Haven, CT: Incentives for Global Health

Hunt, Paul (Special Rapporteur on the right of everyone to the enjoyment of the highest attainable standard of health). (2009). Human Rights Council, Eleventh session, Agenda item 3: Promotion and Protection of all Human Rights, Civil, Political, Economic, Social and Cultural Rights, Including the Right to Development—*Annex: Mission to Glaxosmithkline*. United Nations A/HRC/11/12/Add.2

Institute of Medicine, Committee for the Evaluation of the President's Emergency Plan for AIDS Relief (PEPFAR) Implementation. (2007). *PEPFAR Implementation: Progress and Promise*. Washington, DC: National Academies Press

ISO staff. *ISO—International Organization for Standardization*. www.iso.org/iso/home.htm

JCI staff. *JCI—Joint Commission International*. http://jointcommissioninternational.org.

Kimberley Process staff. *The Kimberley Process*. www.kimberleyprocess.com.

Krugman, Paul. (April 25, 2010). Berating the raters. *New York Times*

Lessig, Lawrence. (Oct. 9, 2009). Against transparency: the perils of openness in government. *The New Republic*

Lev-On, Azi, and Russell Hardin. (2008). Internet-based collaborations and their political significance. *Journal of Information Technology & Politics* 4(2), pp. 5–27

Manson, Neil C., and Onora O'Neill. (2007). *Rethinking Informed Consent in Bioethics*. Cambridge: Cambridge University Press

Margolis, Joshua D., and James P. Walsh. (2003). Misery loves companies: rethinking social initiatives by business. *Administrative Science Quarterly* 48, pp. 268–305

McFarland, Janet. (March 22, 2008). Wal-Mart move 'tipping point' for non-hormone milk. *Toronto Globe and Mail*

McMahan, Jeff. (2009). Humanitarian intervention, consent, and proportionality. In N. Ann Davis, Richard Keshen and Jeff McMahan, eds. *Ethics and Humanity: Themes from the Philosophy of Jonathan Glover*. New York: Oxford University Press

Moore, Don A., Daylian M. Cain, and George Loewenstein. (2005). Coming clean but playing dirtier: the shortcomings of disclosure as a solution to conflicts of interest. In Don A. Moore, Daylian M. Cain, George Loewenstein and Max H. Bazerman, eds. *Conflicts of*

Interest: Challenges and Solutions in Business, Law, Medicine, and Public Policy. Cambridge: Cambridge University Press, pp. 104–125

Nardin, Terry. (2003). The moral basis for humanitarian intervention. In Anthony F. Lang Jr., ed. *Just Intervention*. Washington DC: Georgetown University Press, pp. 11–27

New York Times staff. (Sept. 6, 1995). Greenpeace apologizes to Shell Oil Company. *New York Times*, p. A11

Nozick, Robert. (1986). *Anarchy, State and Utopia*. Oxford: Oxford University Press

O'Connell, Vanessa. (April 24, 2010). 'Green' goods, red flags. *Wall Street Journal*

O'Neill, Onora. (2002). *Autonomy and Trust in Bioethics*. Cambridge: Cambridge University Press

Ord, Toby. *Giving What We Can*. www.givingwhatwecan.org

Overdorf, Jason. (2009). The dark side of India's medical tourism. *Global Post*. www.globalpost.com/dispatch/india/091013/the-dark-side-medical-tourism

PTI. 2010. Medical tourism to become USD 100 billion industry by 2012: Report. *Times of India*, December 11.

Rabkin, Jeremy. (1998). *Why Sovereignty Matters*. Washington DC: American Enterprise Institute

Rajan, T. D. (2005). Private health sector in India—let's not confuse the issues. *British Medical Journal* 331, p. 1339.

(Red)™ Staff. *(Red)™ Results*. www.joinred.com/red/#impact_134.

Resnick, Paul, and Rahul Sami, (2007). The influence-limiter: provably manipulation-resistant recommender systems. *In Proceedings of the ACM Recommender Systems Conference*. Minneapolis, MN: ACM Conference on Recommender Systems, 2007, pp. 25–32. Available at www-personal.umich.edu/~rsami/MRRS/index.html

Resnick, Paul, Richard Zeckhauser, Eric Friedman, and Ko Kuwabara. (2000). Reputation systems. *Communications of the ACM* 43(12), pp. 45–48

Richardson, Matthew, and Lawrence White. (2008). The rating agencies: is regulation the answer? *NYU Stern White Papers*, http://whitepapers.stern.nyu.edu/summaries/ch03.html

Riedl, John, and Joseph Konstan. (2002). *Word of Mouse: The Marketing Power of Collaborative Filtering*. New York: Warner Business Plus

Rosenbloom, Stephanie. (Feb. 25, 2010). Wal-Mart unveils plan to make supply chain greener. *New York Times*, p. B3

Roth, Kenneth. (2004). Defending economic, social and cultural rights: practical issues faced by an international human rights organization. *Human Rights Quarterly* 26(1), pp. 63–73

Rubenstein, Jennifer. (2009). The ethics of INGO advocacy, or why it is OK that no one elected Oxfam. In Democracy Seminar Series. Kennedy School of Government, Harvard University. Cambridge, MA

Ruggie, John (Special Representative of the Secretary-General on the issue of human rights and transnational corporations and other business enterprises). (April 9, 2010). Human Rights Council, Fourteenth session. Agenda item 3: Promotion and protection of all human rights, civil, political, economic, social and cultural rights, including the right to development. *Business and Human Rights: Further steps toward the operationalization of the 'protect, respect and remedy' framework*. United Nations A/HRC/14/27

Sengupta, Amit, and Samiran Nundy. (2005). The private health sector in India. *British Medical Journal* 331, pp. 1157–1158

Sethi, S. Prakash, and Olga Emelianova (2011). The Kimberley Process Certification Scheme (KPCS): a voluntary multigroup initiative to control trade in conflict diamonds. In S. Prakash Sethi, ed. *Globalization and self-regulation: the crucial role that corporate codes of conduct play in global business*. New York City: Palgrave Macmillan, pp. 213–248

Singer, Peter. (2009). *The Life You Can Save: Acting Now to End World Poverty*. New York: Random House

Singh, Nirvikar. (2008). Decentralization and public delivery of health care services in India. *Health Affairs* 27(4), pp. 991–1001

Slomanson, William R. (2007). *Fundamental Perspectives on International Law*, 5th ed. Belmont, CA: Thomson Publishers

Smith, Adam. (1976). *An Inquiry into the Nature and Causes of the Wealth of Nations*. Edited by R. H. Campbell and A. S. Skinner, 2 vols. *Glasgow Edition of the Works and Correspondence of Adam Smith*. Oxford: Clarendon Press

Social Investment Forum staff. *Socially Responsible Investing Facts*. www.socialinvest.org/resources/sriguide/srifacts.cfm

Stockman, Farah. (Oct. 10, 2006). For those excluded, loan program is no success. *Boston Globe*

Timmons, Karen. (2007). The value of accreditation. *Medical Tourism Magazine*, pp. 12–13

Streitfeld, David. (2011). In a Race to Out-Rave Rivals, 5-Star Web Reviews Go for $5. *New York Times*, August 19, A1

UL Certification staff, *About UL Mark Product Certification*. www.ul.com/global/eng/pages/corporate/certifications

UN Sub-Commission for the Promotion and Protection of Human Rights. (2003). *Norms on the Responsibilities of Transnational Corporations and Other Business Enterprises with Regard to Human Rights*. Draft E/Cn.4/Sub.2/2003/12/Rev.2

Vandenbergh, Michael P., and Mark A. Cohen. (2009). Climate change governance: boundaries and leakage. *NYU Law Review*

Visser, Wayne. (2008). Corporate social responsibility in developing countries. In Andrew Crane, Abagail McWilliams, Dirk Matten, and Jeremy Moon, eds. *The Oxford Handbook of Corporate Social Responsibility*. Oxford: Oxford University Press, pp. 473–501

Vogel, David. (2005). *The Market for Virtue*. Washington DC: Brookings Institution Press

Von Mises, Ludwig. (1922/2010). *Socialism: An Economic and Sociological Analysis*. Translated by J. Kahane. 2nd ed. Indianapolis, IN: Liberty Fund Inc.Wal-Mart Stores, Inc., *Standards for Suppliers* [January 2009 Update], http://walmartstores.com/ download/2727.pdf.

Wal-Mart Stores, Inc., *Sustainability* (May 2010 Update). http://instoresnow.walmart.com/Sustainability.aspx?povid=cat14503-env172199-module042610-lLink_wnsus

Walzer, Michael. (1983). *Spheres of Justice: A Defense of Pluralism and Equality*. New York: Basic Books

Wertheimer, A. (1987). *Coercion*. Princeton: Princeton University Press

WHO. (2006). *World Health Report 2006: Working Together for Health*. Geneva: WHO

WHO Secretariat Executive Board,. (Dec. 3, 2009). *International Recruitment of Health Personnel: Draft Global Code of Practice. Provisional Agenda Item 4.5* (Eb126/8), 126th Session

Yamin, Alicia Ely. (2005). The future in the mirror: incorporating strategies for the defense and promotion of economic, social, and cultural rights into the mainstream human rights agenda. *Human Rights Quarterly* 27(4), pp. 1200–1244

CHAPTER 11

་ན⟩

The Obligations of Researchers Amidst Injustice or Deprivation

ALAN WERTHEIMER[*]

What are the obligations of researchers towards subjects within a framework of global injustice or deprivation? It is a fact that hundreds of millions of people living in less-developed countries (LDCs) lack adequate medical care. It is also a fact that only a small proportion of medical research targets diseases such as malaria, dengue fever, rotavirus that primarily affect impoverished people. As is often said, 90% of the research resources are targeted to diseases of 10% of the world. It is not a proven fact—although it may be true—that the burdens of poverty are primarily caused by unjust actions or policies of nations and organizations of the more developed countries. Causal issues aside, it is not a fact—but it may also be a moral truth—that the governments and peoples of more-developed countries are morally required to do more to improve the health of the populations of LDCs through greater support for treatment and research. It is less clear whether those moral requirements are best justified by considerations of justice or rights or beneficence. And it is less clear in what ways these obligations apply to medical researchers.

The editors of this volume maintain that "researchers and their sponsors have to confront the special problems of doing research in an unjust world, with corresponding obligations to *correct injustice* and *avoid exploitation*." (In what follows, "researchers" refers to both researchers *and* their

* The views expressed are those of the author. They do not necessarily represent the views of the National Institutes of Health or the Department of Health and Human Services.

sponsors.) These obligations are distinct. As I understand it, the obligation to *correct injustice* principally concerns macro-level issues and will be mostly fulfilled through research that targets the diseases of impoverished people. By contrast, the obligation to *avoid exploitation* principally concerns micro-level issues of justice and will be mostly fulfilled through proper treatment of research subjects and, perhaps, their communities—setting aside for the moment precisely what that involves. This chapter says relatively little about macro issues of global justice. Instead, I will focus on the micro-level relationship between researchers and subjects. What must researchers do in order to avoid exploitation? Is there a deep relationship between a background of injustice or deprivation and the exploitation of individual subjects? Does avoiding exploitation exhaust the researchers' obligations to subjects? Does a researcher's interaction with or use of subjects create additional obligations? If so, why?

I consider these issues in two steps. The next part of the chapter examines these questions through the lens of exploitation. Then I go on to consider the broader and more general question as to the moral importance and force of interactions.

EXPLOITATION

Research ethics has long been concerned about the exploitation of subjects, particularly when they are members of "vulnerable" groups such as prisoners, children, and the decisionally impaired. Concerns about exploitation in clinical research have increased in recent years as a growing proportion of research is being "off-shored" to LDCs in a context of global injustice or severe deprivation. Those concerns are arguably greater when research is sponsored by pharmaceutical corporations or conducted by "contract research organizations." Unfortunately, the claim that researchers or sponsors exploit their subjects is often used for its rhetorical power and does not provide a helpful moral description of the problem or a basis for its solution. To provide a framework for the discussion that follows, let us consider four cases—two hypothetical, two actual.

Hypertension

Pharma, a major U.S. drug company, proposes to conduct a randomized controlled superiority trial of a new hypertension drug, Q, in Uganda with hypertensive subjects who have never received medication. Half the participants would receive Q and half would receive an existing drug, R, which has proven to be effective. Pharma offers to provide Q or R (whichever is more effective) to all participants when the trial is complete. If the trial is successful, Pharma intends to market Q in developed nations.

The Short-Course Antiretroviral Treatment Trial

Placebo-controlled trials already have established the efficacy of a "long-course" antiretroviral treatment for reducing maternal–fetal transmission of HIV.[2] This proven treatment involves administering the drug zidovudine (AZT) orally to HIV-positive women during pregnancy and during labor, and then to the newborn infant. Unfortunately, the long-course treatment was arguably not a feasible approach in LDCs. It would be too expensive and compliance with the regime would prove virtually impossible for many women. Investigators wanted to determine whether a less-expensive and easier-to-administer "short-course" treatment with AZT would be at least reasonably effective in reducing maternal–fetal transmission of HIV. They conducted a placebo-controlled trial in which they compared a short-course regimen against no treatment. Subjects were informed about the design of the trial and consented to participate.

The Surfaxin Trial

Respiratory distress syndrome (RDS) is a common and potentially fatal disease in premature infants that is caused by insufficient surfactant in the lungs. Several replacement surfactants have been approved by the U.S. Food and Drug Administration since the 1990s. Surfactant replacement therapy is the standard treatment for RDS in the developed world, where it has sharply reduced neonatal mortality from RDS. In LDCs, however, infants do not typically have access to surfactant therapy or ventilator support.

In 2000, a private U.S. drug company, Discovery Labs, proposed to conduct a Phase III clinical trial of a new synthetic surfactant, Surfaxin.[3] The principal target market for Surfaxin was the United States and Europe. Although Discovery could have proposed to conduct (and eventually did conduct) an active-controlled trial in which Surfaxin was compared with an approved surfactant, it preferred to conduct a placebo-controlled trial because that would be quicker and less expensive and would require fewer subjects. In the original trial design, Discovery proposed to conduct a randomized placebo-controlled trial involving 650 premature infants with RDS in Bolivia and three other countries in South America. On this trial design, Discovery would provide endotracheal tubes, ventilators, and antibiotics for all participants. Half the infants would receive air suffused with Surfaxin. Half would receive untreated air. Parents of infants with RDS would be asked to give consent for their infants to participate.

HIV Vaccine

A nonprofit research organization, Vaccine for AIDS, wants to conduct a Phase III randomized placebo-controlled trial of a new HIV vaccine that it hopes to make available in LDCs. The vaccine showed some promise in Phase II testing with a small sample. Because VFA regards quick results as crucial, it prefers that subjects engage in unprotected sexual relations. VFA overstates the likely efficacy of the vaccine, and subjects are given no advice on "safe sex" practices. The vaccine proves to be ineffective, and a higher proportion of subjects get HIV than comparable people outside the study.

SOME DISTINCTIONS

These trials vary on several dimensions. The hypertension and Surfaxin trials are sponsored by pharmaceutical corporations and, if successful, the product will be marketed in more developed countries. The short-course antiretroviral and HIV vaccine trials are sponsored by nonprofit organizations and, if successful, will benefit people in LDCs. Participants in the antiretroviral, Surfaxin, and hypertension trials all stand to benefit from participation as compared with their pre-trial status quo. By contrast, the subjects in the HIV vaccine trial may be more likely to become infected than if they had not engaged in the trial unless (*ex ante*) the vaccine was likely to be effective. Finally, although it is plausible to assume that participants in the antiretroviral, Surfaxin, and hypertension trials did or could give valid informed consent, participants in the HIV vaccine trial did not give valid informed consent because they were deceived. (In these examples, the hypertension and HIV vaccine trials are hypothetical; the Surfaxin and antiretroviral trials are real.)

Despite these differences, it may be thought that participants are exploited in all four trials because researchers are taking advantage of the subjects' poor background conditions to enroll them in research. The hypertension trial is allegedly exploitative because the researchers are testing a product that will benefit a commercial organization and people in more developed countries. Participants in the antiretroviral trial are allegedly exploited because the researchers deliberately withheld proven effective treatment from those in the control group. The Surfaxin trial is arguably more exploitative than the antiretroviral trial, not only because it withholds proven effective treatment from the control group, but also because it tests a product that will benefit a commercial organization and people in more developed countries, not in LDCs. Participants in the HIV vaccine trial are allegedly exploited because they are harmed by participation and they have not given valid informed consent, even though the sponsors may be using them for an (otherwise) admirable purpose.

I have argued elsewhere that it is important to distinguish between *harmful and nonconsensual* exploitation, as in the HIV vaccine trial, and *mutually advantageous and consensual* exploitation, a category that may be exemplified by the other three trials.[4] On that view, a transaction is exploitative if A takes unfair advantage of B. The theoretically important point is that A can take unfair advantage of B even if B benefits from the transaction and consents to its terms. For example, a surgeon takes unfair advantage of a patient if he charges three times the normal price for life-saving surgery even though the patient benefits from the transaction and gives informed consent. Similarly, researchers exploit subjects if the terms of the relationship are unfair even if the subjects benefit from participation and consent to participate.

A mutually advantageous and consensual transaction is exploitative only if its terms are unfair. A surgeon is not wrongfully exploiting a patient if he charges a fair fee, and this is so even if the patient's background situation is unjust or desperate. Unfortunately, it is not easy to specify the criteria by which to evaluate the fairness of a transaction. Although some argue that A exploits B when A gains (much) more from the transaction than B, that comparative view is false. As illustrated by the previous example, a surgeon exploits a patient by charging an unfair fee even if the patient gains *much more* (his life) than the surgeon from the transaction. We need normative criteria by which to assess the fairness of a transaction. And while we have intuitions (albeit differing intuitions) about these matters, I know of no adequate theory of transactional fairness.[5]

In my view, it is an open question as to whether the antiretroviral and Surfaxin trials are examples of exploitation. To withhold proven effective treatment from the placebo group may be unfair, but it is not obviously unfair because the subjects would otherwise receive no treatment, and because they benefit considerably from participation. I am inclined to think that subjects are not exploited in the hypertension trial at all, given that they will be provided with the best available treatment after the trial even though they are being "used" for the benefit of a commercial sponsor and citizens in more developed countries.

The hypertension trial exemplifies an important distinction between *taking unfair advantage* and *taking advantage of unfairness or desperation*. If A hires B, who has been unjustly fired from his previous job by C, then A is taking advantage of C's unfairness to B. But unless the terms of A's offer are unfair, A has not taken unfair advantage of B. If the pipes in B's house are frozen, B may be desperate to have a plumber thaw them out. The plumber is taking advantage of B's desperation, but the plumber does not exploit B unless the terms of the transaction are unfair. The distinction between taking unfair advantage and taking advantage of unfairness has important implications for the ethics of research. The unjust background conditions of prospective subjects may render them particularly vulnerable to exploitation. Their desperation may increase the probability that they will be exploited. Nonetheless, the mere fact that researchers take advantage of background injustice or desperation does not—by itself—show that they exploit their subjects. The latter claim requires its own support.

SHOULD WE PROHIBIT EXPLOITATION?

Let us assume for the sake of argument that a particular trial such as the proposal for the placebo-controlled version of the Surfaxin trial exemplifies mutually advantageous and consensual exploitation. It may be thought that if a trial is wrongfully exploitative, then we are justified in seeking to prevent its occurrence. Not so. For even if a proposed trial is wrongfully exploitative, it is another question as to whether it should be permitted. Better to be allowed to be exploited by the surgeon than allowed to die.

To pursue this issue, it will prove useful to consider the off-shoring of clinical research, within the context of global injustice or deprivation, as an example of the more general phenomenon of the globalization of economic activity. We "trade" when we buy goods that we generally have not produced for ourselves. And so the United States trades with Costa Rica when it exports computers and imports bananas. We "outsource" labor when we pay others to do work that we could do. A university may outsource the maintenance of its grounds to a landscaping company. We "off-shore" labor when we pay people in nations separated by oceans to perform labor that we could or once did perform at home. An American computer manufacturer may off-shore its customer service to India.

When such trade works well, it benefits all parties even if it does not benefit them equally or justly. American consumers benefit from having clothing produced in China and telephone inquiries answered by people in India, given that the Chinese are paid less than Californians and Indians are paid substantially less than Indianans. And the Chinese and Indians arguably benefit by having employment opportunities that would not otherwise be available to them.

Critics frequently raise numerous objections to such off-shoring. Particularly in times of high unemployment, people may object to the loss of jobs at home. But people may also object to what we regard as low wages or poor working conditions for the workers who get the off-shored jobs, even if those jobs are desirable by local standards. Nonetheless, unless interfering with such employment is likely to improve their wages and working conditions, it is hard to justify interfering with such employment on behalf of the *employees* if they can reasonably expect to benefit from employment and if they consent. This is so even if they are being treated unfairly. Of course it is better to be paid a fair wage than an unfair wage. But it is arguably better to allow someone to be paid an unfair wage than not paid at all.

This does not mean that we must be silent. We may object to and work to change the global injustices that render it reasonable for people to consent to work for low wages, and we may press employers to offer better wages and working conditions. But, as a general rule, we do the people of LDCs no favor if we deny them the opportunity to produce goods and provide services on the ground that they are being exploited. Thus a moderate left-wing economist such as Paul Krugman writes that "while fat-cat capitalists might benefit from globalization, the biggest beneficiaries are, yes, Third World Workers."[6]

From this wider perspective, and to put the point in admittedly crude terms, it could be argued that prospective subjects in LDCs may have a "comparative advantage" in serving as research participants as compared with prospective subjects at home. They are more likely to consent. It is cheaper to enroll them. And they are more likely to be treatment-naïve. As an advertisement by iGATE Clinical Research, Inc. puts it: "A huge population with a diversity of diseases that are untreated—yes, that is the 'India Advantage.'"[7] Setting aside our own interests in having research conducted more cheaply and quickly, if we are concerned for the well-being of prospective subjects we should not object to off-shoring medical research if the subjects can expect to benefit and if they give valid consent. Indeed, we should encourage it. Distasteful as it might seem, it may be advantageous for the citizens of LDCs to allow themselves to be used as research subjects, given their background conditions.

There is a connection between the case for *allowing* micro-level exploitation and the *amelioration* of the background global injustice or deprivation on which it feeds. Interestingly, but perhaps not surprisingly, philosophers and ethicists have paid little attention to the benefits of trade or "off-shoring" as a response to global poverty. As Teson and Klick have noted, "None of the major works on global justice draws on the relevant economic literature, the general consensus of which recommends free trade as a way to enhance global and national wealth and thus benefit the poor."[8] It is not clear why this is so, although I suspect that philosophers may lose interest because the benefits of trade rely on egoistic behavior rather than a commitment to justice and the amelioration of deprivation.[9] Perhaps trade is not a sufficiently lofty way to ameliorate poverty and supporting trade is not a way to signal one's concerns for the global poor to fellow academics.

In any case, even if this general argument for the benefits of trade were accepted, it might be thought that it is one thing to employ Indians to answer phones, and quite another to employ them as subjects in a medical trial whether they do it for monetary payment or to get access to medical treatment that they would otherwise not receive. Shamoo and Resnik say that "it is especially important to prevent pharmaceutical companies from using people in developing countries as cheap labor to test drugs that will only be used in the developed world, because this would constitute an egregious form of exploitation."[10] But why is this so? Would they seek to prevent running shoe manufacturers from "using people in developing countries as cheap labor" to manufacture shoes that will only be used in the developed world, when those are among the better jobs available in such societies? If not, why should we seek to prevent Ugandans from serving as research subjects in the hypertension trial? There may be a compelling argument for treating research subjects differently than ordinary workers, but it is not clear what it is.

Of course even if the default position is to permit mutually advantageous and consensual but exploitative research, there nonetheless may be reasons to prevent such research. What I call the *strategic argument* claims that interference with mutually advantageous and consensual transactions is justified if it would serve to advance the welfare of subjects. It may or may not do so.

On the one hand, if researchers are told that they must provide greater benefits in order to conduct the trial, they might abandon the research or shift it to a locale where such treatment would be provided by others. When Discovery Labs was criticized for its placebo-controlled trial design, it redesigned the trial to compare Surfaxin with another surfactant, but shifted the location to more developed countries. As a result, Discovery Labs provided virtually no treatment to infants in LDCs. If that is the likely outcome, then interference is indeed hard to justify. On the other hand, there may well be cases where preventing researchers from conducting research that provides inadequate benefits to subjects will not cause them to abandon the research or shift locations. Rather, preventing such trials will lead researchers to provide greater benefits to subjects in order to conduct the trial anyway. If this is the result, preventing or condemning an unfair mutually advantageous and consensual interaction will actually benefit subjects rather than penalize them. In such cases, interference is much easier to justify.

It may be argued that interfering with exploitative trials may not benefit the people who would have participated in those trials if the interference causes researchers to abandon the trials or to shift locations (as in the Surfaxin trial); but defenders also can argue that a policy of interference or condemnation of exploitative trials will somehow work to the long-term benefit of research subjects in *other* trials, or will help to generate greater amelioration of background injustices or deprivations. I am sympathetic to such consequentialist arguments, but it is important to note that this strategy cannot be justified by appeal to the protection or welfare of the prospective subjects in the trials that do not occur. To the contrary: their interests will be (perhaps justifiably) sacrificed for the well-being of others.

THE INTERACTION PRINCIPLE

Although some of the ethical concerns about the interactions between researchers and subjects can be analyzed through the lens of exploitation, *always or sometimes* I believe there is a more fundamental ethical principle at work here. It is often suggested—or perhaps simply assumed—that the *interaction* between investigators and subjects generates ethical obligations for ✗ researchers beyond those to which the parties might reasonably agree— even if the provision of such additional benefits is not strictly necessary to avoid the charge of exploitation. I believe that such interactions exemplify a deep but problematic principle about the ethics of interactions that takes on additional weight when those interactions occur under background conditions of global injustice or deprivation. This section explores the force of that principle.

The interactions in which I am interested involve what we might call *package deals*. A proposes to provide benefits to B in return for obtaining something from B or being able to do certain things *to* B. Consider this example: A proposes to hire B to mow his lawn for a fee (X). B agrees to mow A's lawn for X. Call (X) "the contractual level" of benefits.

I assume that it is morally permissible for A to mow his own lawn rather than to hire B, even if A knows that hiring B would be to B's advantage. But what I shall call the *Interaction Principle* maintains that if A does choose to interact with B, A acquires an obligation to provide *super-contractual* benefits to B—that is, benefits that go beyond that to which A and B might otherwise have agreed. If the law requires A to pay a minimum wage of $7.50 per hour to B, whereas A and B would otherwise have settled on $5.00 per hour, then the *actual* contract between A and B already incorporates super-contractual benefits because the minimum wage is not itself subject to contract.

Yet even though the interaction principle might seem to be a natural and intuitively attractive moral principle, it has some paradoxical implications and, at a minimum, the rationale, scope, and force of that principle is unclear. The interaction principle has two principal corollaries with which I shall be concerned. First, to accept the interaction principle is to *reject* what I will call the *nonworseness claim*. Bracketing effects on third parties, the nonworseness claim maintains that it *cannot* be morally worse for A to interact with B than not to interact with B at all if the interaction or package deal is beneficial to B and if B consents to the interaction. Second, to accept the interaction principle is to *endorse* what I will call the *greater obligation claim*. This maintains that among the potential beneficiaries of A's actions or resources, A has greater obligations to provide super-contractual benefits to B than to others, even though B has *already* benefited from interaction with A whereas others have not benefited at all.

Consider a (not so) hypothetical example. Two companies produce athletic shoes: Nike and Hike. Nike sets up factories in LDCs in which it employs people for wages that are at or above typical local wages, but that do not provide what could reasonably be called a "living wage" at local prices, much less a living wage by American standards. The employees work long hours under "sweatshop" conditions that would be unacceptable in the United States, but working for Nike is better than the employees' other alternatives, and they work voluntarily. Hike believes that it cannot compete successfully with Nike if it has to pay higher wages or provide (much) better working conditions in LDCs, but it can compete by making big capital investments in automation and producing athletic shoes in the United States. Observing that Nike has been subject to considerable criticism for its employment practices, or, perhaps because it believes that it is immoral to employ workers in sweatshops at low wages, Hike builds a highly automated plant in the United States.

Let us assume for the sake of argument that Nike exploits its workers (which has not been shown) because the terms of the transaction are unfair. Even so, it seems that Nike is doing more than Hike to advance the interests of people in LDCs. How should we compare the moral quality of the decisions made by Nike and Hike? If one uses Hike as one's anchor, and if Hike is doing nothing wrong by producing shoes without sweatshop labor, and if Nike's practices are better for poor people in LDCs than are Hike's, then perhaps Nike's practices are not morally objectionable, first appearances to the contrary notwithstanding. On the other hand, if one uses Nike as one's anchor, and if Nike exploits its workers, and if Hike does less for poor people than Nike, then perhaps Hike's practices are morally objectionable. Exploiting workers à la Nike may be bad, but not providing jobs for them à la Hike may be worse.

The nonworseness claim is consistent with the two views just described. It maintains that it *cannot* be morally worse for A (Nike) to interact with B (the LDC workers) than not to interact with B if: (1) the overall interaction or package deal is better for B than non-interaction, (2) B consents to the interaction, and (3) such interaction has no negative effects on others. By contrast, the interaction principle implies that Nike's practices might actually be morally worse than Hike's because Nike benefits from the labors of its employees on unfair terms whereas, *ex hypothesi*, Hike is already treating its relatively few workers fairly.[11]

Let us now consider the greater obligation claim. The interaction principle claims that A may have a greater special obligation to provide super-contractual benefits to those who have already benefited from interaction with A than to provide benefits to those with whom A has not (yet) interacted. Suppose that Nike earns greater profits than it anticipated and that it is considering three options: (1) giving a bonus to its workers; (2) contributing to a fund for those who applied for jobs with Nike but who were not hired; (3) investing its profits in new factories that will expand the number of persons it can employ in LDCs. A commitment to the interaction principle and the greater obligation claim suggests that Nike has reason (of undetermined weight) to choose (1) because it has a greater obligation to the workers from whose efforts it has made its profits than to those that it excluded or to those that it might hire in the future, even though its current workers have already benefited from their employment and are demonstrably better off than those (2) who applied but were not hired or (3) would be hired if Nike built additional factories.

The greater obligation claim is ethically problematic because there are opportunity costs to any super-contractual resources that Nike provides to its workers. Although this is obvious and should not need stating, it does need to be stated in the present context because many of the canonical statements of research ethics appear to be quite indifferent to the opportunity costs of various principles. Much bioethics literature agrees with the spirit although perhaps not the letter of the Declaration of Helsinki's statement that "In medical research involving human subjects, the well-being of the individual research subject must take precedence over all other interests."[12] It is certainly plausible to claim that research should ordinarily not impose *nonconsensual* burdens on subjects in order to promote the interests of others, but it is not clear why the well-being of *subjects* should trump the well-being of others who might be helped by research if participation in research is already beneficial to the subjects and if they consent to participate on the proposed terms.

Yet this is precisely what the greater obligation claim seems to require. If we frame the choice as one between greater profits for a corporation and greater benefits for the research subjects, we may think that morality should push us towards the latter and towards minimizing the former. Better to benefit the poor than the rich. But that would be too quick: it is a complex empirical question as to who does or would ultimately pay the price of requiring researchers to provide super-contractual benefits to subjects. Would it inhibit the pace of innovation and the cure of disease? Would it reduce the number of studies that a company can conduct, thereby depriving others of the opportunity to benefit by participating in research? It is hard to know.

It should be noted that while many critics take special aim at pharmaceutical corporations that seek to profit financially by the use of poor persons in LDCs, the opportunity costs entailed by a commitment to the greater obligation claim will be felt whoever sponsors the research. If NIH-sponsored research provides super-contractual benefits to research subjects or their communities, there will be fewer research projects or those projects will have smaller samples, and so forth. Here, as elsewhere, there are no free lunches to be had.

PRINCIPLES OF RESEARCH ETHICS

Although I believe that there are numerous accepted or advocated principles of research ethics that seem to be motivated by the interaction principle, I shall focus on but a few. The reader is invited to apply the general structure of the argument to other principles that I have not considered in detail.

Researchers Have Greater Obligations Than the Rest of Us

If one doesn't conduct research in LDCs, then one is subject to the general demands of justice and beneficence, about which views vary from the very demanding to the moderately demanding to the non-demanding.[13] But when we ask what *researchers* owe subjects or their communities, we tacitly assume that they have special obligations to correct injustice or ameliorate deprivation—*qua* researchers—that do not apply to the governments and citizens of more developed countries. It may be thought that because researchers benefit directly from the use of research subjects, they have special obligations of reciprocity that go beyond the general obligations (large or small) of justice, respect, and beneficence that apply to everyone else. In addition, researchers are *there* and in a position to help, whereas others are not. As Alex London puts it, researchers may "have the knowledge and training to prevent some of the harms they encounter as a result of [the subject's] vast, unmet healthcare needs."[14] And so they may also have a duty to rescue those they encounter whereas the rest of us do not.

Responsiveness

It is accepted wisdom that clinical research in developing societies should be "responsive" to the health care needs of the community in which research takes place. Exemplifying the views of many canonical statements, Guideline 10 of the Council for International Organizations of Medical Sciences (CIOMS) ethical guidelines for biomedical research states: "Before undertaking research in a population or community *with limited resources*, the sponsor and the researcher must make every effort to ensure that the research is responsive to the health needs and the priorities of the population or community in which it is to be carried out" (emphasis added).[15] Call this the "responsiveness principle."

On most formulations, the responsiveness principle does not claim that researchers have a positive obligation to conduct responsive research. It is generally not claimed, for example, that it would be wrong for Pharma to conduct the hypertension trial in the United States rather than to conduct malaria research in Uganda because treatment for hypertension is not a priority health need in Uganda. The main thrust of the responsiveness principle is negative: it is wrong to conduct research in LDCs that does *not* satisfy the responsiveness principle. If the people of LDCs are to be used for research, then that research must generate super-contractual benefits for the host society by addressing a research question that is responsive to its needs.

Despite its wide acceptance, the responsiveness principle is deeply puzzling. No one argues that Nike should locate production plants only where there is a special need for athletic shoes. We may criticize Nike if the workers are not treated fairly, but we are not concerned that the workers are producing goods which neither they nor their fellow citizens have any prospect of using. It is surely paradoxical to claim that it would be permissible to conduct hypertension research in the United States, in which case *no* Ugandans benefit, but that it would be wrong to conduct this research in Uganda, where *some* Ugandans benefit, simply because hypertension is not a health priority in Uganda. How can it be worse that *some* Ugandans benefit than that *no* Ugandans benefit?

One might defend the responsiveness principle on consequentialist grounds by arguing that nonresponsive research such as the hypertension trial typically inflicts hidden negative effects on the host societies. For example, it might be argued that nonresponsive research tends to displace responsive research. And so by putting roadblocks in the way of non-responsive research, the responsiveness principle might motivate or incentivize researchers to conduct research that will benefit the populations of LDCs and not just the research subjects.

This blocking strategy provides a plausible defense of the responsiveness principle, although it is an empirical question as to whether it will actually generate more responsive research. It does not give us reason to disallow nonresponsive research such as the hypertension trial if allowing such research does not reduce the amount of more responsive research. But, despite appearances, this argument is not a counterexample to the nonworseness claim because the nonworseness claim assumes that the interaction has no adverse effect on others. And even if such externalities did give us reason to reject such research, it would not be because conducting such research is a wrong to the *subjects* of nonresponsive research.

Standard Care Principle

It is often claimed that it is ethically impermissible to withhold a proven effective therapy as part of a randomized controlled trial, particularly if the subject has a serious disease. The Declaration of Helsinki states: "The benefits, risks, burdens and effectiveness of a new intervention must be tested against those of the best current proven intervention"[16] And this is true even if subjects would otherwise receive no effective therapy were no trial to be conducted.

The standard care principle might be defended on the grounds that research subjects should be regarded as "patients" toward whom investigators owe specific obligations *qua* physicians. That view must be rejected because it would preclude many procedures that are designed to measure study outcomes but that provide no benefit to the subject. It is more promising to assume that the standard care principle assumes that researchers acquire special obligations to those subjects with whom they interact—obligations that they do not have towards those who are not research subjects. In condemning the Surfaxin trial, Lurie and Wolfe argued that "the provision of placebo . . . to the 325 infants in the control group will result in the preventable deaths of 16 infants"—that is, those infants in the study who would die because they received sham air, but would not have died if they had received traditional surfactant therapy.[17] Fair enough. But note that a decision to include only 650 subjects in the study, as contrasted, say, with 1,000, also results in preventable deaths— this time to infants who were not included in the study because of a decision about sample size, and who therefore were not in the Surfaxin treatment arm of the trial. Lurie and Wolfe assume without argument that Discovery Labs must provide surfactant therapy to the subjects *in* the research but need not provide it to infants with RDS who are not a part of the study, even though *all* of the infants in the study would have been prospectively better off than infants not in the study. Why? Because Discovery Labs interacts with and seeks to benefit from its subjects but does not interact with or benefit from the non-subjects.

Post-Trial Obligations to Research Subjects

It is frequently argued that if researchers use subjects to test the effectiveness of an intervention, then those subjects should be provided with the intervention at the conclusion of a successful trial. For example, the Declaration of Helsinki states: "At the conclusion of the study, patients entered into the study are entitled to be informed about the outcome of the study and to share any benefits that result from it, for example, access to interventions identified as beneficial in the study or to other appropriate care or benefits."[18]

There are two related questions here: (1) Do researchers have super-contractual obligations of post-trial (or extra-trial) treatment or other benefits to anyone, as the nonworseness claim might deny? (2) If researchers have super-contractual obligations, does the obligation extend only or primarily to their particular subjects, as the greater obligation claim would imply?

In discussing an antiretroviral study, David Rothman argues, "As *compensation* to their subjects for enrolling in the research, investigators who come to Uganda should be required to leave their subjects better off" (emphasis added).[19] But compensation for what? If the subjects are *harmed* by participation, then they may be owed compensation for the harm. But if the subjects can expect to benefit from and consent to participation, then compensation for *harm* seems not to apply.

It might be argued that researchers have obligations of reciprocity to provide treatment to subjects in exchange for the subjects' willingness to allow themselves to be used by the researcher. Despite frequent appeals to the notion of reciprocity in bioethics, there are precious few theoretical accounts on offer. Henry Sidgwick more than a century ago complained that he could find "no clear accepted principle" by which to determine how much one party owes another.[20] There's been little progress since then. Although we lack a systematic account of reciprocity, its demands appear to reflect at least three factors: (1) the degree to which A benefits from the interaction with B; (2) the degree to which B has sacrificed or benefited (relative to A's benefit) from the interaction; (3) A's capacity to reciprocate. If this is right, then the issue as to how much investigators owe to participants will depend on the extent to which researchers gain from the participants' efforts and how much participants *already* benefit from participation.

If researchers have a reciprocity-based obligation to provide post-trial treatment to those who are research subjects, then researchers have a strong incentive to prefer to conduct trials in locations where subjects have other sources of post-trial treatment. Once again, the nonworseness claim rears its head. If it is ethical not to conduct research in locations where people have little or no access to post-trial medical services, and thereby to provide *no* treatment to subjects, it seems odd to claim that it is worse to provide only intra-trial treatment to subjects if inclusion in research is otherwise beneficial to the subjects.

At least one group of prospective subjects made precisely this claim. In 1997, a multinational drug company wanted to conduct research in South Africa because of its combination of a "large infected population and proven medical expertise." The sponsors said that they would provide the "cocktail" of antiretroviral drugs free for 2 to 3 years but would not guarantee that the treatment would be continued at the conclusion of the trial, in part because providing the "cocktail" would require them to buy drugs produced by other companies.[21] A Research Ethics Committee (the equivalent of an institutional review board) whose approval was necessary maintained that "it is not ethical" to do such research unless post-trial treatment is assured and that it had the responsibility "to ensure that patients are not exploited."[22]

In response to these worries, AIDS activists argued that, *for them,* the trials "are seen as treatment rather than research—and are often the only way in which [they] have any access to treatment." From their perspective, "access to limited and potentially beneficial treatment is better than no treatment at all."[23] If the researchers have no obligation to conduct the research, and would not do so if they were required to provide post-trial treatment, then it seems problematic to argue that it is worse for researchers to conduct the study without providing post-trial treatment.

From a broader perspective, the obligation to provide post-trial treatment seems strange. We do not believe that employers have an obligation to ensure that employees are no worse off when their employment ends than during the period of employment. This is particularly so when the employment is task-specific, as when one hires someone to paint one's house, but this is also true of other temporary jobs, such as when retailers hire extra clerks for the holiday shopping season. It hardly seems worse to hire them for a short while and then stop their wages than not to hire them at all. On that view, it is hard to see why researchers acquire post-trial obligations to the subjects that they "employ."

Moreover, and to pursue the employment model, let us assume that there are more applicants than jobs for these temporary positions. Suppose that an employer is considering a choice between the following options after a particularly profitable season:

1. Keeping all the profits
2. Sharing some of the profits with its temporary employees
3. Sharing the profits with those who were not hired

Although it might be morally desirable that the employer choose (2) over (1), it is simply not clear that there is greater moral reason for employers to choose (2) over (3), thereby giving special weight to the persons who were lucky enough to be hired as contrasted with those who were less fortunate and remain unemployed.

We can make a similar point about research subjects. There can be significant opportunity costs in providing for the needs of subjects when those resources could be devoted to additional or higher-quality research. The desirability of using those resources for other research projects or other persons must gain weight when subjects have, *ex hypothesi, already benefited* from participation.

THE INTERACTION PRINCIPLE RECONSIDERED

In the previous section, I suggested that a commitment to the interaction principle underlies some of the common wisdom about the ethics of conducting research in LDCs. I also suggested that we have some reason to be skeptical about the merits of two corollaries of the interaction principle: the rejection of the nonworseness claim and the endorsement of the greater obligation claim. Despite the concerns that I have raised, it is entirely possible that (some version of) the interaction principle is basically correct. To properly evaluate the viability of these principles of ethical research that receive support from the interaction principle, we need to consider the principle at a more theoretical level.

Of course even if the interaction principle is defensible, there is no reason to suppose that *every* mutually beneficial interaction gives rise to super-contractual obligations. There must be considerable space in our lives for parties to reach agreement on whatever terms they please (perhaps within broad limits) without being morally required to do more than fulfill the terms of the agreement. But some interactions may strike us as cases in which the terms are subject to moral evaluation—depending, perhaps, on the duration and intensity of the interaction, the level of benefits the parties have received, the pre-transaction levels of well-being of the parties, and the surplus available for providing additional benefits.

Consider these cases.

1. A, a bank customer, transacts with B, a bank teller, thereby benefiting from her services. Although B is better off working as a bank teller than being unemployed, B's salary does not provide for a decent living.
2. A employs B as a full-time housekeeper for a minimum wage. B does not earn enough to live adequately. B has no other job opportunities and would be worse off if she were not employed by A.
3. B performs life-saving surgery on A for a normal fee.

Although I believe that the interaction in the bank situation is so low in duration and depth as not to raise moral worries, Ruth Sample suggests that A exploits or wrongfully uses B because A effectively "tolerates" B's inadequate wages while benefiting from the interaction.[24] Interestingly, if we view the surgery example as a case in which the *patient* employs the surgeon, few would say that patient A owes surgeon B more than a normal fee out of gratitude or reciprocity, even though this is a high-intensity interaction from which A has benefited much more than B. It is not clear why; perhaps it is because we assume that B is already well off and our moral evaluation of the transaction takes place against this background. In any case, and for present purposes, I will focus on the housekeeper situation because the intensity, proximity, and inequality of the interaction is most analogous to the interaction between researcher and subject and because it is the case in which the interaction principle is most plausible.

Suppose that A has two applicants for the housekeeper job: B and C. A judges them to be equally qualified, so she flips a coin, and B wins the lottery. Suppose as well that A has four options: (1) not hire B (or C) and do the housework for herself; (2) hire B for the minimum wage of

$200 per week, which is not sufficient to provide for B's needs; (3) pay B $240 per week, which is sufficient to provide for B's needs; (4) pay $200 per week to B and give $40 per week to C, who is worse off than B and who would gladly have taken B's job if it had been offered to her (Table 11.1).

The nonworseness claim maintains that it cannot be worse for A to offer (2) than (1) because being hired is better for B. By contrast, those who endorse the interaction principle may claim that it could be worse for A to employ B but not to provide adequately for her needs (2) than for A not to employ B at all (1).

When I reflect on this case, I feel the pull of the intuition that A's action in (2) is (or might be) morally worse than (1). At the same time, when I reflect on the *principles* involved, I am sympathetic to the view that (2) cannot be worse than (1) if it's better for B and if B consents to it. All that said, the intuition that (2) is worse than (1) is sufficiently strong that we should consider what might be said in its favor.

It might be thought that the nonworseness claim is a consequentialist principle and that it is therefore open to the standard deontological objections to consequentialism. Not quite. The claim is consequentialist in that it maintains that if the consequences of A's choosing (2) are better for *everyone* than A's choosing (1), then it cannot be morally worse for A to choose (2) rather than (1) if B consents to (2). But the nonworseness claim is not consequentialist in the sense that it claims that that it is permissible to harm B if doing so has better *aggregate* consequences for others. The nonworseness claim does not allow us to kill one person to redistribute her organs to save five others. Indeed, appeals to deontology seem somewhat out of place here. Deontological principles are normally advanced as *side-constraints* on agents that serve to protect the targets of actions against nonconsensual harms or violations of rights that could otherwise be justified on consequentialist grounds. But such constraints would seem to have no direct bearing on interactions from which the target benefits and to which she consents.

Table 11.1. HOUSEKEEPER SCENARIO

Option	B	C
1	$0	$0
2	$200	$0
3	$240	$0
4	$200	$40

It might be objected, however, that fairness is also a moral constraint. If A does not hire B in situation (1), there is no issue as to the *fairness* of B's compensation or whether A exploits B or wrongfully uses B. So if it is wrong to treat people unfairly or to exploit them, there is a moral reason to criticize (2) that does not and could not occur in (1). And it might be argued that because (2) involves a wrongful unfair action whereas (1) does not, it follows that (2) can be worse than (1).

But the question is not whether A acts wrongly when A treats B unfairly as in (2). A does act wrongly. The question is whether A's action in (2) is *morally worse* than A's action in (1) when evaluated by *the reasons that we care about fairness*. After all, we do not think that we should be constrained by the principle of fairness for the *constraint's* sake. We observe it for the *person's* sake. Indeed, to observe or enforce a constraint for the sake of the constraint itself "would arguably fail to treat the person as an end in himself" and so violate the deontologist's most fundamental commitment.[25]

Taking a slightly different tack, Derek Parfit argues that "some things are worth doing for their own sake," such as "acts that express respect for people, or some act of loyalty to some dead friend."[26] Similarly, some acts may be bad to do because they express callousness or indifference towards others. So it might be thought that A's decision not to hire anyone as a housekeeper does not express indifference towards B, whereas paying B an inadequate wage does express indifference or callousness.

Even if this is so, this argument has little relevance to the principles of research ethics that are concerned with the protection or interests of research subjects. Those principles are not advanced as a way to "express respect" for subjects. Rather, their advocates believe that compliance with those principles will be *better* for subjects. So this line of argument does not support rejecting the nonworseness claim in the present context.

There may be agent-oriented or integrity-based arguments that would support such principles. It might be argued that one should not violate certain principles even if violating them would be better for subjects: "I know you'd be better off if I engaged in this research, but I can't do it. And so I'll eschew doing such research. For my sake, not for yours." Such "integrity-based" arguments may be plausible, but they would not justify these principles by appeal to considerations of the interests of subjects.

Where does all this leave us? Despite the intuitive appeal of the interaction principle, I have not found a persuasive argument for rejecting the nonworseness claim. Assume, for the sake of argument, that the nonworseness claim is roughly correct. What follows? There are at least three possibilities. First, if non-interaction is not (seriously) wrong, it is possible that some interactions such as Nike or (2) in the housekeeper scenario or some cases of allegedly unethical clinical research are indeed *less* wrong than some ethicists are inclined to think. Second, if an interaction is wrong and if non-interaction is worse than the interaction, then some non-interactions such as Hike or (1) in the housekeeper scenario may be *more* wrong than we were otherwise inclined to think. On the first view, Discovery Labs' proposal to conduct a placebo-controlled trial in South America was not wrong. On the second view, Discovery Labs' decision to shift the locations of the trial away from South America was wrong.

There is a third alternative. It might be argued that the nonworseness claim is correct with respect to the moral evaluation of "states of affairs," but that it is incorrect with respect to the moral evaluation of "actions." Recall the housekeeper scenario. It might be argued that whereas (2) is a morally superior state of affairs to (1), A acts wrongly in (2) but not in (1). Here principles of moral action may be justified by something like the "strategic argument" for intervention that we considered above. Faced with a choice between (1), (2), and (3), A would choose (2), which is better for B than (1). But if the relevant set of moral principles presents A with a choice that is limited to (1) and (3), A is likely to choose (3), which is better for B than (2). By taking (2) off the moral table, the expected result is *not* worse for B, given a sufficient likelihood that A will choose (3) rather than (1). Thus it is arguably better to regard certain types of interaction, such as (2), as wrong even when they yield a morally superior state of affairs vis-à-vis (1). And the same might apply with respect to the principles of ethical research. It may be best to regard certain research interactions as wrong even if those interactions yield a better state of affairs than non-interaction in a particular case. Whether this is so depends upon the likely consequences of compliance with or acceptance of the relevant principles of ethical research.

It is, of course, an empirical question as to whether acceptance of or compliance with particular norms of research ethics tend to advance the well-being of subjects. If they do so, then that is a reason in their favor. If, however, the regulations or norms deter research from which subjects will benefit and that does not compromise their autonomy, then we should probably reject the norm or regulation. Of course, a norm or regulation could have *both* effects. Minimum wage laws may raise the income of some employees while (unintentionally) deterring the employment of others to the detriment of the non-employed. What then? We have to choose. If minimum wage laws cause a significant increase in non-employment, then such laws would be hard to justify even if they help some workers. On the other hand, if the principal effect of minimum wage laws is to raise the income of employees without significantly reducing employment, then such laws are much easier to justify.

A similar point applies to research. If insisting on the standard care principle that would bar placebo-controlled trials advances the interests of research subjects, then we have at least a very strong reason to defend it. But if the principle has the opposite effect, as was arguably the case in the Surfaxin trial, then there is reason to be skeptical. Much the same can be said for the principle that researchers should provide post-trial treatment or ancillary care to research subjects or provide benefits to the host community.

But even if requiring researchers to devote greater resources to research subjects will not deter them from doing research, there is still a question as to whether we *should* push them in that direction, given the opportunity costs of providing those resources to subjects as opposed to others. Suppose, for example, that we favor principles that require Pfizer and Merck to devote greater resources to research subjects in the form of post-trial treatment. Because the requirement affects both companies, it will not weaken either company's competitive position, and so it will not deter them from conducting research. Even so, compliance with the requirements will mean that they will have to devote more resources to each research project, leaving fewer resources for additional projects or basic research. Or perhaps the costs will be passed on to consumers in the form of higher prices. Here we face the issue raised by the greater obligation claim.

THE GREATER OBLIGATION CLAIM

The greater obligation claim says that if A benefits from interacting with B, A may have a special obligation to provide super-contractual benefits to B as opposed to other persons. The issue is not whether A has obligations that arise from *commitments* that A has made to B. If A borrows $10 from B, then, *ceteris paribus*, A has greater reason to give $10 to B than to C, even if C would derive a greater benefit from the $10. The question is whether A has special obligations to B simply in virtue of the interaction between A and B.

The issue here is not whether A may have obligations of justice or beneficence to those who live under non-ideal conditions that go beyond his contractual obligations. For on the assumption that A has some such obligations, the issue here is the *direction* of any such obligations, whether A should provide greater resources to B who has already benefited from A or to others who have not.

Consider the housekeeper situation once again (Table 11.2).

Table 11.2. HOUSEKEEPER SCENARIO

Option	B	C
1	$0	$0
2	$200	$0
3	$240	$0
4	$200	$40

Even if A is prepared to pay a total of $240 for someone to clean her house (3), if it were wrong for her to pay $200 to B (2), that leaves open the question as to whether A's super-contractual obligations should take the form of paying $240 to B (3) rather than distributing the $240 between B and C (4).

The greater obligation claim maintains that because A derives benefits from her interaction with B, A has greater moral reason to provide super-contractual benefits to *B*, as in (3), than to use those resources to benefit others with whom A has not interacted, as in (4). If A employs B to clean her house, then A has greater moral reason to provide for B's needs than to provide for C's needs, even if A is aware of C's needs, even though C is *worse* off than B, even though C applied for and would have taken B's position if it had been offered, and even if it was simply a matter of the flip of a coin that B got the job rather than C. For, according to the greater obligation claim, the fact is that A benefits from B's work, not C's.

As with the nonworseness claim, we have principles and intuitions that pull us in different directions. While proximity, feelings of compassion, and a commitment to reciprocity pull us towards (3), our concerns for impartiality and for assisting the worse off pull us towards (4). The greater obligation claim is less paradoxical than the nonworseness claim. Whereas in the housekeeper scenario it is paradoxical to claim that it is worse or wrong for A to choose (2) than (1), given that (2) is better for both A and B, there is nothing paradoxical about thinking that (3) is morally superior to (4), given that neither alternative is better for everyone. Indeed, the general thrust of the greater obligation claim is quite familiar. The notion that we are entitled or indeed obligated to be partial to family and friends is central to our moral psychology. As Scheffler puts it: "the importance of special obligations in common-sense moral thought seems undeniable ... the demands of morality, as ordinarily interpreted, have less to do with abstractions like the overall good than with the specific web of roles and relationships that serve to situate a person in social space."[27] So the question is not whether it is reasonable to think that we may have special obligations. The question is whether and why such obligations emerge from interactions between researchers and subjects and how they compare with competing uses of those resources.

Suppose that the Bill and Melinda Gates Foundation is supporting clinical research on AIDS in sub-Saharan Africa. Even a very wealthy foundation has limited resources. The Gates Foundation is considering the following possible uses of a pool of funds:

1. Provide antiretrovirals as post-trial treatment to HIV-positive participants in a successful antiretroviral trial
2. Provide antiretrovirals to those (previously) HIV-negative subjects who are included in an AIDS *vaccine* trial, but who seroconvert during the trial

I do not have a firm idea as to which option is morally superior. I would note, however, that the participants in (1) have already benefited from participation, whereas the participants in (2) have not benefited and may be worse off because they participated. It is no doubt more difficult to stop therapy to those who have been treated, as (2) would require, than not to begin therapy with others, as (1) would require, even if the therapy will produce comparable benefits in both cases. Nonetheless, the fact that the scientific benefits of the antiretroviral trial come from the use of the participants in (1) does not necessarily give them a (much) weightier claim to resources that could otherwise be provided to the participants in (2).

After all this, I find that I am unable to reach a firm conclusion about the force of the greater obligation claim with regard to issues such as the standard of care or the provision of post-trial treatment or ancillary care. The case for special obligations is strong enough to render it permissible or reasonable for researchers or sponsors to use their resources in a way that is consistent with the greater obligation claim, even if it would be morally preferable from an impartial perspective if they were to do less for those who have already benefited from participation and more for those who have not. On the other hand, I do not think it unreasonable not to use available resources on them if those resources would otherwise be used for non-subjects in ways that would do more to reduce human suffering subject to the demands of immediacy and urgency of suffering that no person should be expected to refuse.

That said, as a practical principle the case for the greater obligation claim and the principles of research ethics that it supports may be stronger than would be warranted on its (pure) philosophical merits. Research resources may be saved if they are not used for the benefit of the research subjects; but the saved resources often may not be directed to first best uses—in pursuit of generalizable knowledge for the benefit of others in LDCs. The greater obligation claim might be a reasonable accommodation to this fact. And this might be especially so when research is sponsored by private corporations. It might well be ethically better if researchers were to devote available resources to the amelioration of unmet medical needs without special regard for research subjects, but it would be arguably *un*ethical for policymakers and advocates to recommend principles on the assumption that individuals will make ethical decisions when the evidence suggests that they will not.

CONCLUSION

A cautionary reminder: the argument in this chapter is limited to medical research to which subjects give morally transformative consent and from which subjects can expect to benefit (setting aside what counts *as* a benefit). I have no doubt but that there is *much* research—such as the AIDS vaccine trial described earlier, tainted by lack of counseling and deceptive consent procedures—that does not fit this description, even if that research might be justified on utilitarian grounds. But as morally objectionable as such research may be, it does not present difficult theoretical issues. By contrast, I hope to have shown that it is much less clear that researchers have an obligation to provide super-contractual benefits to subjects who give morally transformative consent and can reasonably expect to benefit from participation. At a minimum, I hope to have shown that what I have called the interaction principle needs a more sustained defense than has been provided, and that the principles of research ethics that receive support from the interaction principle may need to be abandoned, substantially revised, or supported on other grounds. And, if I am right, this is particularly so in the context of global injustice or deprivation.

NOTES

1. Quoted from the dust-jacket of this book.
2. Lurie P and Wolfe S. (1997). Unethical trials of interventions to reduce perinatal transmission of the human immunodeficiency virus in developing countries. *New England Journal of Medicine* 337:853–856.
3. See the description of the Surfaxin trial in Lavery J, et al., eds. (2007). *Ethical Issues in International Biomedical Research: A Casebook.* New York: Oxford University Press, pp. 151–159.
4. See Wertheimer A. (1996). *Exploitation.* Princeton, NJ: Princeton University Press.

5. I try to develop such an account in Chapter 7 of *Exploitation,* but I am not sure that it works or that it is applicable to all the contexts in which it makes sense to refer to exploitation.
6. Krugman P. (March 21, 1997). In praise of cheap labor. *Slate.*
7. http://www.igatecr.com/company/indiaadvantage/index.php
8. Teson F and Klick J. *Global Justice and Trade: A Puzzling Omission.* http://papers.ssrn.com/sol3/papers.cfm?abstract_id=1022996, p. 54.
9. *Ibid.,* 76.
10. Shamoo A and Resnik D. (2006). Strategies to minimize risks and exploitation in phase one trials on healthy subjects. *American Journal of Bioethics* 6(3):W1–W13.
11. The contrast between the obligations of Nike and Hike is mirrored by arguments about the obligations of consumers who indirectly benefit from workers. Those who think they should boycott Nike products are typically more concerned about the plight of the Nike laborer who works for $1 per hour than with the peasant who works for $0.50 per hour because the consumer would otherwise benefit from the Nike laborer whereas she does not benefit from the peasant's labor.
12. Adopted by the 18th World Medical Association General Assembly, Helsinki, Finland, June 1964. Amended in 2008. Paragraph 6.
13. For the very demanding, see Singer P. (1972). Famine, affluence, and morality. *Philosophy and Public Affairs* 1:229–243. For the moderately demanding, see Pogge T. (2002). *World Poverty and Human Rights: Cosmopolitan Responsibilities and Reforms.* Cambridge, UK: Polity Press. For somewhat but less demanding, see Rawls J. (1999). *The Law of Peoples.* Cambridge, MA: Harvard University Press.
14. London A. (2000). The ambiguity and the exigency: clarifying 'standard of care' arguments in international research. *Journal of Medicine and Philosophy* 25:379–397.
15. International Ethical Guidelines for Biomedical Research Involving Human Subjects, The Council for International Organizations of Medical Sciences. http://www.cioms.ch/publications/frame_printable_publications.htm
16. World Medical Association Declaration of Helsinki, Paragraph 32. The document also states that a placebo-controlled trial may be permissible "[w]here for compelling and scientifically sound methodological reasons its use is necessary to determine the efficacy or safety of a prophylactic, diagnostic or therapeutic method."
17. Lurie P and Wolfe S. (Feb. 22, 2001). Letter to Tommy Thompson. *The Public Citizen,* http://www.citizen.org/publications/publicationredirect.cfm?ID=6761
18. Declaration of Helsinki, Paragraph 33.
19. Rothman D. (Nov. 30, 2000). The shame of medical research. *New York Review of Books.*
20. Sidgwick H. (1907/1982). *The Methods of Ethics,* 7th ed. Indianapolis: Hackett, p. 261.
21. Cleaton-Jones PE. (1997). An ethical dilemma: availability of antiretroviral therapy after clinical trials with HIV-infected patients are ended. *BMJ* 314:887.
22. *Ibid.*
23. Busse P. (1997). Strident but essential: the voices of people with Aids. *BMJ* 314:888.
24. Sample R. (2003). *Exploitation: What It Is and Why It's Wrong.* Lanham, MD: Rowman and Littlefield, p. 69.
25. Applbaum A. (1999). *Ethics for Adversaries.* Princeton: Princeton University Press.
26. Parfit D (unpublished mss). *Climbing the Mountain,* p. 59.
27. Scheffler S. (2001). *Boundaries and Allegiances.* Oxford: Oxford University Press, pp. 36–37.

INDEX